SCIENCE OF EVERYDAY THINGS

SCIENCE OF EVERYDAY THINGS

VOLUME 2: REAL-LIFE PHYSICS

EDITED BY NEIL SCHLAGER
WRITTEN BY JUDSON KNIGHT

A SCHLAGER INFORMATION GROUP BOOK

GALE GROUP

THOMSON LEARNING

Detroit • New York • San Diego • San Francisco
Boston • New Haven, Conn. • Waterville, Maine
London • Munich

A Schlager Information Group Book
Neil Schlager, Editor
Written by Judson Knight

Gale Group Staff

Kimberley A. McGrath, *Senior Editor*

Maria Franklin, *Permissions Manager*

Margaret A. Chamberlain, *Permissions Specialist*

Shalice Shah-Caldwell, *Permissions Associate*

Mary Beth Trimper, *Manager, Composition and Electronic Prepress*

Evi Seoud, *Assistant Manager, Composition and Electronic Prepress*

Dorothy Maki, *Manufacturing Manager*

Rita Wimberley, *Buyer*

Michelle DiMercurio, *Senior Art Director*

Barbara J. Yarrow, *Manager, Imaging and Multimedia Content*

Robyn V. Young, *Project Manager, Imaging and Multimedia Content*

Leitha Etheridge-Sims, Mary K. Grimes, and David G. Oblender, *Image Catalogers*

Pam A. Reed, *Imaging Coordinator*

Randy Bassett, *Imaging Supervisor*

Robert Duncan, *Senior Imaging Specialist*

Dan Newell, *Imaging Specialist*

ISBN   0-7876-5631-3 (set)
            0-7876-5632-1 (vol. 1)   0-7876-5634-8 (vol. 3)
            0-7876-5633-X (vol. 2)   0-7876-5635-6 (vol. 4)

Printed in the United States of America
10 9 8 7 6 5 4 3 2 1

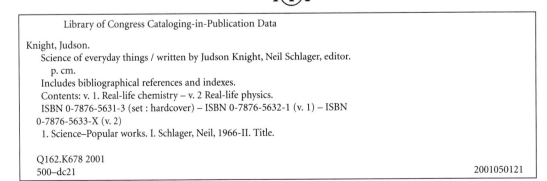

Library of Congress Cataloging-in-Publication Data

Knight, Judson.
    Science of everyday things / written by Judson Knight, Neil Schlager, editor.
      p. cm.
    Includes bibliographical references and indexes.
    Contents: v. 1. Real-life chemistry – v. 2 Real-life physics.
    ISBN 0-7876-5631-3 (set : hardcover) – ISBN 0-7876-5632-1 (v. 1) – ISBN
0-7876-5633-X (v. 2)
    1. Science–Popular works. I. Schlager, Neil, 1966-II. Title.

Q162.K678 2001
500–dc21
                                                               2001050121

CONTENTS

INTRODUCTION

Overview of the Series

Welcome to *Science of Everyday Things*. Our aim is to explain how scientific phenomena can be understood by observing common, real-world events. From luminescence to echolocation to buoyancy, the series will illustrate the chief principles that underlay these phenomena and explore their application in everyday life. To encourage cross-disciplinary study, the entries will draw on applications from a wide variety of fields and endeavors.

Science of Everyday Things initially comprises four volumes:

Volume 1: *Real-Life Chemistry*
Volume 2: *Real-Life Physics*
Volume 3: *Real-Life Biology*
Volume 4: *Real-Life Earth Science*

Future supplements to the series will expand coverage of these four areas and explore new areas, such as mathematics.

Arrangement of Real Life Physics

This volume contains 40 entries, each covering a different scientific phenomenon or principle. The entries are grouped together under common categories, with the categories arranged, in general, from the most basic to the most complex. Readers searching for a specific topic should consult the table of contents or the general subject index.

Within each entry, readers will find the following rubrics:

- **Concept** Defines the scientific principle or theory around which the entry is focused.

- **How It Works** Explains the principle or theory in straightforward, step-by-step language.
- **Real-Life Applications** Describes how the phenomenon can be seen in everyday events.
- **Where to Learn More** Includes books, articles, and Internet sites that contain further information about the topic.

Each entry also includes a "Key Terms" section that defines important concepts discussed in the text. Finally, each volume includes numerous illustrations, graphs, tables, and photographs.

In addition, readers will find the comprehensive general subject index valuable in accessing the data.

About the Editor, Author, and Advisory Board

Neil Schlager and Judson Knight would like to thank the members of the advisory board for their assistance with this volume. The advisors were instrumental in defining the list of topics, and reviewed each entry in the volume for scientific accuracy and reading level. The advisors include university-level academics as well as high school teachers; their names and affiliations are listed elsewhere in the volume.

NEIL SCHLAGER is the president of Schlager Information Group Inc., an editorial services company. Among his publications are *When Technology Fails* (Gale, 1994); *How Products Are Made* (Gale, 1994); the *St. James Press Gay and Lesbian Almanac* (St. James Press, 1998); *Best Literature By and About Blacks* (Gale,

2000); *Contemporary Novelists, 7th ed.* (St. James Press, 2000); and *Science and Its Times* (7 vols., Gale, 2000-2001). His publications have won numerous awards, including three RUSA awards from the American Library Association, two Reference Books Bulletin/Booklist Editors' Choice awards, two New York Public Library Outstanding Reference awards, and a CHOICE award for best academic book.

Judson Knight is a freelance writer, and author of numerous books on subjects ranging from science to history to music. His work on science titles includes *Science, Technology, and Society, 2000 B.C.-A.D. 1799* (U*X*L, 2002), as well as extensive contributions to Gale's seven-volume *Science and Its Times* (2000-2001). As a writer on history, Knight has published *Middle Ages Reference Library* (2000), *Ancient Civilizations* (1999), and a volume in U*X*L's *African American Biography* series (1998). Knight's publications in the realm of music include *Parents Aren't Supposed to Like It* (2001), an overview of contemporary performers and genres, as well as *Abbey Road to Zapple Records: A Beatles Encyclopedia* (Taylor, 1999). His wife, Deidre Knight, is a literary agent and president of the Knight Agency. They live in Atlanta with their daughter Tyler, born in November 1998.

COMMENTS AND SUGGESTIONS

Your comments on this series and suggestions for future editions are welcome. Please write: The Editor, *Science of Everyday Things*, Gale Group, 27500 Drake Road, Farmington Hills, MI 48331.

ADVISORY BOARD

William E. Acree, Jr.
Professor of Chemistry, University of North Texas

Russell J. Clark
Research Physicist, Carnegie Mellon University

Maura C. Flannery
Professor of Biology, St. John's University, New
York

John Goudie
Science Instructor, Kalamazoo (MI) Area
Mathematics and Science Center

Cheryl Hach
Science Instructor, Kalamazoo (MI) Area
Mathematics and Science Center

Michael Sinclair
Physics instructor, Kalamazoo (MI) Area
Mathematics and Science Center

Rashmi Venkateswaran
Senior Instructor and Lab Coordinator,
University of Ottawa
Ottawa, Ontario, Canada

GENERAL CONCEPTS

FRAME OF REFERENCE

KINEMATICS AND DYNAMICS

DENSITY AND VOLUME

CONSERVATION LAWS

FRAME OF REFERENCE

CONCEPT

Among the many specific concepts the student of physics must learn, perhaps none is so deceptively simple as frame of reference. On the surface, it seems obvious that in order to make observations, one must do so from a certain point in space and time. Yet, when the implications of this idea are explored, the fuller complexities begin to reveal themselves. Hence the topic occurs at least twice in most physics textbooks: early on, when the simplest principles are explained—and near the end, at the frontiers of the most intellectually challenging discoveries in science.

HOW IT WORKS

There is an old story from India that aptly illustrates how frame of reference affects an understanding of physical properties, and indeed of the larger setting in which those properties are manifested. It is said that six blind men were presented with an elephant, a creature of which they had no previous knowledge, and each explained what he thought the elephant was.

The first felt of the elephant's side, and told the others that the elephant was like a wall. The second, however, grabbed the elephant's trunk, and concluded that an elephant was like a snake. The third blind man touched the smooth surface of its tusk, and was impressed to discover that the elephant was a hard, spear-like creature. Fourth came a man who touched the elephant's legs, and therefore decided that it was like a tree trunk. However, the fifth man, after feeling of its tail, disdainfully announced that the elephant was nothing but a frayed piece of rope. Last of all, the sixth blind man, standing beside the elephant's slowly flapping ear, felt of the ear itself and

determined that the elephant was a sort of living fan.

These six blind men went back to their city, and each acquired followers after the manner of religious teachers. Their devotees would then argue with one another, the snake school of thought competing with adherents of the fan doctrine, the rope philosophy in conflict with the tree trunk faction, and so on. The only person who did not join in these debates was a seventh blind man, much older than the others, who had visited the elephant after the other six.

While the others rushed off with their separate conclusions, the seventh blind man had taken the time to pet the elephant, to walk all around it, to smell it, to feed it, and to listen to the sounds it made. When he returned to the city and found the populace in a state of uproar between the six factions, the old man laughed to himself: he was the only person in the city who was not convinced he knew exactly what an elephant was like.

UNDERSTANDING FRAME OF REFERENCE

The story of the blind men and the elephant, within the framework of Indian philosophy and spiritual beliefs, illustrates the principle of syadvada. This is a concept in the Jain religion related to the Sanskrit word *syat*, which means "may be." According to the doctrine of syadvada, no judgment is universal; it is merely a function of the circumstances in which the judgment is made.

On a complex level, syadvada is an illustration of relativity, a topic that will be discussed later; more immediately, however, both syadvada and the story of the blind men beautifully illustrate the ways that frame of reference affects perceptions. These are concerns of fundamental importance both in physics and philosophy, disciplines that once were closely allied until each became more fully defined and developed. Even in the modern era, long after the split between the two, each in its own way has been concerned with the relationship between subject and object.

These two terms, of course, have numerous definitions. Throughout this book, for instance, the word "object" is used in a very basic sense, meaning simply "a physical object" or "a thing." Here, however, an object may be defined as something that is perceived or observed. As soon as that definition is made, however, a flaw becomes apparent: nothing is just perceived or observed in and of itself—there has to be someone or something that actually perceives or observes. That something or someone is the subject, and the perspective from which the subject perceives or observes the object is the subject's frame of reference.

AMERICA AND CHINA: FRAME OF REFERENCE IN PRACTICE. An old joke—though not as old as the story of the blind men—goes something like this: "I'm glad I wasn't born in China, because I don't speak Chinese." Obviously, the humor revolves around the fact that if the speaker were born in China, then he or she would have grown up speaking Chinese, and English would be the foreign language.

The difference between being born in America and speaking English on the one hand—even if one is of Chinese descent—or of being born in China and speaking Chinese on the other, is not just a contrast of countries or languages. Rather, it is a difference of worlds—a difference, that is, in frame of reference.

Indeed, most people would see a huge distinction between an English-speaking American and a Chinese-speaking Chinese. Yet to a visitor from another planet—someone whose frame of reference would be, quite literally, otherworldly—the American and Chinese would have much more in common with each other than either would with the visitor.

THE VIEW FROM OUTSIDE AND INSIDE

Now imagine that the visitor from outer space (a handy example of someone with no preconceived ideas) were to land in the United States. If the visitor landed in New York City, Chicago, or Los Angeles, he or she would conclude that America is a very crowded, fast-paced country in which a number of ethnic groups live in close proximity. But if the visitor first arrived in Iowa or Nebraska, he or she might well decide that the United States is a sparsely populated land, economically dependent on agriculture and composed almost entirely of Caucasians.

A landing in San Francisco would create a falsely inflated impression regarding the number of Asian Americans or Americans of Pacific Island descent, who actually make up only a small portion of the national population. The same would be true if one first arrived in Arizona or New Mexico, where the Native American population is much higher than for the nation as a whole. There are numerous other examples to be made in the same vein, all relating to the visitors' impressions of the population, economy, climate, physical features, and other aspects of a specific place. Without consulting some outside reference point—say, an almanac or an atlas—it would be impossible to get an accurate picture of the entire country.

The principle is the same as that in the story of the blind men, but with an important distinction: an elephant is an example of an identifiable species, whereas the United States is a unique entity, not representative of some larger class of thing. (Perhaps the only nation remotely comparable is Brazil, also a vast land settled by outsiders and later populated by a number of groups.) Another important distinction between the blind men story and the United States example is the fact that the blind men were viewing the elephant from outside, whereas the visitor to America views it from inside. This in turn reflects a difference in frame of reference relevant to the work of a scientist: often it is possible to view a process, event, or phenomenon from outside; but sometimes one must view it from inside—which is more challenging.

FRAME OF REFERENCE IN SCIENCE

Philosophy (literally, "love of knowledge") is the most fundamental of all disciplines: hence, most persons who complete the work for a doctorate receive a "doctor of philosophy" (Ph.D.) degree. Among the sciences, physics—a direct offspring of philosophy, as noted earlier—is the most fundamental, and frame of reference is among its most basic concepts.

Hence, it is necessary to take a seemingly backward approach in explaining how frame of reference works, examining first the broad applications of the principle and then drawing upon its specific relation to physics. It makes little sense to discuss first the ways that physicists apply frame of reference, and only then to explain the concept in terms of everyday life. It is more meaningful to relate frame of reference first to familiar, or at least easily comprehensible, experiences—as has been done.

At this point, however, it is appropriate to discuss how the concept is applied to the sciences. People use frame of reference every day—indeed, virtually every moment—of their lives, without thinking about it. Rare indeed is the person who "walks a mile in another person's shoes"—that is, someone who tries to see events from the viewpoint of another. Physicists, on the other hand, have to be acutely aware of their frame of reference. Moreover, they must "rise above" their frame of reference in the sense that they have to take it into account in making calculations. For physicists in particular, and scientists in general, frame of reference has abundant "real-life applications."

LINES OF LONGITUDE ON EARTH ARE MEASURED AGAINST THE LINE PICTURED HERE: THE "PRIME MERIDIAN" RUNNING THROUGH GREENWICH, ENGLAND. AN IMAGINARY LINE DRAWN THROUGH THAT SPOT MARKS THE Y-AXIS FOR ALL VERTICAL COORDINATES ON EARTH, WITH A VALUE OF 0° ALONG THE X-AXIS, WHICH IS THE EQUATOR. THE PRIME MERIDIAN, HOWEVER, IS AN ARBITRARY STANDARD THAT DEPENDS ON ONE'S FRAME OF REFERENCE. *(Photograph by Dennis di Cicco/Corbis. Reproduced by permission.)*

REAL-LIFE APPLICATIONS

POINTS AND GRAPHS

There is no such thing as an absolute frame of reference—that is, a frame of reference that is fixed, and not dependent on anything else. If the entire universe consisted of just two points, it would be impossible (and indeed irrelevant) to say which was to the right of the other. There would be no right and left: in order to have such a distinction, it is necessary to have a third point from which to evaluate the other two points.

As long as there are just two points, there is only one dimension. The addition of a third point—as long as it does not lie along a straight line drawn through the first two points—creates two dimensions, length and width. From the frame of reference of any one point, then, it is possible to say which of the other two points is to the right.

Clearly, the judgment of right or left is relative, since it changes from point to point. A more absolute judgment (but still not a completely absolute one) would only be possible from the frame of reference of a fourth point. But to constitute a new dimension, that fourth point could not lie on the same plane as the other three points—more specifically, it should not be possible to create a single plane that encompasses all four points.

Assuming that condition is met, however, it then becomes easier to judge right and left. Yet right and left are never fully absolute, a fact easily illustrated by substituting people for points. One may look at two objects and judge which is to the right of the other, but if one stands on one's head, then of course right and left become reversed.

Of course, when someone is upside-down, the correct orientation of left and right is still fairly obvious. In certain situations observed by physicists and other scientists, however, orientation is not so simple. It then becomes necessary to assign values to various points, and for this, scientists use tools such as the Cartesian coordinate system.

COORDINATES AND AXES. Though it is named after the French mathematician and philosopher René Descartes (1596-1650), who first described its principles, the Cartesian system owes at least as much to Pierre de Fermat (1601-1665). Fermat, a brilliant French amateur mathematician—amateur in the sense that he was not trained in mathematics, nor did he earn a living from that discipline—greatly developed the Cartesian system.

A coordinate is a number or set of numbers used to specify the location of a point on a line, on a surface such as a plane, or in space. In the Cartesian system, the x-axis is the horizontal line of reference, and the y-axis the vertical line of reference. Hence, the coordinate (0, 0) designates the point where the x- and y-axes meet. All numbers to the right of 0 on the x-axis, and above 0 on the y-axis, have a positive value, while those to the left of 0 on the x-axis, or below 0 on the y-axis have a negative value.

This version of the Cartesian system only accounts for two dimensions, however; therefore, a z-axis, which constitutes a line of reference for the third dimension, is necessary in three-dimensional graphs. The z-axis, too, meets the x- and y-axes at (0, 0), only now that point is designated as (0, 0, 0).

In the two-dimensional Cartesian system, the x-axis equates to "width" and the y-axis to "height." The introduction of a z-axis adds the dimension of "depth"—though in fact, length, width, and height are all relative to the observer's frame of reference. (Most representations of the three-axis system set the x- and y-axes along a horizontal plane, with the z-axis perpendicular to them.) Basic studies in physics, however, typically involve only the x- and y-axes, essential to plotting graphs, which, in turn, are integral to illustrating the behavior of physical processes.

THE TRIPLE POINT. For instance, there is a phenomenon known as the "triple point," which is difficult to comprehend unless one sees it on a graph. For a chemical compound such as water or carbon dioxide, there is a point at which it is simultaneously a liquid, a solid, and a vapor. This, of course, seems to go against common sense, yet a graph makes it clear how this is possible.

Using the x-axis to measure temperature and the y-axis pressure, a number of surprises become apparent. For instance, most people associate water as a vapor (that is, steam) with very high temperatures. Yet water can also be a vapor—for example, the mist on a winter morning—at relatively low temperatures and pressures, as the graph shows.

The graph also shows that the higher the temperature of water vapor, the higher the pressure will be. This is represented by a line that curves upward to the right. Note that it is not a straight line along a 45° angle: up to about 68°F (20°C), temperature increases at a somewhat greater rate than pressure does, but as temperature gets higher, pressure increases dramatically.

As everyone knows, at relatively low temperatures water is a solid—ice. Pressure, however, is relatively high: thus on a graph, the values of temperatures and pressure for ice lie above the vaporization curve, but do not extend to the right of 32°F (0°C) along the x-axis. To the right of 32°F, but above the vaporization curve, are the coordinates representing the temperature and pressure for water in its liquid state.

Water has a number of unusual properties, one of which is its response to high pressures and low temperatures. If enough pressure is applied, it is possible to melt ice—thus transforming it from a solid to a liquid—at temperatures below the normal freezing point of 32°F. Thus, the line that divides solid on the left from liquid on the right is not exactly parallel to the y-axis: it slopes gradually toward the y-axis, meaning that at ultra-high pressures, water remains liquid even though it is well below the freezing point.

Nonetheless, the line between solid and liquid has to intersect the vaporization curve somewhere, and it does—at a coordinate slightly above freezing, but well below normal atmospheric pressure. This is the triple point, and though "common sense" might dictate that a thing cannot possibly be solid, liquid, and vapor all at once, a graph illustrating the triple point makes it clear how this can happen.

NUMBERS

In the above discussion—and indeed throughout this book—the existence of the decimal, or base-

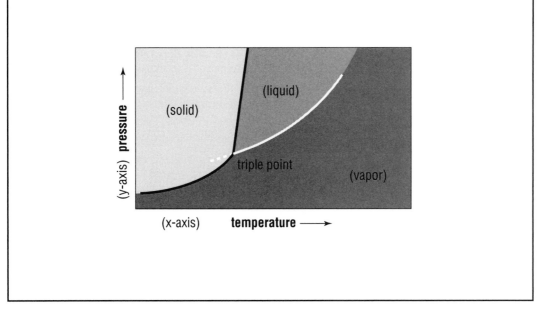

THIS CARTESIAN COORDINATE GRAPH SHOWS HOW A SUBSTANCE SUCH AS WATER COULD EXPERIENCE A TRIPLE POINT—A POINT AT WHICH IT IS SIMULTANEOUSLY A LIQUID, A SOLID, AND A VAPOR.

10, numeration system is taken for granted. Yet that system is a wonder unto itself, involving a complicated interplay of arbitrary and real values. Though the value of the number 10 is absolute, the expression of it (and its use with other numbers) is relative to a frame of reference: one could just as easily use a base-12 system.

Each numeration system has its own frame of reference, which is typically related to aspects of the human body. Thus throughout the course of history, some societies have developed a base-2 system based on the two hands or arms of a person. Others have used the fingers on one hand (base-5) as their reference point, or all the fingers and toes (base-20). The system in use throughout most of the world today takes as its frame of reference the ten fingers used for basic counting.

COEFFICIENTS. Numbers, of course, provide a means of assigning relative values to a variety of physical characteristics: length, mass, force, density, volume, electrical charge, and so on. In an expression such as "10 meters," the numeral 10 is a coefficient, a number that serves as a measure for some characteristic or property. A coefficient may also be a factor against which other values are multiplied to provide a desired result.

For instance, the figure 3.141592, better known as pi (π), is a well-known coefficient used in formulae for measuring the circumference or area of a circle. Important examples of coefficients in physics include those for static and sliding friction for any two given materials. A coefficient is simply a number—not a value, as would be the case if the coefficient were a measure of something.

STANDARDS OF MEASUREMENT

Numbers and coefficients provide a convenient lead-in to the subject of measurement, a practical example of frame of reference in all sciences—and indeed, in daily life. Measurement always requires a standard of comparison: something that is fixed, against which the value of other things can be compared. A standard may be arbitrary in its origins, but once it becomes fixed, it provides a frame of reference.

Lines of longitude, for instance, are measured against an arbitrary standard: the "Prime Meridian" running through Greenwich, England. An imaginary line drawn through that spot marks the line of reference for all longitudinal measures on Earth, with a value of 0°. There is nothing special about Greenwich in any profound scientific sense; rather, its place of importance reflects that of England itself, which ruled the seas and indeed much of the world at the time the Prime Meridian was established.

The Equator, on the other hand, has a firm scientific basis as the standard against which all lines of latitude are measured. Yet today, the coordinates of a spot on Earth's surface are given in relation to both the Equator and the Prime Meridian.

CALIBRATION. Calibration is the process of checking and correcting the performance of a measuring instrument or device against the accepted standard. America's preeminent standard for the exact time of day, for instance, is the United States Naval Observatory in Washington, D.C. Thanks to the Internet, people all over the country can easily check the exact time, and correct their clocks accordingly.

There are independent scientific laboratories responsible for the calibration of certain instruments ranging from clocks to torque wrenches, and from thermometers to laser beam power analyzers. In the United States, instruments or devices with high-precision applications—that is, those used in scientific studies, or by high-tech industries—are calibrated according to standards established by the National Institute of Standards and Technology (NIST).

THE VALUE OF STANDARD-IZATION TO A SOCIETY. Standardization of weights and measures has always been an important function of government. When Ch'in Shih-huang-ti (259-210 B.C.) united China for the first time, becoming its first emperor, he set about standardizing units of measure as a means of providing greater unity to the country—thus making it easier to rule.

More than 2,000 years later, another empire—Russia—was negatively affected by its failure to adjust to the standards of technologically advanced nations. The time was the early twentieth century, when Western Europe was moving forward at a rapid pace of industrialization. Russia, by contrast, lagged behind—in part because its failure to adopt Western standards put it at a disadvantage.

Train travel between the West and Russia was highly problematic, because the width of railroad tracks in Russia was different than in Western Europe. Thus, adjustments had to be performed on trains making a border crossing, and this created difficulties for passenger travel.

More importantly, it increased the cost of transporting freight from East to West.

Russia also used the old Julian calendar, as opposed to the Gregorian calendar adopted throughout much of Western Europe after 1582. Thus October 25, 1917, in the Julian calendar of old Russia translated to November 7, 1917 in the Gregorian calendar used in the West. That date was not chosen arbitrarily: it was then that Communists, led by V. I. Lenin, seized power in the weakened former Russian Empire.

METHODS OF DETERMINING STANDARDS. It is easy to understand, then, why governments want to standardize weights and measures—as the U.S. Congress did in 1901, when it established the Bureau of Standards (now NIST) as a nonregulatory agency within the Commerce Department. Today, NIST maintains a wide variety of standard definitions regarding mass, length, temperature, and so forth, against which other devices can be calibrated.

Note that NIST keeps on hand definitions rather than, say, a meter stick or other physical model. When the French government established the metric system in 1799, it calibrated the value of a kilogram according to what is now known as the International Prototype Kilogram, a platinum-iridium cylinder housed near Sèvres in France. In the years since then, the trend has moved away from such physical expressions of standards, and toward standards based on a constant figure. Hence, the meter is defined as the distance light travels in a vacuum (an area of space devoid of air or other matter) during the interval of 1/299,792,458 of a second.

METRIC VS. BRITISH. Scientists almost always use the metric system, not because it is necessarily any less arbitrary than the British or English system (pounds, feet, and so on), but because it is easier to use. So universal is the metric system within the scientific community that it is typically referred to simply as SI, an abbreviation of the French *Système International d'Unités*—that is, "International System of Units."

The British system lacks any clear frame of reference for organizing units: there are 12 inches in a foot, but 3 feet in a yard, and 1,760 yards in a mile. Water freezes at 32°F instead of 0°, as it does in the Celsius scale associated with the metric system. In contrast to the English system, the metric system is neatly arranged according to the base-10 numerical framework: 10 millimeters to a centimeter, 100 centimeters to a meter, 1,000 meters to kilometer, and so on.

The difference between the pound and the kilogram aptly illustrates the reason scientists in general, and physicists in particular, prefer the metric system. A pound is a unit of weight, meaning that its value is entirely relative to the gravitational pull of the planet on which it is measured. A kilogram, on the other hand, is a unit of mass, and does not change throughout the universe. Though the basis for a kilogram may not ultimately be any more fundamental than that for a pound, it measures a quality that—unlike weight—does not vary according to frame of reference.

FRAME OF REFERENCE IN CLASSICAL PHYSICS AND ASTRONOMY

Mass is a measure of inertia, the tendency of a body to maintain constant velocity. If an object is at rest, it tends to remain at rest, or if in motion, it tends to remain in motion unless acted upon by some outside force. This, as identified by the first law of motion, is inertia—and the greater the inertia, the greater the mass.

Physicists sometimes speak of an "inertial frame of reference," or one that has a constant velocity—that is, an unchanging speed and direction. Imagine if one were on a moving bus at constant velocity, regularly tossing a ball in the air and catching it. It would be no more difficult to catch the ball than if the bus were standing still, and indeed, there would be no way of determining, simply from the motion of the ball itself, that the bus was moving.

But what if the inertial frame of reference suddenly became a non-inertial frame of reference—in other words, what if the bus slammed on its brakes, thus changing its velocity? While the bus was moving forward, the ball was moving along with it, and hence, there was no relative motion between them. By stopping, the bus responded to an "outside" force—that is, its brakes. The ball, on the other hand, experienced that force indirectly. Hence, it would continue to move forward as before, in accordance with its own inertia—only now it would be in motion relative to the bus.

ASTRONOMY AND RELATIVE MOTION. The idea of relative motion plays a powerful role in astronomy. At every moment, Earth is turning on its axis at about 1,000 MPH (1,600 km/h) and hurtling along its orbital path around the Sun at the rate of 67,000 MPH (107,826 km/h.) The fastest any human being—that is, the astronauts taking part in the Apollo missions during the late 1960s—has traveled is about 30% of Earth's speed around the Sun.

Yet no one senses the speed of Earth's movement in the way that one senses the movement of a car—or indeed the way the astronauts perceived their speed, which was relative to the Moon and Earth. Of course, everyone experiences the results of Earth's movement—the change from night to day, the precession of the seasons—but no one experiences it directly. It is simply impossible, from the human frame of reference, to feel the movement of a body as large as Earth—not to mention larger progressions on the part of the Solar System and the universe.

FROM ASTRONOMY TO PHYSICS. The human body is in an inertial frame of reference with regard to Earth, and hence experiences no relative motion when Earth rotates or moves through space. In the same way, if one were traveling in a train alongside another train at constant velocity, it would be impossible to perceive that either train was actually moving—unless one referred to some fixed point, such as the trees or mountains in the background. Likewise, if two trains were sitting side by side, and one of them started to move, the relative motion might cause a person in the stationary train to believe that his or her train was the one moving.

For any measurement of velocity, and hence, of acceleration (a change in velocity), it is essential to establish a frame of reference. Velocity and acceleration, as well as inertia and mass, figured heavily in the work of Galileo Galilei (1564-1642) and Sir Isaac Newton (1642-1727), both of whom may be regarded as "founding fathers" of modern physics. Before Galileo, however, had come Nicholas Copernicus (1473-1543), the first modern astronomer to show that the Sun, and not Earth, is at the center of "the universe"—by which people of that time meant the Solar System.

In effect, Copernicus was saying that the frame of reference used by astronomers for millennia was incorrect: as long as they believed Earth to be the center, their calculations were bound to be wrong. Galileo and later Newton, through their studies in gravitation, were able to prove Copernicus's claim in terms of physics.

At the same time, without the understanding of a heliocentric (Sun-centered) universe that he inherited from Copernicus, it is doubtful that Newton could have developed his universal law of gravitation. If he had used Earth as the center-point for his calculations, the results would have been highly erratic, and no universal law would have emerged.

RELATIVITY

For centuries, the model of the universe developed by Newton stood unchallenged, and even today it identifies the basic forces at work when speeds are well below that of the speed of light. However, with regard to the behavior of light itself—which travels at 186,000 mi (299,339 km) a second—Albert Einstein (1879-1955) began to observe phenomena that did not fit with Newtonian mechanics. The result of his studies was the Special Theory of Relativity, published in 1905, and the General Theory of Relativity, published a decade later. Together these altered humanity's view of the universe, and ultimately, of reality itself.

Einstein himself once offered this charming explanation of his epochal theory: "Put your hand on a hot stove for a minute, and it seems like an hour. Sit with a pretty girl for an hour, and it seems like a minute. That's relativity." Of course, relativity is not quite as simple as that—though the mathematics involved is no more challenging than that of a high-school algebra class. The difficulty lies in comprehending how things that seem impossible in the Newtonian universe become realities near the speed of light.

PLAYING TRICKS WITH TIME. An exhaustive explanation of relativity is far beyond the scope of the present discussion. What is important is the central precept: that no measurement of space or time is absolute, but depends on the relative motion of the observer (that is, the subject) and the observed (the object). Einstein further established that the movement of time itself is relative rather than absolute, a fact that would become apparent at speeds close to that of light. (His theory also showed that it is impossible to surpass that speed.)

Imagine traveling on a spaceship at nearly the speed of light while a friend remains station-ary on Earth. Both on the spaceship and at the friend's house on Earth, there is a TV camera trained on a clock, and a signal relays the image from space to a TV monitor on Earth, and vice versa. What the TV monitor reveals is surprising: from your frame of reference on the spaceship, it seems that time is moving more slowly for your friend on Earth than for you. Your friend thinks exactly the same thing—only, from the friend's perspective, time on the spaceship is moving more slowly than time on Earth. How can this happen?

Again, a full explanation—requiring reference to formulae regarding time dilation, and so on—would be a rather involved undertaking. The short answer, however, is that which was stated above: no measurement of space or time is absolute, but each depends on the relative motion of the observer and the observed. Put another way, there is no such thing as absolute motion, either in the three dimensions of space, or in the fourth dimension identified by Einstein, time. All motion is relative to a frame of reference.

RELATIVITY AND ITS IMPLICATIONS. The ideas involved in relativity have been verified numerous times, and indeed the only reason why they seem so utterly foreign to most people is that humans are accustomed to living within the Newtonian framework. Einstein simply showed that there is no universal frame of reference, and like a true scientist, he drew his conclusions entirely from what the data suggested. He did not form an opinion, and only then seek the evidence to confirm it, nor did he seek to extend the laws of relativity into any realm beyond that which they described.

Yet British historian Paul Johnson, in his unorthodox history of the twentieth century, *Modern Times* (1983; revised 1992), maintained that a world disillusioned by World War I saw a moral dimension to relativity. Describing a set of tests regarding the behavior of the Sun's rays around the planet Mercury during an eclipse, the book begins with the sentence: "The modern world began on 29 May 1919, when photographs of a solar eclipse, taken on the Island of Principe off West Africa and at Sobral in Brazil, confirmed the truth of a new theory of the universe."

As Johnson went on to note, "...for most people, to whom Newtonian physics... were perfectly

KEY TERMS

ABSOLUTE: Fixed; not dependent on anything else. The value of 10 is absolute, relating to unchanging numerical principles; on the other hand, the value of 10 dollars is relative, reflecting the economy, inflation, buying power, exchange rates with other currencies, etc.

CALIBRATION: The process of checking and correcting the performance of a measuring instrument or device against a commonly accepted standard.

CARTESIAN COORDINATE SYSTEM: A method of specifying coordinates in relation to an x-axis, y-axis, and z-axis. The system is named after the French mathematician and philosopher René Descartes (1596-1650), who first described its principles, but it was developed greatly by French mathematician and philosopher Pierre de Fermat (1601-1665).

COEFFICIENT: A number that serves as a measure for some characteristic or property. A coefficient may also be a factor against which other values are multiplied to provide a desired result.

COORDINATE: A number or set of numbers used to specify the location of a point on a line, on a surface such as a plane, or in space.

FRAME OF REFERENCE: The perspective of a subject in observing an object.

OBJECT: Something that is perceived or observed by a subject.

RELATIVE: Dependent on something else for its value or for other identifying qualities. The fact that the United States has a constitution is an absolute, but the fact that it was ratified in 1787 is relative: that date has meaning only within the Western calendar.

SUBJECT: Something (usually a person) that perceives or observes an object and/or its behavior.

X-AXIS: The horizontal line of reference for points in the Cartesian coordinate system.

Y-AXIS: The vertical line of reference for points in the Cartesian coordinate system.

Z-AXIS: In a three-dimensional version of the Cartesian coordinate system, the z-axis is the line of reference for points in the third dimension. Typically the x-axis equates to "width," the y-axis to "height," and the z-axis to "depth"—though in fact length, width, and height are all relative to the observer's frame of reference.

comprehensible, relativity never became more than a vague source of unease. It was grasped that absolute time and absolute length had been dethroned.... All at once, nothing seemed certain in the spheres....At the beginning of the 1920s the belief began to circulate, for the first time at a popular level, that there were no longer any absolutes: of time and space, of good and evil, of knowledge, above all of value. Mistakenly but perhaps inevitably, relativity became confused with relativism."

Certainly many people agree that the twentieth century—an age that saw unprecedented mass murder under the dictatorships of Adolf Hitler and Josef Stalin, among others—was characterized by moral relativism, or the belief that there is no right or wrong. And just as Newton's discoveries helped usher in the Age of Reason, when thinkers believed it was possible to solve any problem through intellectual effort, it is quite plausible that Einstein's theory may have had this negative moral effect.

If so, this was certainly not Einstein's intention. Aside from the fact that, as stated, he did not set out to describe anything other than the physical behavior of objects, he continued to believe that there was no conflict between his ideas and a belief in an ordered universe: "Relativity," he once said, "teaches us the connection between the different descriptions of one and the same reality."

WHERE TO LEARN MORE

Beiser, Arthur. *Physics,* 5th ed. Reading, MA: Addison-Wesley, 1991.

Fleisher, Paul. *Relativity and Quantum Mechanics: Principles of Modern Physics.* Minneapolis, MN: Lerner Publications, 2002.

"Frame of Reference" (Web site). <http://www.physics.reading.ac.uk/units/flap/glossary/ff/frameref.html> (March 21, 2001).

"Inertial Frame of Reference" (Web site). <http://id.mind.net/~zona/mstm/physics/mechanics/framesOfReference /inertialFrame.html> (March 21, 2001).

Johnson, Paul. *Modern Times: The World from the Twenties to the Nineties.* Revised edition. New York: HarperPerennial, 1992.

King, Andrew. *Plotting Points and Position.* Illustrated by Tony Kenyon. Brookfield, CT: Copper Beech Books, 1998.

Parker, Steve. *Albert Einstein and Relativity.* New York: Chelsea House, 1995.

Robson, Pam. *Clocks, Scales, and Measurements.* New York: Gloucester Press, 1993.

Rutherford, F. James; Gerald Holton; and Fletcher G. Watson. *Project Physics.* New York: Holt, Rinehart, and Winston, 1981.

Swisher, Clarice. *Relativity: Opposing Viewpoints.* San Diego, CA: Greenhaven Press, 1990.

KINEMATICS AND DYNAMICS

CONCEPT

Webster's defines physics as "a science that deals with matter and energy and their interactions." Alternatively, physics can be described as the study of matter and motion, or of matter inmotion. Whatever the particulars of the definition, physics is among the most fundamental of disciplines, and hence, the rudiments of physics are among the most basic building blocks for thinking about the world. Foundational to an understanding of physics are kinematics, the explanation of how objects move, and dynamics, the study of why they move. Both are part of a larger branch of physics called mechanics, the study of bodies in motion. These are subjects that may sound abstract, but in fact, are limitless in their applications to real life.

HOW IT WORKS

THE PLACE OF PHYSICS IN THE SCIENCES

Physics may be regarded as the queen of the sciences, not because it is "better" than chemistry or astronomy, but because it is the foundation on which all others are built. The internal and interpersonal behaviors that are the subject of the social sciences (psychology, anthropology, sociology, and so forth) could not exist without the biological framework that houses the human consciousness. Yet the human body and other elements studied by the biological and medical sciences exist within a larger environment, the framework for earth sciences, such as geology.

Earth sciences belong to a larger grouping of physical sciences, each more fundamental in concerns and broader in scope. Earth, after all, is but one corner of the realm studied by astronomy; and before a universe can even exist, there must be interactions of elements, the subject of chemistry. Yet even before chemicals can react, they have to do so within a physical framework—the realm of the most basic science—physics.

THE BIRTH OF PHYSICS IN GREECE

THE FIRST HYPOTHESIS. Indeed, physics stands in relation to the sciences as philosophy does to thought itself: without philosophy to provide the concept of concepts, it would be impossible to develop a consistent worldview in which to test ideas. It is no accident, then, that the founder of the physical sciences was also the world's first philosopher, Thales (c. 625?-547? B.C.) of Miletus in Greek Asia Minor (now part of Turkey.) Prior to Thales's time, religious figures and mystics had made statements regarding ethics or the nature of deity, but none had attempted statements concerning the fundamental nature of reality.

For instance, the Bible offers a story of Earth's creation in the Book of Genesis which was well-suited to the understanding of people in the first millennium before Christ. But the writer of the biblical creation story made no attempt to explain how things came into being. He was concerned, rather, with showing that God had willed the existence of all physical reality by calling things into being—for example, by saying, "Let there be light."

Thales, on the other hand, made a genuine philosophical and scientific statement when he said that "Everything is water." This was the first hypothesis, a statement capable of being scientifically tested for accuracy. Thales's pronouncement did not mean he believed all things were necessarily made of water, literally. Rather, he appears to have been referring to a general tendency of movement: that the whole world is in a fluid state.

ATTEMPTING TO UNDERSTAND PHYSICAL REALITY.

While we can respect Thales's statement for its truly earth-shattering implications, we may be tempted to read too much into it. Nonetheless, it is striking that he compared physical reality to water. On the one hand, there is the fact that water is essential to all life, and pervades Earth—but that is a subject more properly addressed by the realms of chemistry and the biological sciences. Perhaps of more interest to the physicist is the allusion to a fluid nature underlying all physical reality.

The physical realm is made of matter, which appears in four states: solid, liquid, gas, and plasma. The last of these is not the same as blood plasma: containing many ionized atoms or molecules which exhibit collective behavior, plasma is the substance from which stars, for instance, are composed. Though not plentiful on Earth, within the universe it may be the most common of all four states. Plasma is akin to gas, but different in molecular structure; the other three states differ at the molecular level as well.

Nonetheless, it is possible for a substance such as water—genuine H_2O, not the figurative water of Thales—to exist in liquid, gas, or solid form, and the dividing line between these is not always fixed. In fact, physicists have identified a phenomenon known as the triple point: at a certain temperature and pressure, a substance can be solid, liquid, and gas all at once!

The above statement shows just how challenging the study of physical reality can be, and indeed, these concepts would be far beyond the scope of Thales's imagination, had he been presented with them. Though he almost certainly deserves to be called a "genius," he lived in a world that viewed physical processes as a product of the gods' sometimes capricious will. The behavior of the tides, for instance, was attributed to Poseidon. Though Thales's statement began the process of digging humanity out from under the burden of superstition that had impeded scientific progress for centuries, the road forward would be a long one.

MATHEMATICS, MEASUREMENT, AND MATTER.

In the two centuries after Thales's death, several other thinkers advanced understanding of physical reality in one way or another. Pythagoras (c. 580-c. 500 B.C.) taught that everything could be quantified, or related to numbers. Though he entangled this idea with mysticism and numerology, the concept itself influenced the idea that physical processes could be measured. Likewise, there were flaws at the heart of the paradoxes put forth by Zeno of Elea (c. 495-c. 430 B.C.), who set out to prove that motion was impossible—yet he was also the first thinker to analyze motion seriously.

In one of Zeno's paradoxes, he referred to an arrow being shot from a bow. At every moment of its flight, it could be said that the arrow was at rest within a space equal to its length. Though it would be some 2,500 years before slow-motion photography, in effect he was asking his listeners to imagine a snapshot of the arrow in flight. If it was at rest in that "snapshot," he asked, so to speak, and in every other possible "snapshot," when did the arrow actually move? These paradoxes were among the most perplexing questions of premodern times, and remain a subject of inquiry even today.

In fact, it seems that Zeno unwittingly (for there is no reason to believe that he deliberately deceived his listeners) inserted an error in his paradoxes by treating physical space as though it were composed of an infinite number of points. In the ideal world of geometric theory, a point takes up no space, and therefore it is correct to say that a line contains an infinite number of points; but this is not the case in the real world, where a "point" has some actual length. Hence, if the number of points on Earth were limitless, so too would be Earth itself.

Zeno's contemporary Leucippus (c. 480-c. 420 B.C.) and his student Democritus (c. 460-370 B.C.) proposed a new and highly advanced model for the tiniest point of physical space: the atom. It would be some 2,300 years, however, before physicists returned to the atomic model.

ARISTOTLE'S FLAWED PHYSICS

The study of matter and motion began to take shape with Aristotle (384-322 B.C.); yet, though his *Physics* helped establish a framework for the discipline, his errors are so profound that any praise must be qualified. Certainly, Aristotle was one of the world's greatest thinkers, who originated a set of formalized realms of study. However, in *Physics* he put forth an erroneous explanation of matter and motion that still prevailed in Europe twenty centuries later.

Actually, Aristotle's ideas disappeared in the late ancient period, as learning in general came to a virtual halt in Europe. That his writings—which on the whole did much more to advance the progress of science than to impede it—survived at all is a tribute to the brilliance of Arab, rather than European, civilization. Indeed, it was in the Arab world that the most important scientific work of the medieval period took place. Only after about 1200 did Aristotelian thinking once again enter Europe, where it replaced a crude jumble of superstitions that had been substituted for learning.

THE FOUR ELEMENTS. According to Aristotelian physics, all objects consisted, in varying degrees, of one or more elements: air, fire, water, and earth. In a tradition that went back to Thales, these elements were not necessarily pure: water in the everyday world was composed primarily of the element water, but also contained smaller amounts of the other elements. The planets beyond Earth were said to be made up of a "fifth element," or quintessence, of which little could be known.

The differing weights and behaviors of the elements governed the behavior of physical objects. Thus, water was lighter than earth, for instance, but heavier than air or fire. It was due to this difference in weight, Aristotle reasoned, that certain objects fall faster than others: a stone, for instance, because it is composed primarily of earth, will fall much faster than a leaf, which has much less earth in it.

Aristotle further defined "natural" motion as that which moved an object toward the center of the Earth, and "violent" motion as anything that propelled an object toward anything other than its "natural" destination. Hence, all horizontal or upward motion was "violent," and must be the direct result of a force. When the force was removed, the movement would end.

ARISTOTLE'S MODEL OF THE UNIVERSE. From the fact that Earth's center is the destination of all "natural" motion, it is easy to comprehend the Aristotelian cosmology, or model of the universe. Earth itself was in the center, with all other bodies (including the Sun)

ARISTOTLE. *(The Bettmann Archive. Reproduced by permission.)*

revolving around it. Though in constant movement, these heavenly bodies were always in their "natural" place, because they could only move on the firmly established—almost groove-like—paths of their orbits around Earth. This in turn meant that the physical properties of matter and motion on other planets were completely different from the laws that prevailed on Earth.

Of course, virtually every precept within the Aristotelian system is incorrect, and Aristotle compounded the influence of his errors by promoting a disdain for quantification. Specifically, he believed that mathematics had little value for describing physical processes in the real world, and relied instead on pure observation without attempts at measurement.

MOVING BEYOND ARISTOTLE

Faulty as Aristotle's system was, however, it possessed great appeal because much of it seemed to fit with the evidence of the senses. It is not at all immediately apparent that Earth and the other planets revolve around the Sun, nor is it obvious that a stone and a leaf experience the same acceleration as they fall toward the ground. In fact, quite the opposite appears to be the case: as everyone knows, a stone falls faster than a leaf. Therefore, it would seem reasonable—on the

GALILEO. *(Archive Photos, Inc. Reproduced by permission.)*

surface of it, at least—to accept Aristotle's conclusion that this difference results purely from a difference in weight.

Today, of course, scientists—and indeed, even people without any specialized scientific knowledge—recognize the lack of merit in the Aristotelian system. The stone does fall faster than the leaf, but only because of air resistance, not weight. Hence, if they fell in a vacuum (a space otherwise entirely devoid of matter, including air), the two objects would fall at exactly the same rate.

As with a number of truths about matter and motion, this is not one that appears obvious, yet it has been demonstrated. To prove this highly nonintuitive hypothesis, however, required an approach quite different from Aristotle's—an approach that involved quantification and the separation of matter and motion into various components. This was the beginning of real progress in physics, and in a sense may be regarded as the true birth of the discipline. In the years that followed, understanding of physics would grow rapidly, thanks to advancements of many individuals; but their studies could not have been possible without the work of one extraordinary thinker who dared to question the Aristotelian model.

REAL-LIFE APPLICATIONS

KINEMATICS: HOW OBJECTS MOVE

By the sixteenth century, the Aristotelian worldview had become so deeply ingrained that few European thinkers would have considered the possibility that it could be challenged. Professors all over Europe taught Aristotle's precepts to their students, and in this regard the University of Pisa in Italy was no different. Yet from its classrooms would emerge a young man who not only questioned, but ultimately overturned the Aristotelian model: Galileo Galilei (1564-1642.)

Challenges to Aristotle had been slowly growing within the scientific communities of the Arab and later the European worlds during the preceding millennium. Yet the ideas that most influenced Galileo in his break with Aristotle came not from a physicist but from an astronomer, Nicolaus Copernicus (1473-1543.) It was Copernicus who made a case, based purely on astronomical observation, that the Sun and not Earth was at the center of the universe.

Galileo embraced this model of the cosmos, but was later forced to renounce it on orders from the pope in Rome. At that time, of course, the Catholic Church remained the single most powerful political entity in Europe, and its endorsement of Aristotelian views—which philosophers had long since reconciled with Christian ideas—is a measure of Aristotle's impact on thinking.

GALILEO'S REVOLUTION IN PHYSICS. After his censure by the Church, Galileo was placed under house arrest and was forbidden to study astronomy. Instead he turned to physics—where, ironically, he struck the blow that would destroy the bankrupt scientific system endorsed by Rome. In 1638, he published *Discourses and Mathematical Demonstrations Concerning Two New Sciences Pertaining to Mathematics and Local Motion,* a work usually referred to as *Two New Sciences.* In it, he laid the groundwork for physics by emphasizing a new method that included experimentation, demonstration, and quantification of results.

In this book—highly readable for a work of physics written in the seventeenth century—Galileo used a dialogue, an established format among philosophers and scientists of the past.

The character of Salviati argued for Galileo's ideas and Simplicio for those of Aristotle, while the genial Sagredo sat by and made occasional comments. Through Salviati, Galileo chose to challenge Aristotle on an issue that to most people at the time seemed relatively settled: the claim that objects fall at differing speeds according to their weight.

In order to proceed with his aim, Galileo had to introduce a number of innovations, and indeed, he established the subdiscipline of kinematics, or how objects move. Aristotle had indicated that when objects fall, they fall at the same rate from the moment they begin to fall until they reach their "natural" position. Galileo, on the other hand, suggested an aspect of motion, unknown at the time, that became an integral part of studies in physics: acceleration.

SCALARS AND VECTORS

Even today, many people remain confused as to what acceleration is. Most assume that acceleration means only an increase in speed, but in fact this represents only one of several examples of acceleration. Acceleration is directly related to velocity, often mistakenly identified with speed.

In fact, speed is what scientists today would call a scalar quantity, or one that possesses magnitude but no specific direction. Speed is the rate at which the position of an object changes over a given period of time; thus people say "miles (or kilometers) per hour." A story problem concerning speed might state that "A train leaves New York City at a rate of 60 miles (96.6 km/h). How far will it have traveled in 73 minutes?"

Note that there is no reference to direction, whereas if the story problem concerned velocity—a vector, that is, a quantity involving both magnitude and direction—it would include some crucial qualifying phrase after "New York City": "for Boston," perhaps, or "northward." In practice, the difference between speed and velocity is nearly as large as that between a math problem and real life: few people think in terms of driving 60 miles, for instance, without also considering the direction they are traveling.

RESULTANTS. One can apply the same formula with velocity, though the process is more complicated. To obtain change in distance, one must add vectors, and this is best done by means of a diagram. You can draw each vector as an arrow on a graph, with the tail of each vector at the head of the previous one. Then it is possible to draw a vector from the tail of the first to the head of the last. This is the sum of the vectors, known as a resultant, which measures the net change.

Suppose, for instance, that a car travels east 4 mi (6.44 km), then due north 3 mi (4.83 km). This may be drawn on a graph with four units along the x axis, then 3 units along the y axis, making two sides of a triangle. The number of sides to the resulting shape is always one more than the number of vectors being added; the final side is the resultant. From the tail of the first segment, a diagonal line drawn to the head of the last will yield a measurement of 5 units—the resultant, which in this case would be equal to 5 mi (8 km) in a northeasterly direction.

VELOCITY AND ACCELERATION. The directional component of velocity makes it possible to consider forms of motion other than linear, or straight-line, movement. Principal among these is circular, or rotational motion, in which an object continually changes direction and thus, velocity. Also significant is projectile motion, in which an object is thrown, shot, or hurled, describing a path that is a combination of horizontal and vertical components.

Furthermore, velocity is a key component in acceleration, which is defined as a change in velocity. Hence, acceleration can mean one of five things: an increase in speed with no change in direction (the popular, but incorrect, definition of the overall concept); a decrease in speed with no change in direction; a decrease or increase of speed with a change in direction; or a change in direction with no change in speed. If a car speeds up or slows down while traveling in a straight line, it experiences acceleration. So too does an object moving in rotational motion, even if its speed does not change, because its direction will change continuously.

DYNAMICS: WHY OBJECTS MOVE

GALILEO'S TEST. To return to Galileo, he was concerned primarily with a specific form of acceleration, that which occurs due to the force of gravity. Aristotle had provided an explanation of gravity—if a highly flawed one—with his claim that objects fall to their "natural" position; Galileo set out to develop the first truly scientific explanation concerning how objects fall to the ground.

According to Galileo's predictions, two metal balls of differing sizes would fall with the same rate of acceleration. To test his hypotheses, however, he could not simply drop two balls from a rooftop—or have someone else do so while he stood on the ground—and measure their rate of fall. Objects fall too fast, and lacking sophisticated equipment available to scientists today, he had to find another means of showing the rate at which they fell.

This he did by resorting to a method Aristotle had shunned: the use of mathematics as a means of modeling the behavior of objects. This is such a deeply ingrained aspect of science today that it is hard to imagine a time when anyone would have questioned it, and that very fact is a tribute to Galileo's achievement. Since he could not measure speed, he set out to find an equation relating total distance to total time. Through a detailed series of steps, Galileo discovered that in uniform or constant acceleration from rest—that is, the acceleration he believed an object experiences due to gravity—there is a proportional relationship between distance and time.

With this mathematical model, Galileo could demonstrate uniform acceleration. He did this by using an experimental model for which observation was easier than in the case of two falling bodies: an inclined plane, down which he rolled a perfectly round ball. This allowed him to extrapolate that in free fall, though velocity was greater, the same proportions still applied and therefore, acceleration was constant.

POINTING THE WAY TOWARD NEWTON. The effects of Galileo's system were enormous: he demonstrated mathematically that acceleration is constant, and established a method of hypothesis and experiment that became the basis of subsequent scientific investigation. He did not, however, attempt to calculate a figure for the acceleration of bodies in free fall; nor did he attempt to explain the overall principle of gravity, or indeed why objects move as they do—the focus of a subdiscipline known as dynamics.

At the end of *Two New Sciences,* Sagredo offered a hopeful prediction: "I really believe that... the principles which are set forth in this little treatise will, when taken up by speculative minds, lead to another more remarkable result...." This prediction would come true with the work of a man who, because he lived in a somewhat more enlightened time—and because

he lived in England, where the pope had no power—was free to explore the implications of his physical studies without fear of Rome's intervention. Born in the very year Galileo died, his name was Sir Isaac Newton (1642-1727.)

NEWTON'S THREE LAWS OF MOTION. In discussing the movement of the planets, Galileo had coined the term inertia to describe the tendency of an object in motion to remain in motion, and an object at rest to remain at rest. This idea would be the starting point of Newton's three laws of motion, and Newton would greatly expand on the concept of inertia.

The three laws themselves are so significant to the understanding of physics that they are treated separately elsewhere in this volume; here they are considered primarily in terms of their implications regarding the larger topic of matter and motion.

Introduced by Newton in his *Principia* (1687), the three laws are:

- First law of motion: An object at rest will remain at rest, and an object in motion will remain in motion, at a constant velocity unless or until outside forces act upon it.
- Second law of motion: The net force acting upon an object is a product of its mass multiplied by its acceleration.
- Third law of motion: When one object exerts a force on another, the second object exerts on the first a force equal in magnitude but opposite in direction.

These laws made final the break with Aristotle's system. In place of "natural" motion, Newton presented the concept of motion at a uniform velocity—whether that velocity be a state of rest or of uniform motion. Indeed, the closest thing to "natural" motion (that is, true "natural" motion) is the behavior of objects in outer space. There, free from friction and away from the gravitational pull of Earth or other bodies, an object set in motion will remain in motion forever due to its own inertia. It follows from this observation, incidentally, that Newton's laws were and are universal, thus debunking the old myth that the physical properties of realms outside Earth are fundamentally different from those of Earth itself.

MASS AND GRAVITATIONAL ACCELERATION. The first law establishes the principle of inertia, and the second law makes reference to the means by which inertia is measured: mass, or the resistance of an object to a

KEY TERMS

ACCELERATION: A change in velocity.

DYNAMICS: The study of why objects move as they do; compare with kinematics.

FORCE: The product of mass multiplied by acceleration.

HYPOTHESIS: A statement capable of being scientifically tested for accuracy.

INERTIA: The tendency of an object in motion to remain in motion, and of an object at rest to remain at rest.

KINEMATICS: The study of how objects move; compare with dynamics.

MASS: A measure of inertia, indicating the resistance of an object to a change in its motion—including a change in velocity.

MATTER: The material of physical reality. There are four basic states of matter: solid, liquid, gas, and plasma.

MECHANICS: The study of bodies in motion.

RESULTANT: The sum of two or more vectors, which measures the net change in distance and direction.

SCALAR: A quantity that possesses only magnitude, with no specific direction. Mass, time, and speed are all scalars. The opposite of a scalar is a vector.

SPEED: The rate at which the position of an object changes over a given period of time.

VACUUM: Space entirely devoid of matter, including air.

VECTOR: A quantity that possesses both magnitude and direction. Velocity, acceleration, and weight (which involves the downward acceleration due to gravity) are examples of vectors. Its opposite is a scalar.

VELOCITY: The speed of an object in a particular direction.

WEIGHT: A measure of the gravitational force on an object; the product of mass multiplied by the acceleration due to gravity. (The latter is equal to 32 ft or 9.8 m per second per second, or 32 ft/9.8 m per second squared.)

change in its motion—including a change in velocity. Mass is one of the most fundamental notions in the world of physics, and it too is the subject of a popular misconception—one which confuses it with weight. In fact, weight is a force, equal to mass multiplied by the acceleration due to gravity.

It was Newton, through a complicated series of steps he explained in his *Principia,* who made possible the calculation of that acceleration—an act of quantification that had eluded Galileo. The figure most often used for gravitational acceleration at sea level is 32 ft (9.8 m) per second squared. This means that in the first second, an object falls at a velocity of 32 ft per second, but its

velocity is also increasing at a rate of 32 ft per second per second. Hence, after 2 seconds, its velocity will be 64 ft (per second; after 3 seconds 96 ft per second, and so on.

Mass does not vary anywhere in the universe, whereas weight changes with any change in the gravitational field. When United States astronaut Neil Armstrong planted the American flag on the Moon in 1969, the flagpole (and indeed Armstrong himself) weighed much less than on Earth. Yet it would have required exactly the same amount of force to move the pole (or, again, Armstrong) from side to side as it would have on Earth, because their mass and therefore their inertia had not changed.

BEYOND MECHANICS

The implications of Newton's three laws go far beyond what has been described here; but again, these laws, as well as gravity itself, receive a much more thorough treatment elsewhere in this volume. What is important in this context is the gradually unfolding understanding of matter and motion that formed the basis for the study of physics today.

After Newton came the Swiss mathematician and physicist Daniel Bernoulli (1700-1782), who pioneered another subdiscipline, fluid dynamics, which encompasses the behavior of liquids and gases in contact with solid objects. Air itself is an example of a fluid, in the scientific sense of the term. Through studies in fluid dynamics, it became possible to explain the principles of air resistance that cause a leaf to fall more slowly than a stone—even though the two are subject to exactly the same gravitational acceleration, and would fall at the same speed in a vacuum.

EXTENDING THE REALM OF PHYSICAL STUDY. The work of Galileo, Newton, and Bernoulli fit within one of five major divisions of classical physics: mechanics, or the study of matter, motion, and forces. The other principal divisions are acoustics, or studies in sound; optics, the study of light; thermodynamics, or investigations regarding the relationships between heat and other varieties of energy; and electricity and magnetism. (These subjects, and subdivisions within them, also receive extensive treatment elsewhere in this book.)

Newton identified one type of force, gravitation, but in the period leading up to the time of Scottish physicist James Clerk Maxwell (1831-1879), scientists gradually became aware of a new fundamental interaction in the universe. Building on studies of numerous scientists, Maxwell hypothesized that electricity and magnetism are in fact differing manifestations of a second variety of force, electromagnetism.

MODERN PHYSICS. The term classical physics, used above, refers to the subjects of study from Galileo's time through the end of the nineteenth century. Classical physics deals primarily with subjects that can be discerned by the senses, and addressed processes that could be observed on a large scale. By contrast, modern physics, which had its beginnings with the work of Max Planck (1858-1947), Albert Einstein (1879-1955), Niels Bohr (1885-1962), and others at the beginning of the twentieth century, addresses quite a different set of topics.

Modern physics is concerned primarily with the behavior of matter at the molecular, atomic, or subatomic level, and thus its truths cannot be grasped with the aid of the senses. Nor is classical physics much help in understanding modern physics. The latter, in fact, recognizes two forces unknown to classical physicists: weak nuclear force, which causes the decay of some subatomic particles, and strong nuclear force, which binds the nuclei of atoms with a force 1 trillion (10^{12}) times as great as that of the weak nuclear force.

Things happen in the realm of modern physics that would have been inconceivable to classical physicists. For instance, according to quantum mechanics—first developed by Planck—it is not possible to make a measurement without affecting the object (e.g., an electron) being measured. Yet even atomic, nuclear, and particle physics can be understood in terms of their effects on the world of experience: challenging as these subjects are, they still concern—though within a much more complex framework—the physical fundamentals of matter and motion.

WHERE TO LEARN MORE

Ballard, Carol. *How Do We Move?* Austin, TX: Raintree Steck-Vaughn, 1998.

Beiser, Arthur. *Physics*, 5th ed. Reading, MA: Addison-Wesley, 1991.

Fleisher, Paul. *Objects in Motion: Principles of Classical Mechanics.* Minneapolis, MN: Lerner Publications, 2002.

Hewitt, Sally. *Forces Around Us.* New York: Children's Press, 1998.

Measure for Measure: Sites That Do the Work for You (Web site). <http://www.wolinskyweb.com/measure.html> (March 7, 2001).

Motion, Energy, and Simple Machines (Web site). <http://www.necc.mass.edu/MRVIS/MR3_13/start.html> (March 7, 2001).

Physlink.com (Web site). <http://www.physlink.com> (March 7, 2001).

Rutherford, F. James; Gerald Holton; and Fletcher G. Watson. *Project Physics.* New York: Holt, Rinehart, and Winston, 1981.

Wilson, Jerry D. *Physics: Concepts and Applications,* second edition. Lexington, MA: D. C. Heath, 1981.

DENSITY AND VOLUME

CONCEPT

Density and volume are simple topics, yet in order to work within any of the hard sciences, it is essential to understand these two types of measurement, as well as the fundamental quantity involved in conversions between them—mass. Measuring density makes it possible to distinguish between real gold and fake gold, and may also give an astronomer an important clue regarding the internal composition of a planet.

HOW IT WORKS

There are four fundamental standards by which most qualities in the physical world can be measured: length, mass, time, and electric current. The volume of a cube, for instance, is a unit of length cubed: the length is multiplied by the width and multiplied by the height. Width and height, however, are not distinct standards of measurement: they are simply versions of length, distinguished by their orientation. Whereas length is typically understood as a distance along an *x*-axis in one-dimensional space, width adds a second dimension, and height a third.

Of particular concern within this essay are length and mass, since volume is measured in terms of length, and density in terms of the ratio between mass and volume. Elsewhere in this book, the distinction between mass and weight has been presented numerous times from the standpoint of a person whose mass and weight are measured on Earth, and again on the Moon. Mass, of course, does not change, whereas weight does, due to the difference in gravitational force exerted by Earth as compared with that of its satellite, the Moon. But consider instead the role of the funda-

mental quality, mass, in determining this significantly less fundamental property of weight.

According to the second law of motion, weight is a force equal to mass multiplied by acceleration. Acceleration, in turn, is equal to change in velocity divided by change in time. Velocity, in turn, is equal to distance (a form of length) divided by time. If one were to express weight in terms of *l*, *t*, and *m*, with these representing, respectively, the fundamental properties of length, time, and mass, it would be expressed as

$$\frac{M \cdot D}{t^2}$$

—clearly, a much more complicated formula than that of mass!

MASS

So what is mass? Again, the second law of motion, derived by Sir Isaac Newton (1642-1727), is the key: mass is the ratio of force to acceleration. This topic, too, is discussed in numerous places throughout this book; what is actually of interest here is a less precise identification of mass, also made by Newton.

Before formulating his laws of motion, Newton had used a working definition of mass as the quantity of matter an object possesses. This is not of much value for making calculations or measurements, unlike the definition in the second law. Nonetheless, it serves as a useful reminder of matter's role in the formula for density.

Matter can be defined as a physical substance not only having mass, but occupying space. It is composed of atoms (or in the case of subatomic particles, it is part of an atom), and is

HOW DOES A GIGANTIC STEEL SHIP, SUCH AS THE SUPERTANKER PICTURED HERE, STAY AFLOAT, EVEN THOUGH IT HAS A WEIGHT DENSITY FAR GREATER THAN THE WATER BELOW IT? THE ANSWER LIES IN ITS CURVED HULL, WHICH CONTAINS A LARGE AMOUNT OF OPEN SPACE AND ALLOWS THE SHIP TO SPREAD ITS AVERAGE DENSITY TO A LOWER LEVEL THAN THE WATER. (*Photograph by Vince Streano/Corbis. Reproduced by permission.*)

convertible with energy. The form or state of matter itself is not important: on Earth it is primarily observed as a solid, liquid, or gas, but it can also be found (particularly in other parts of the universe) in a fourth state, plasma.

Yet there are considerable differences among types of matter—among various elements and states of matter. This is apparent if one imagines three gallon jugs, one containing water, the second containing helium, and the third containing iron filings. The volume of each is the same, but obviously, the mass is quite different.

The reason, of course, is that at a molecular level, there is a difference in mass between the compound H_2O and the elements helium and iron. In the case of helium, the second-lightest of all elements after hydrogen, it would take a great deal of helium for its mass to equal that of iron. In fact, it would take more than 43,000 gallons of helium to equal the mass of the iron in one gallon jug!

DENSITY

Rather than comparing differences in molecular mass among the three substances, it is easier to analyze them in terms of density, or mass divid-

ed by volume. It so happens that the three items represent the three states of matter on Earth: liquid (water), solid (iron), and gas (helium). For the most part, solids tend to be denser than liquids, and liquids denser than gasses.

One of the interesting things about density, as distinguished from mass and volume, is that it has nothing to do with the amount of material. A kilogram of iron differs from 10 kilograms of iron both in mass and volume, but the density of both samples is the same. Indeed, as discussed below, the known densities of various materials make it possible to determine whether a sample of that material is genuine.

VOLUME

Mass, because of its fundamental nature, is sometimes hard to comprehend, and density requires an explanation in terms of mass and volume. Volume, on the other hand, appears to be quite straightforward—and it is, when one is describing a solid of regular shape. In other situations, however, volume is more complicated.

As noted earlier, the volume of a cube can be obtained simply by multiplying length by width by height. There are other means for measuring

SINCE SCIENTISTS KNOW EARTH'S MASS AS WELL AS ITS VOLUME, THEY ARE EASILY ABLE TO COMPUTE ITS DENSITY—APPROXIMATELY 5 G/CM^3. *(Corbis. Reproduced by permission.)*

the volume of other straight-sided objects, such as a pyramid. That formula applies, indeed, for any polyhedron (a three-dimensional closed solid bounded by a set number of plane figures) that constitutes a modified cube in which the lengths of the three dimensions are unequal—that is, an oblong shape.

For a cylinder or sphere, volume measurements can be obtained by applying formulae involving radius (r) and the constant π, roughly equal to 3.14. The formula for volume of a cylinder is $V = \pi r^2 h$, where h is the height. A sphere's volume can be obtained by the formula $(4/3)\pi r^3$. Even the volume of a cone can be easily calculated: it is one-third that of a cylinder of equal base and height.

REAL-LIFE APPLICATIONS

MEASURING VOLUME

What about the volume of a solid that is irregular in shape? Some irregularly shaped objects, such as a scooter, which consists primarily of one round wheel and a number of oblong shapes, can be measured by separating them into regular shapes. Calculus may be employed with more complex problems to obtain the volume of an irregular shape—but the most basic method is simply to immerse the object in water. This procedure involves measuring the volume of the water before and after immersion, and calculating the difference. Of course, the object being measured cannot be water-soluble; if it is, its volume must be measured in a non-water-based liquid such as alcohol.

Measuring liquid volumes is easy, given the fact that liquids have no definite shape, and will simply take the shape of the container in which they are placed. Gases are similar to liquids in the sense that they expand to fit their container; however, measurement of gas volume is a more involved process than that used to measure either liquids or solids, because gases are highly responsive to changes in temperature and pressure.

If the temperature of water is raised from its freezing point to its boiling point (32° to 212°F or 0 to 100°C), its volume will increase by only 2%. If its pressure is doubled from 1 atm (defined as normal air pressure at sea level—14.7 pounds-per-square-inch or 1.013×10^5 Pa) to 2 atm, volume will decrease by only 0.01%.

Yet, if air were heated from 32° to 212°F, its volume would increase by 37%; and if its pressure were doubled from 1 atm to 2, its volume would decrease by 50%. Not only do gases respond dramatically to changes in temperature and pressure, but also, gas molecules tend to be non-attractive toward one another—that is, they do not tend to stick together. Hence, the concept of "volume" involving gas is essentially meaningless, unless its temperature and pressure are known.

BUOYANCY: VOLUME AND DENSITY

Consider again the description above, of an object with irregular shape whose volume is measured by immersion in water. This is not the only interesting use of water and solids when dealing with volume and density. Particularly intriguing is the concept of buoyancy expressed in Archimedes's principle.

More than twenty-two centuries ago, the Greek mathematician, physicist, and inventor Archimedes (c. 287-212 B.C.) received orders from the king of his hometown—Syracuse, a Greek colony in Sicily—to weigh the gold in the royal crown. According to legend, it was while bathing that Archimedes discovered the principle that is today named after him. He was so excited, legend maintains, that he jumped out of his bath and ran naked through the streets of Syracuse shouting "Eureka!" (I have found it).

What Archimedes had discovered was, in short, the reason why ships float: because the buoyant, or lifting, force of an object immersed in fluid is equal to the weight of the fluid displaced by the object.

HOW A STEEL SHIP FLOATS ON WATER. Today most ships are made of steel, and therefore, it is even harder to understand why an aircraft carrier weighing many thousands of tons can float. After all, steel has a weight density (the preferred method for measuring density according to the British system of measures) of 480 pounds per cubic foot, and a density of 7,800 kilograms-per-cubic-meter. By contrast, sea water has a weight density of 64 pounds per cubic foot, and a density of 1,030 kilograms-per-cubic-meter.

This difference in density should mean that the carrier would sink like a stone—and indeed it would, if all the steel in it were hammered flat. As it is, the hull of the carrier (or indeed of any seaworthy ship) is designed to displace or move a quantity of water whose weight is greater than that of the vessel itself. The weight of the displaced water—that is, its mass multiplied by the downward acceleration due to gravity—is equal to the buoyant force that the ocean exerts on the ship. If the ship weighs less than the water it displaces, it will float; but if it weighs more, it will sink.

Put another way, when the ship is placed in the water, it displaces a certain quantity of water whose weight can be expressed in terms of Vdg—volume multiplied by density multiplied by the downward acceleration due to gravity. The density of sea water is a known figure, as is g (32 ft or 9.8 m/sec^2); thus the only variable for the water displaced is its volume.

For the buoyant force on the ship, g will of course be the same, and the value of V will be the same as for the water. In order for the ship to float, then, its density must be much less than that of the water it has displaced. This can be achieved by designing the ship in order to maximize displacement. The steel is spread over as large an area as possible, and the curved hull, when seen in cross section, contains a relatively large area of open space. Obviously, the density of this space is much less than that of water; thus, the average density of the ship is greatly reduced, which enables it to float.

KEY TERMS

ARCHIMEDES'S PRINCIPLE: A rule of physics which holds that the buoyant force of an object immersed in fluid is equal to the weight of the fluid displaced by the object. It is named after the Greek mathematician, physicist, and inventor Archimedes (c. 287-212 B.C.), who first identified it.

BUOYANCY: The tendency of an object immersed in a fluid to float. This can be explained by Archimedes's principle.

DENSITY: The ratio of mass to volume—in other words, the amount of matter within a given area.

MASS: According to the second law of motion, mass is the ratio of force to acceleration. Mass may likewise be defined, though much less precisely, as the amount of matter an object contains. Mass is also the product of volume multiplied by density.

MATTER: Physical substance that occupies space, has mass, is composed of atoms (or in the case of subatomic particles, is part of an atom), and is convertible into energy.

SPECIFIC GRAVITY: The density of an object or substance relative to the density of water; or more generally, the ratio between the densities of two objects or substances.

VOLUME: The amount of three-dimensional space an object occupies. Volume is usually expressed in cubic units of length.

WEIGHT DENSITY: The proper term for density within the British system of weights and measures. The pound is a unit of weight rather than of mass, and thus British units of density are usually rendered in terms of weight density—that is, pounds-per-cubic-foot. By contrast, the metric or international units measure mass density (referred to simply as "density"), which is rendered in terms of kilograms-per-cubic-meter, or grams-per-cubic-centimeter.

COMPARING DENSITIES

As noted several times, the densities of numerous materials are known quantities, and can be easily compared. Some examples of density, all expressed in terms of kilograms per cubic meter, are:

- Hydrogen: 0.09 kg/m^3
- Air: 1.3 kg/m^3
- Oak: 720 kg/m^3
- Ethyl alcohol: 790 kg/m^3
- Ice: 920 kg/m^3
- Pure water: 1,000 kg/m^3
- Concrete: 2,300 kg/m^3
- Iron and steel: 7,800 kg/m^3
- Lead: 11,000 kg/m^3
- Gold: 19,000 kg/m^3

Note that pure water (as opposed to sea water, which is 3% denser) has a density of 1,000 kilograms per cubic meter, or 1 gram per cubic centimeter. This value is approximate; however, at a temperature of 39.2°F (4°C) and under normal atmospheric pressure, it is exact, and so, water is a useful standard for measuring the specific gravity of other substances.

SPECIFIC GRAVITY AND THE DENSITIES OF PLANETS.   Specific gravity is the ratio between the densities of two objects or substances, and it is expressed as a number without units of measure. Due to the value of 1 g/cm^3 for water, it is easy to determine the specific gravity of a given substance, which will have the same number value as its density. For example, the specific gravity of concrete, which has a density of 2.3 g/cm^3, is 2.3. The specific gravities of gases are usually determined in comparison to the specific gravity of dry air.

Most rocks near the surface of Earth have a specific gravity of somewhere between 2 and 3, while the specific gravity of the planet itself is about 5. How do scientists know that the density of Earth is around 5 g/cm^3? The computation is fairly simple, given the fact that the mass and volume of the planet are known. And given the fact that most of what lies close to Earth's surface—sea water, soil, rocks—has a specific gravity well below 5, it is clear that Earth's interior must contain high-density materials, such as nickel or iron. In the same way, calculations regarding the density of other objects in the Solar System provide a clue as to their interior composition.

ALL THAT GLITTERS. Closer to home, a comparison of density makes it possible to determine whether a piece of jewelry alleged to be solid gold is really genuine. To determine the answer, one must drop it in a beaker of water with graduated units of measure clearly marked. (Here, figures are given in cubic centimeters, since these are easiest to use in this context.)

Suppose the item has a mass of 10 grams. The density of gold is 19 g/cm^3, and since $V = m/d = 10/19$, the volume of water displaced by the gold should be 0.53 cm^3. Suppose that instead, the item displaced 0.91 cm^3 of water. Clearly, it is not gold, but what is it?

Given the figures for mass and volume, its density would be equal to $m/V = 10/0.91 = 11$ g/cm^3—which happens to be the density of lead. If on the other hand the amount of water displaced were somewhere between the values for

pure gold and pure lead, one could calculate what portion of the item was gold and which lead. It is possible, of course, that it could contain some other metal, but given the high specific gravity of lead, and the fact that its density is relatively close to that of gold, lead is a favorite gold substitute among jewelry counterfeiters.

WHERE TO LEARN MORE

Beiser, Arthur. *Physics,* 5th ed. Reading, MA: Addison-Wesley, 1991.

Chahrour, Janet. *Flash! Bang! Pop! Fizz!: Exciting Science for Curious Minds.* Illustrated by Ann Humphrey Williams. Hauppauge, N.Y.: Barron's, 2000.

"Density and Specific Gravity" (Web site). <http://www.tpub.com/fluid/ch1e.htm> (March 27, 2001).

"Density, Volume, and Cola" (Web site). <http://student.biology.arizona.edu/sciconn/density/density_coke.html> (March 27, 2001).

"The Mass Volume Density Challenge" (Web site). <http://science-math-technology.com/mass_volume_density.html> (March 27, 2001).

"Metric Density and Specific Gravity" (Web site). <http://www.essex1.com/people/speer/density.html> (March 27, 2001).

"Mineral Properties: Specific Gravity" The Mineral and Gemstone Kingdom (Web site). <http://www.minerals.net/resource/property/sg.htm> (March 27, 2001).

Robson, Pam. *Clocks, Scales and Measurements.* New York: Gloucester Press, 1993.

"Volume, Mass, and Density" (Web site). <http://www.nyu.edu/pages/mathmol/modules/water/density_intro.html> (March 27, 2001).

Willis, Shirley. *Tell Me How Ships Float.* Illustrated by the author. New York: Franklin Watts, 1999.

CONSERVATION LAWS

CONCEPT

The term "conservation laws" might sound at first like a body of legal statutes geared toward protecting the environment. In physics, however, the term refers to a set of principles describing certain aspects of the physical universe that are preserved throughout any number of reactions and interactions. Among the properties conserved are energy, linear momentum, angular momentum, and electrical charge. (Mass, too, is conserved, though only in situations well below the speed of light.) The conservation of these properties can be illustrated by examples as diverse as dropping a ball (energy); the motion of a skater spinning on ice (angular momentum); and the recoil of a rifle (linear momentum).

HOW IT WORKS

The conservation laws describe physical properties that remain constant throughout the various processes that occur in the physical world. In physics, "to conserve" something means "to result in no net loss of" that particular component. For each such component, the input is the same as the output: if one puts a certain amount of energy into a physical system, the energy that results from that system will be the same as the energy put into it.

The energy may, however, change forms. In addition, the operations of the conservation laws are—on Earth, at least—usually affected by a number of other forces, such as gravity, friction, and air resistance. The effects of these forces, combined with the changes in form that take place within a given conserved property, sometimes make it difficult to perceive the working of the conservation laws. It was stated above that

the resulting energy of a physical system will be the same as the energy that was introduced to it. Note, however, that the usable energy output of a system will not be equal to the energy input. This is simply impossible, due to the factors mentioned above—particularly friction.

When one puts gasoline into a motor, for instance, the energy that the motor puts out will never be as great as the energy contained in the gasoline, because part of the input energy is expended in the operation of the motor itself. Similarly, the angular momentum of a skater on ice will ultimately be dissipated by the resistant force of friction, just as that of a Frisbee thrown through the air is opposed both by gravity and air resistance—itself a specific form of friction.

In each of these cases, however, the property is still conserved, even if it does not seem so to the unaided senses of the observer. Because the motor has a usable energy output less than the input, it seems as though energy has been lost. In fact, however, the energy has only changed forms, and some of it has been diverted to areas other than the desired output. (Both the noise and the heat of the motor, for instance, represent uses of energy that are typically considered undesirable.) Thus, upon closer study of the motor—itself an example of a system—it becomes clear that the resulting energy, if not the desired usable output, is the same as the energy input.

As for the angular momentum examples in which friction, or air resistance, plays a part, here too (despite all apparent evidence to the contrary) the property is conserved. This is easier to understand if one imagines an object spinning in outer space, free from the opposing force of friction. Thanks to the conservation of angular

AS THIS HUNTER FIRES HIS RIFLE, THE RIFLE PRODUCES A BACKWARD "KICK" AGAINST HIS SHOULDER. THIS KICK, WITH A VELOCITY IN THE OPPOSITE DIRECTION OF THE BULLET'S TRAJECTORY, HAS A MOMENTUM EXACTLY THE SAME AS THAT OF THE BULLET ITSELF: HENCE MOMENTUM IS CONSERVED. *(Photograph by Tony Arruza/Corbis. Reproduced by permission.)*

momentum, an object set into rotation in space will continue to spin indefinitely. Thus, if an astronaut in the 1960s, on a spacewalk from his capsule, had set a screwdriver spinning in the emptiness of the exosphere, the screwdriver would still be spinning today!

ENERGY AND MASS

Among the most fundamental statements in all of science is the conservation of energy: a system isolated from all outside factors will maintain the same total amount of energy, even though energy transformations from one form or another take place.

Energy is manifested in many varieties, including thermal, electromagnetic, sound, chemical, and nuclear energy, but all these are merely reflections of three basic types of energy. There is potential energy, which an object possesses by virtue of its position; kinetic energy, which it possesses by virtue of its motion; and rest energy, which it possesses by virtue of its mass.

The last of these three will be discussed in the context of the relationship between energy and mass; at present the concern is with potential and kinetic energy. Every system possesses a certain quantity of both, and the sum of its potential and kinetic energy is known as mechanical energy. The mechanical energy within a system does not change, but the relative values of potential and kinetic energy may be altered.

A SIMPLE EXAMPLE OF MECHANICAL ENERGY. If one held a baseball at the top of a tall building, it would have a certain amount of potential energy. Once it was dropped, it would immediately begin losing potential energy and gaining kinetic energy proportional to the potential energy it lost. The relationship between the two forms, in fact, is inverse: as the value of one variable decreases, that of the other increases in exact proportion.

The ball cannot keep falling forever, losing potential energy and gaining kinetic energy. In fact, it can never gain an amount of kinetic energy greater than the potential energy it possessed in the first place. At the moment before it hits the ground, the ball's kinetic energy is equal to the potential energy it possessed at the top of the building. Correspondingly, its potential energy is zero—the same amount of kinetic energy it possessed before it was dropped.

Then, as the ball hits the ground, the energy is dispersed. Most of it goes into the ground, and depending on the rigidity of the ball and the ground, this energy may cause the ball to bounce. Some of the energy may appear in the form of sound, produced as the ball hits bottom, and some will manifest as heat. The total energy, however, will not be lost: it will simply have changed form.

REST ENERGY. The values for mechanical energy in the above illustration would most likely be very small; on the other hand, the rest or mass energy of the baseball would be staggering. Given the weight of 0.333 pounds for a regulation baseball, which on Earth converts to 0.15 kg in mass, it would possess enough energy by virtue of its mass to provide a year's worth of electrical power to more than 150,000 American homes. This leads to two obvious questions: how can a mere baseball possess all that energy? And if it does, how can the energy be extracted and put to use?

The answer to the second question is, "By accelerating it to something close to the speed of light"—which is more than 27,000 times faster than the fastest speed ever achieved by humans. (The astronauts on *Apollo 10* in May 1969 reached nearly 25,000 MPH (40,000 km/h), which is more than 33 times the speed of sound but still insignificant when compared to the speed of light.) The answer to the first question lies in the most well-known physics formula of all time: $E = mc^2$

In 1905, Albert Einstein (1879-1955) published his Special Theory of Relativity, which he followed a decade later with his General Theory of Relativity. These works introduced the world to the above-mentioned formula, which holds that energy is equal to mass multiplied by the squared speed of light. This formula gained its widespread prominence due to the many implications of Einstein's Relativity, which quite literally changed humanity's perspective on the universe. Most concrete among those implications was the atom bomb, made possible by the understanding of mass and energy achieved by Einstein.

In fact, $E = mc^2$ is the formula for rest energy, sometimes called mass energy. Though rest energy is "outside" of kinetic and potential energy in the sense that it is not defined by the above-described interactions within the larger system of

As Surya Bonaly goes into a spin on the ice, she draws in her arms and leg, reducing the moment of inertia. Because of the conservation of angular momentum, her angular velocity will increase, meaning that she will spin much faster. *(Bolemian Nomad Picturemakers/Corbis. Reproduced by permission.)*

mechanical energy, its relation to the other forms can be easily shown. All three are defined in terms of mass. Potential energy is equal to *mgh*, where *m* is mass, *g* is gravity, and *h* is height. Kinetic energy is equal to $\frac{1}{2} mv^2$, where *v* is velocity. In fact—using a series of steps that will not be demonstrated here—it is possible to directly relate the kinetic and rest energy formulae.

The kinetic energy formula describes the behavior of objects at speeds well below the speed of light, which is 186,000 mi (297,600 km) per second. But at speeds close to that of the speed of light, $\frac{1}{2} mv^2$ does not accurately reflect the energy possessed by the object. For instance, if *v* were equal to 0.999*c* (where *c* represents the speed of light), then the application of the formula $\frac{1}{2} mv^2$ would yield a value equal to less than 3% of the object's real energy. In order to calculate the true energy of an object at 0.999*c*, it would be necessary to apply a different formula for total energy, one that takes into account the fact that, at such a speed, mass itself becomes energy.

CONSERVATION OF MASS.
Mass itself is relative at speeds approaching *c*, and, in fact, becomes greater and greater the closer an object comes to the speed of light. This may seem strange in light of the fact that there is, after all, a law stating that mass is conserved. But mass is only conserved at speeds well below c: as an object approaches 186,000 mi (297,600 km) per second, the rules are altered.

The conservation of mass states that total mass is constant, and is unaffected by factors such as position, velocity, or temperature, in any system that does not exchange any matter with its environment. Yet, at speeds close to *c*, the mass of an object increases dramatically.

In such a situation, the mass would be equal to the rest, or starting mass, of the object divided by $\sqrt{1 - (v^2/c^2)}$, where *v* is the object's speed of relative motion. The denominator of this equation will always be less than one, and the greater the value of *v*, the smaller the value of the denominator. This means that at a speed of *c*, the denominator is zero—in other words, the object's mass is infinite! Obviously, this is not possible, and indeed, what the formula actually shows is that no object can travel faster than the speed of light.

Of particular interest to the present discussion, however, is the fact, established by relativity theory, that mass can be converted into energy. Hence, as noted earlier, a baseball or indeed any object can be converted into energy—and since the formula for rest energy requires that the mass be multiplied by c^2, clearly, even an object of virtually negligible mass can generate a staggering amount of energy. This conversion of mass to energy happens well below the speed of light, in a very small way, when a stick of dynamite explodes. A portion of that stick becomes energy, and the fact that this portion is equal to just 6 parts out of 100 billion indicates the vast proportions of energy available from converted mass.

OTHER CONSERVATION LAWS

In addition to the conservation of energy, as well as the limited conservation of mass, there are laws governing the conservation of momentum, both for an object in linear (straight-line) motion, and for one in angular (rotational) motion. Momentum is a property that a moving body possesses by virtue of its mass and velocity, which determines the amount of force and time required to stop it. Linear momentum is equal to mass multiplied by velocity, and the conservation of linear momentum law states that when the sum of the external force vectors acting on a physical system is equal to zero, the total linear momentum of the system remains unchanged, or conserved.

Angular momentum, or the momentum of an object in rotational motion, is equal to $mr^2\omega$, where *m* is mass, *r* is the radius of rotation, and ω (the Greek letter omega) stands for angular velocity. According to the conservation of angular momentum law, when the sum of the external torques acting on a physical system is equal to zero, the total angular momentum of the system remains unchanged. Torque is a force applied around an axis of rotation. When playing the old game of "spin the bottle," for instance, one is applying torque to the bottle and causing it to rotate.

ELECTRIC CHARGE. The conservation of both linear and angular momentum are best explained in the context of real-life examples, provided below. Before going on to those examples, however, it is appropriate here to discuss a conservation law that is outside the realm of everyday experience: the conservation of electric charge, which holds that for an isolated system, the net electric charge is constant.

This law is "outside the realm of everyday experience" such that one cannot experience it through the senses, but at every moment, it is happening everywhere. Every atom has positively charged protons, negatively charged electrons, and uncharged neutrons. Most atoms are neutral, possessing equal numbers of protons and electrons; but, as a result of some disruption, an atom may have more protons than electrons, and thus, become positively charged. Conversely, it may end up with a net negative charge due to a greater number of electrons. But the protons or electrons it released or gained did not simply appear or disappear: they moved from one part of the system to another—that is, from one atom to another atom, or to several other atoms.

Throughout these changes, the charge of each proton and electron remains the same, and the net charge of the system is always the sum of its positive and negative charges. Thus, it is impossible for any electrical charge in the universe to be smaller than that of a proton or electron. Likewise, throughout the universe, there is always the same number of negative and positive electrical charges: just as energy changes form, the charges simply change position.

There are also conservation laws describing the behavior of subatomic particles, such as the positron and the neutrino. However, the most significant of the conservation laws are those involving energy (and mass, though with the limitations discussed above), linear momentum, angular momentum, and electrical charge.

REAL-LIFE APPLICATIONS

CONSERVATION OF LINEAR MOMENTUM: RIFLES AND ROCKETS

FIRING A RIFLE. The conservation of linear momentum is reflected in operations as simple as the recoil of a rifle when it is fired, and in those as complex as the propulsion of a rocket through space. In accordance with the conservation of momentum, the momentum of a system must be the same after it undergoes an operation as it was before the process began. Before firing, the momentum of a rifle and bullet is zero, and therefore, the rifle-bullet system must return to that same zero-level of momentum after it is fired. Thus, the momentum of the bullet must be matched—and "cancelled" within the system under study—by a corresponding backward momentum.

When a person shooting a gun pulls the trigger, it releases the bullet, which flies out of the barrel toward the target. The bullet has mass and velocity, and it clearly has momentum; but this is only half of the story. At the same time it is fired, the rifle produces a "kick," or sharp jolt, against the shoulder of the person who fired it. This backward kick, with a velocity in the opposite direction of the bullet's trajectory, has a momentum exactly the same as that of the bullet itself: hence, momentum is conserved.

But how can the rearward kick have the same momentum as that of the bullet? After all, the bullet can kill a person, whereas, if one holds the rifle correctly, the kick will not even cause any injury. The answer lies in several properties of linear momentum. First of all, as noted earlier, momentum is equal to mass multiplied by velocity; the actual proportions of mass and velocity,

however, are not important as long as the backward momentum is the same as the forward momentum. The bullet is an object of relatively small mass and high velocity, whereas the rifle is much larger in mass, and hence, its rearward velocity is correspondingly small.

In addition, there is the element of impulse, or change in momentum. Impulse is the product of force multiplied by change or interval in time. Again, the proportions of force and time interval do not matter, as long as they are equal to the momentum change—that is, the difference in momentum that occurs when the rifle is fired. To avoid injury to one's shoulder, clearly force must be minimized, and for this to happen, time interval must be extended.

If one were to fire the rifle with the stock (the rear end of the rifle) held at some distance from one's shoulder, it would kick back and could very well produce a serious injury. This is because the force was delivered over a very short time interval—in other words, force was maximized and time interval minimized. However, if one holds the rifle stock firmly against one's shoulder, this slows down the delivery of the kick, thus maximizing time interval and minimizing force.

ROCKETING THROUGH SPACE. Contrary to popular belief, rockets do not move by pushing against a surface such as a launchpad. If that were the case, then a rocket would have nothing to propel it once it had been launched, and certainly there would be no way for a rocket to move through the vacuum of outer space. Instead, what propels a rocket is the conservation of momentum.

Upon ignition, the rocket sends exhaust gases shooting downward at a high rate of velocity. The gases themselves have mass, and thus, they have momentum. To balance this downward momentum, the rocket moves upward—though, because its mass is greater than that of the gases it expels, it will not move at a velocity as high as that of the gases. Once again, the upward or forward momentum is exactly the same as the downward or backward momentum, and linear momentum is conserved.

Rather than needing something to push against, a rocket in fact performs best in outer space, where there is nothing—neither launchpad nor even air—against which to push. Not only is "pushing" irrelevant to the operation of

CONSERVATION LAWS: A set of principles describing physical properties that remain constant—that is, are conserved—throughout the various processes that occur in the physical world. The most significant of these laws concerns the conservation of energy (as well as, with qualifications, the conservation of mass); conservation of linear momentum; conservation of angular momentum; and conservation of electrical charge.

CONSERVATION OF ANGULAR MOMENTUM: A physical law stating that when the sum of the external torques acting on a physical system is equal to zero, the total angular momentum of the system remains unchanged. Angular momentum is the momentum of an object in rotational motion, and torque is a force applied around an axis of rotation.

CONSERVATION OF ELECTRICAL CHARGE: A physical law which holds that for an isolated system, the net electrical charge is constant.

CONSERVATION OF ENERGY: A law of physics stating that within a system isolated from all other outside factors, the total amount of energy remains the same, though transformations of energy from one form to another take place.

CONSERVATION OF LINEAR MOMENTUM: A physical law stating that when the sum of the external force vectors acting on a physical system is equal to zero, the total linear momentum of the system remains unchanged—or is conserved.

CONSERVATION OF MASS: A physical principle stating that total mass is constant, and is unaffected by factors such as position, velocity, or temperature, in any system that does not exchange any matter with its environment. Unlike the other conservation laws, however, conservation of mass is not universally applicable, but applies only at speeds significant lower than that of light—186,000 mi (297,600 km) per second. Close to the speed of light, mass begins converting to energy.

CONSERVE: In physics, "to conserve" something means "to result in no net loss of" that particular component. It is possible that within a given system, the component may change form or position, but as long as the net value of the component remains the same, it has been conserved.

FRICTION: The force that resists motion when the surface of one object comes into contact with the surface of another.

MOMENTUM: A property that a moving body possesses by virtue of its mass and velocity, which determines the amount of force and time required to stop it.

SYSTEM: In physics, the term "system" usually refers to any set of physical interactions isolated from the rest of the universe. Anything outside of the system, including all factors and forces irrelevant to a discussion of that system, is known as the environment.

the rocket, but the rocket moves much more efficiently without the presence of air resistance. In the same way, on the relatively frictionless surface of an ice-skating rink, conservation of linear momentum (and hence, the process that makes possible the flight of a rocket through space) is easy to demonstrate.

If, while standing on the ice, one throws an object in one direction, one will be pushed in the opposite direction with a corresponding level of momentum. However, since a person's mass is presumably greater than that of the object thrown, the rearward velocity (and, therefore, distance) will be smaller.

Friction, as noted earlier, is not the only force that counters conservation of linear momentum on Earth: so too does gravity, and thus, once again, a rocket operates much better in space than it does when under the influence of Earth's gravitational field. If a bullet is fired at a bottle thrown into the air, the linear momentum of the spent bullet and the shattered pieces of glass in the infinitesimal moment just after the collision will be the same as that of the bullet and the bottle a moment before impact. An instant later, however, gravity will accelerate the bullet and the pieces downward, thus leading to a change in total momentum.

CONSERVATION OF ANGULAR MOMENTUM: SKATERS AND OTHER SPINNERS

As noted earlier, angular momentum is equal to $mr^2\omega$, where m is mass, r is the radius of rotation, and ω stands for angular velocity. In fact, the first two quantities, mr^2, are together known as moment of inertia. For an object in rotation, moment of inertia is the property whereby objects further from the axis of rotation move faster, and thus, contribute a greater share to the overall kinetic energy of the body.

One of the most oft-cited examples of angular momentum—and of its conservation—involves a skater or ballet dancer executing a spin. As the skater begins the spin, she has one leg planted on the ice, with the other stretched behind her. Likewise, her arms are outstretched, thus creating a large moment of inertia. But when she goes into the spin, she draws in her arms and leg, reducing the moment of inertia. In accordance with conservation of angular momentum, $mr^2\omega$ will remain constant, and therefore, her angular velocity will increase, meaning that she will spin much faster.

CONSTANT ORIENTATION. The motion of a spinning top and a Frisbee in flight also illustrate the conservation of angular momentum. Particularly interesting is the tendency of such an object to maintain a constant orientation. Thus, a top remains perfectly vertical while it spins, and only loses its orientation once friction from the floor dissipates its velocity and brings it to a stop. On a frictionless surface, however, it would remain spinning—and therefore upright—forever.

A Frisbee thrown without spin does not provide much entertainment; it will simply fall to the ground like any other object. But if it is tossed with the proper spin, delivered from the wrist, conservation of angular momentum will keep it in a horizontal position as it flies through the air. Once again, the Frisbee will eventually be brought to ground by the forces of air resistance and gravity, but a Frisbee hurled through empty space would keep spinning for eternity.

WHERE TO LEARN MORE

Beiser, Arthur. *Physics,* 5th ed. Reading, MA: Addison-Wesley, 1991.

"*Conservation Laws: An Online Physics Textbook*" (Web site).
<http://www.lightandmatter.com/area1book2.html> (March 12, 2001).

"*Conservation Laws: The Most Powerful Laws of Physics*" (Web site).
<http://webug.physics.uiuc.edu/courses/phys150/fall 99/slides/lect0 7/> (March 12, 2001).

"*Conservation of Energy.*" NASA (Web site).
<http://www.grc.nasa.gov/WWW/K-12/airplane/thermo1f.html> (March 12, 2001).

Elkana, Yehuda. *The Discovery of the Conservation of Energy.* With a foreword by I. Bernard Cohen. Cambridge, MA: Harvard University Press, 1974.

"*Momentum and Its Conservation*" (Web site).
<http://www.glenbrook.k12.il.us/gbssci/phys/Class/momentum/momtoc. html> (March 12, 2001).

Rutherford, F. James; Gerald Holton; and Fletcher G. Watson. *Project Physics.* New York: Holt, Rinehart, and Winston, 1981.

Suplee, Curt. *Everyday Science Explained.* Washington, D.C.: National Geographic Society, 1996.

KINEMATICS AND PARTICLE DYNAMICS

MOMENTUM

CENTRIPETAL FORCE

FRICTION

LAWS OF MOTION

GRAVITY

PROJECTILE MOTION

TORQUE

MOMENTUM

CONCEPT

The faster an object is moving—whether it be a baseball, an automobile, or a particle of matter—the harder it is to stop. This is a reflection of momentum, or specifically, linear momentum, which is equal to mass multiplied by velocity. Like other aspects of matter and motion, momentum is conserved, meaning that when the vector sum of outside forces equals zero, no net linear momentum within a system is ever lost or gained. A third important concept is impulse, the product of force multiplied by length in time. Impulse, also defined as a change in momentum, is reflected in the proper methods for hitting a baseball with force or surviving a car crash.

HOW IT WORKS

Like many other aspects of physics, the word "momentum" is a part of everyday life. The common meaning of momentum, however, unlike many other physics terms, is relatively consistent with its scientific meaning. In terms of formula, momentum is equal to the product of mass and velocity, and the greater the value of that product, the greater the momentum.

Consider the term "momentum" outside the world of physics, as applied, for example, in the realm of politics. If a presidential candidate sees a gain in public-opinion polls, then wins a debate and embarks on a whirlwind speaking tour, the media comments that he has "gained momentum." As with momentum in the framework of physics, what these commentators mean is that the candidate will be hard to stop—or to carry the analogy further, that he is doing enough of the right things (thus gaining "mass"), and doing them quickly enough, thereby gaining velocity.

MOMENTUM AND INERTIA

It might be tempting to confuse momentum with another physical concept, inertia. Inertia, as defined by the second law of motion, is the tendency of an object in motion to remain in motion, and of an object at rest to remain at rest. Momentum, by definition, involves a body in motion, and can be defined as the tendency of a body in motion to continue moving at a constant velocity.

Not only does momentum differ from inertia in that it relates exclusively to objects in motion, but (as will be discussed below) the component of velocity in the formula for momentum makes it a vector—that is, a quantity that possesses both magnitude and direction. There is at least one factor that momentum very clearly has in common with inertia: mass, a measure of inertia indicating the resistance of an object to a change in its motion.

MASS AND WEIGHT

Unlike velocity, mass is a scalar, a quantity that possesses magnitude without direction. Mass is often confused with weight, a vector quantity equal to its mass multiplied by the downward acceleration due to gravity. The weight of an object changes according to the gravitational force of the planet or other celestial body on which it is measured. Hence, the mass of a person on the Moon would be the same as it is on Earth, whereas the person's weight would be considerably less, due to the smaller gravitational pull of the Moon.

Given the unchanging quality of mass as opposed to weight, as well as the fact that scientists themselves prefer the much simpler metric

WHEN BILLIARD BALLS COLLIDE, THEIR HARDNESS RESULTS IN AN ELASTIC COLLISION—ONE IN WHICH KINETIC ENERGY IS CONSERVED. *(Photograph by John-Marshall Mantel/Corbis. Reproduced by permission.)*

system, metric units will generally be used in the following discussion. Where warranted, of course, conversion to English or British units (for example, the pound, a unit of weight) will be provided. However, since the English unit of mass, the slug, is even more unfamiliar to most Americans than its metric equivalent, the kilogram, there is little point in converting kilos into slugs.

VELOCITY AND SPEED

Not only is momentum often confused with inertia, and mass with weight, but in the everyday world the concepts of velocity and speed tend to be blurred. Speed is the rate at which the position of an object changes over a given period of time, expressed in terms such as "50 MPH." It is a scalar quantity.

Velocity, by contrast, is a vector. If one were to say "50 miles per hour toward the northeast," this would be an expression of velocity. Vectors are typically designated in bold, without italics; thus velocity is typically abbreviated **v**. Scalars, on the other hand, are rendered in italics. Hence, the formula for momentum is usually shown as *m***v**.

LINEAR MOMENTUM AND ITS CONSERVATION

Momentum itself is sometimes designated as *p*. It should be stressed that the form of momentum discussed here is strictly linear, or straight-line, momentum, in contrast to angular momentum, more properly discussed within the framework of rotational motion.

Both angular and linear momentum abide by what are known as conservation laws. These are statements concerning quantities that, under certain conditions, remain constant or unchanging. The conservation of linear momentum law states that when the sum of the external force vectors acting on a physical system is equal to zero, the total linear momentum of the system remains unchanged—or conserved.

The conservation of linear momentum is reflected both in the recoil of a rifle and in the propulsion of a rocket through space. When a rifle is fired, it produces a "kick"—that is, a sharp jolt to the shoulder of the person who has fired it—corresponding to the momentum of the bullet. Why, then, does the "kick" not knock a person's shoulder off the way a bullet would? Because the rifle's mass is much greater than that of the bullet, meaning that its velocity is much smaller.

As for rockets, they do not—contrary to popular belief—move by pushing against a surface, such as a launch pad. If that were the case, then a rocket would have nothing to propel it once it is launched, and certainly there would be no way for a rocket to move through the vacuum of outer space. Instead, as it burns fuel, the rocket expels exhaust gases that exert a backward momentum, and the rocket itself travels forward with a corresponding degree of momentum.

SYSTEMS

Here, "system" refers to any set of physical interactions isolated from the rest of the universe. Anything outside of the system, including all factors and forces irrelevant to a discussion of that system, is known as the environment. In the pool-table illustration shown earlier, the interaction of the billiard balls in terms of momentum is the system under discussion.

It is possible to reduce a system even further for purposes of clarity: hence, one might specify that the system consists only of the pool balls, the force applied to them, and the resulting momentum interactions. Thus, we will ignore the friction of the pool table's surface, and the assumption will be that the balls are rolling across a frictionless plane.

IMPULSE

For an object to have momentum, some force must have set it in motion, and that force must have been applied over a period of time. Likewise, when the object experiences a collision or any other event that changes its momentum, that change may be described in terms of a certain amount of force applied over a certain period of time. Force multiplied by time interval is impulse, expressed in the formula $F \cdot \delta t$, where F is force, δ (the Greek letter delta) means "a change" or "change in..."; and t is time.

As with momentum itself, impulse is a vector quantity. Whereas the vector component of momentum is velocity, the vector quantity in impulse is force. The force component of impulse can be used to derive the relationship between impulse and change in momentum. According to the second law of motion, $F = m\,a$; that is, force is equal to mass multiplied by acceleration. Acceleration can be defined as a change

WHEN PARACHUTISTS LAND, THEY KEEP THEIR KNEES BENT AND OFTEN ROLL OVER—ALL IN AN EFFORT TO LENGTHEN THE PERIOD OF THE FORCE OF IMPACT, THUS REDUCING ITS EFFECTS. *(Photograph by James A. Sugar/Corbis. Reproduced by permission.)*

in velocity over a change or interval in time. Expressed as a formula, this is

$$a = \frac{\Delta v}{\Delta t}.$$

Thus, force is equal to

$$m\left(\frac{\Delta v}{\Delta t}\right),$$

an equation that can be rewritten as $F\delta t = m\delta v$. In other words, impulse is equal to change in momentum.

This relationship between impulse and momentum change, derived here in mathematical terms, will be discussed below in light of several well-known examples from the real world. Note that the metric units of impulse and momentum are actually interchangeable, though they are typically expressed in different forms, for the purpose of convenience. Hence, momentum is usually rendered in terms of kilogram-meters-per-second (kg • m/s), whereas impulse is typically shown as newton-seconds (N • s). In the English system, momentum is shown in units of slug-feet per-

As Sammy Sosa's bat hits this ball, it applies a tremendous momentum change to the ball. After contact with the ball, Sosa will continue his swing, thereby contributing to the momentum change and allowing the ball to travel farther. *(AFP/Corbis. Reproduced by permission.)*

second, and impulse in terms of the pound-second.

REAL-LIFE APPLICATIONS

When Two Objects Collide

Two moving objects, both possessing momentum by virtue of their mass and velocity, collide with one another. Within the system created by their collision, there is a total momentum $M\mathbf{V}$ that is equal to their combined mass and the vector sum of their velocity.

This is the case with any system: the total momentum is the sum of the various individual momentum products. In terms of a formula, this is expressed as $M\mathbf{V} = m_1\mathbf{v}_1 + m_2\mathbf{v}_2 + m_3\mathbf{v}_3 +\dots$ and so on. As noted earlier, the total momentum will be conserved; however, the actual distribution of momentum within the system may change.

TWO LUMPS OF CLAY. Consider the behavior of two lumps of clay, thrown at one another so that they collide head-on. Due to the properties of clay as a substance, the two lumps will tend to stick. Assuming the lumps are not of equal mass, they will continue traveling in the same direction as the lump with greater momentum.

As they meet, the two lumps form a larger mass $M\mathbf{V}$ that is equal to the sum of their two individual masses. Once again, $M\mathbf{V} = m_1\mathbf{v}_1 + m_2\mathbf{v}_2$. The M in $M\mathbf{V}$ is the sum of the smaller values m, and the $\mathbf{V}$ is the *vector sum* of velocity. Whereas M is larger than m_1 or m_2—the reason being that scalars are simply added like ordinary numbers—V is smaller than $\mathbf{v}_1$ or $\mathbf{v}_2$. This lower number for net velocity as compared to particle velocity will always occur when two objects are moving in opposite directions. (If the objects are moving in the same direction, V will have a value between that of $\mathbf{v}_1$ and $\mathbf{v}_2$.)

To add the vector sum of the two lumps in collision, it is best to make a diagram showing the bodies moving toward one another, with arrows illustrating the direction of velocity. By convention, in such diagrams the velocity of an object moving to the right is rendered as a positive number, and that of an object moving to the left is shown with a negative number. It is therefore easier to interpret the results if the object with the larger momentum is shown moving to the right.

The value of $\mathbf{V}$ will move in the same direction as the lump with greater momentum. But since the two lumps are moving in opposite directions, the momentum of the smaller lump will cancel out a portion of the greater lump's momentum—much as a negative number, when added to a positive number of greater magnitude, cancels out part of the positive number's value. They will continue traveling in the direction of the lump with greater momentum, now with a combined mass equal to the arithmetic sum of their masses, but with a velocity much smaller than either had before impact.

BILLIARD BALLS. The game of pool provides an example of a collision in which one object, the cue ball, is moving, while the other—known as the object ball—is stationary. Due to the hardness of pool balls, and their tendency not to stick to one another, this is also an example of an almost perfectly elastic collision—one in which kinetic energy is conserved.

The colliding lumps of clay, on the other hand, are an excellent example of an inelastic collision, or one in which kinetic energy is not conserved. The total energy in a given system, such as that created by the two lumps of clay in collision, is conserved; however, kinetic energy may be transformed, for instance, into heat energy and/or sound energy as a result of collision. Whereas inelastic collisions involve soft, sticky objects, elastic collisions involve rigid, non-sticky objects.

Kinetic energy and momentum both involve components of velocity and mass: p (momentum) is equal to mv, and KE (kinetic energy) equals $\frac{1}{2} mv^2$. Due to the elastic nature of pool-ball collisions, when the cue ball strikes the object ball, it transfers its velocity to the latter. Their masses are the same, and therefore the resulting momentum and kinetic energy of the object ball will be the same as that possessed by the cue ball prior to impact.

If the cue ball has transferred all of its velocity to the object ball, does that mean it has stopped moving? It does. Assuming that the interaction between the cue ball and the object ball constitutes a closed system, there is no other source from which the cue ball can acquire velocity, so its velocity must be zero.

It should be noted that this illustration treats pool-ball collisions as though they were 100% elastic, though in fact, a portion of kinetic energy in these collisions is transformed into heat and sound. Also, for a cue ball to transfer all of its velocity to the object ball, it must hit it straight-on. If the balls hit off-center, not only will the object ball move after impact, but the cue ball will continue to move—roughly at 90° to a line drawn through the centers of the two balls at the moment of impact.

IMPULSE: BREAKING OR BUILDING THE IMPACT

When a cue ball hits an object ball in pool, it is safe to assume that a powerful impact is desired. The same is true of a bat hitting a baseball. But what about situations in which a powerful impact is not desired—as for instance when cars are crashing? There is, in fact, a relationship between impulse, momentum change, transfer of kinetic energy, and the impact—desirable or undesirable—experienced as a result.

Impulse, again, is equal to momentum change—and also equal to force multiplied by time interval (or change in time). This means that the greater the force and the greater the amount of time over which it is applied, the greater the momentum change. Even more interesting is the fact that one can achieve the same momentum change with differing levels of force and time interval. In other words, a relatively low degree of force applied over a relatively long period of time would produce the same momentum change as a relatively high amount of force over a relatively short period of time.

The conservation of kinetic energy in a collision is, as noted earlier, a function of the relative elasticity of that collision. The question of whether KE is transferred has nothing to do with impulse. On the other hand, the question of how KE is transferred—or, even more specifically, the interval over which the transfer takes place—is very much related to impulse.

Kinetic energy, again, is equal to $\frac{1}{2} mv^2$. If a moving car were to hit a stationary car head-on, it would transfer a quantity of kinetic energy to the stationary car equal to one-half its own mass multiplied by the square of its velocity. (This, of course, assumes that the collision is perfectly elastic, and that the mass of the cars is exactly equal.) A transfer of KE would also occur if two moving cars hit one another head-on, especially in a highly elastic collision. Assuming one car had considerably greater mass and velocity than the other, a high degree of kinetic energy would be transferred—which could have deadly consequences for the people in the car with less mass and velocity. Even with cars of equal mass, however, a high rate of acceleration can bring about a potentially lethal degree of force.

CRUMPLE ZONES IN CARS. In a highly elastic car crash, two automobiles would bounce or rebound off one another. This would mean a dramatic change in direction—a reversal, in fact—hence, a sudden change in velocity and therefore momentum. In other words, the figure for $m\delta v$ would be high, and so would that for impulse, $F\delta t$.

On the other hand, it is possible to have a highly inelastic car crash, accompanied by a small change in momentum. It may seem logical to think that, in a crash situation, it would be better for two cars to bounce off one another than

KEY TERMS

ACCELERATION: A change velocity. Acceleration can be expressed as a formula $\delta v/\delta t$—that is, change in velocity divided by change, or interval, in time.

CONSERVATION OF LINEAR MOMENTUM: A physical law, which states that when the sum of the external force vectors acting on a physical system is equal to zero, the total linear momentum of the system remains unchanged—or is conserved.

CONSERVE: In physics, "to conserve" something (for example, momentum or kinetic energy) means "to result in no net loss of" that particular component. It is possible that within a given system, one type of energy may be transformed into another type, but the net energy in the system will remain the same.

ELASTIC COLLISION: A collision in which kinetic energy is conserved. Typically elastic collisions involve rigid, non-sticky objects such as pool balls. At the other extreme is an inelastic collision.

IMPULSE: The amount of force and time required to cause a change in momentum. Impulse is the product of force multiplied by a change, or interval, in time ($F\delta t$): the greater the momentum, the greater the force needed to change it, and the longer the period of time over which it must be applied.

INELASTIC COLLISION: A collision in which kinetic energy is not conserved. (The total energy is conserved: kinetic energy itself, however, may be transformed into heat energy or sound energy.) Typically, inelastic collisions involve non-rigid, sticky objects—for instance, lumps of clay. At the other extreme is an elastic collision.

INERTIA: The tendency of an object in motion to remain in motion, and of an object at rest to remain at rest.

KINETIC ENERGY: The energy an object possesses by virtue of its motion.

MASS: A measure of inertia, indicating the resistance of an object to a change in its

---

for them to crumple together. In fact, however, the latter option is preferable. When the cars crumple rather than rebounding, they do not experience a reversal in direction. They do experience a change in speed, of course, but the momentum change is far less than it would be if they rebounded.

Furthermore, crumpling lengthens the amount of time during which the change in velocity occurs, and thus reduces impulse. But even with the reduced impulse of this momentum change, it is possible to further reduce the effect of force, another aspect of impact. Remember that $m\delta v = F\delta t$: the value of force and time interval do not matter, as long as their product is equal to the momentum change. Because F and

δt are inversely proportional, an increase in impact time will reduce the effects of force.

For this reason, car manufacturers actually design and build into their cars a feature known as a crumple zone. A crumple zone—and there are usually several in a single automobile—is a section in which the materials are put together in such a way as to ensure that they will crumple when the car experiences a collision. Of course, the entire car cannot be one big crumple zone—this would be fatal for the driver and riders; however, the incorporation of crumple zones at key points can greatly reduce the effect of the force a car and its occupants must endure in a crash.

Another major reason for crumple zones is to keep the passenger compartment of the car

KEY TERMS CONTINUED

motion—including a change in velocity. A kilogram is a unit of mass, whereas a pound is a unit of weight.

MOMENTUM: A property that a moving body possesses by virtue of its mass and velocity, which determines the amount of force and time (impulse) required to stop it. Momentum—actually linear momentum, as opposed to the angular momentum of an object in rotational motion—is equal to mass multiplied by velocity.

SCALAR: A quantity that possesses only magnitude, with no specific direction—as contrasted with a vector, which possesses both magnitude and direction. Scalar quantities are usually expressed in italicized letters, thus: *m* (mass).

SPEED: The rate at which the position of an object changes over a given period of time.

SYSTEM: In physics, the term "system" usually refers to any set of physical interactions isolated from the rest of the universe. Anything outside of the system, including all factors and forces irrelevant to a discussion of that system, is known as the environment.

VECTOR: A quantity that possesses both magnitude and direction—as contrasted with a scalar, which possesses magnitude without direction. Vector quantities are usually expressed in bold, non-italicized letters, thus: **F** (force). They may also be shown by placing an arrow over the letter designating the specific property, as for instance v for velocity.

VECTOR SUM: A calculation that yields the net result of all the vectors applied in a particular situation. In the case of momentum, the vector component is velocity. The best method is to make a diagram showing bodies in collision, with arrows illustrating the direction of velocity. On such a diagram, motion to the right is assigned a positive value, and to the left a negative value.

VELOCITY: The speed of an object in a particular direction.

---

intact. Many injuries are caused when the body of the car intrudes on the space of the occupants—as, for instance, when the floor buckles, or when the dashboard is pushed deep into the passenger compartment. Obviously, it is preferable to avoid this by allowing the fender to collapse.

REDUCING IMPULSE: SAVING LIVES, BONES, AND WATER BALLOONS. An airbag is another way of minimizing force in a car accident, in this case by reducing the time over which the occupants move forward toward the dashboard or windshield. The airbag rapidly inflates, and just as rapidly begins to deflate, within the split-second that separates the car's collision and a person's collision with part of the car. As it deflates, it is receding toward the dashboard even as the driver's or passenger's body is being hurled toward the dashboard. It slows down impact, extending the amount of time during which the force is distributed.

By the same token, a skydiver or paratrooper does not hit the ground with legs outstretched: he or she would be likely to suffer a broken bone or worse from such a foolish stunt. Rather, as a parachutist prepares to land, he or she keeps knees bent, and upon impact immediately rolls over to the side. Thus, instead of experiencing the force of impact over a short period of time, the parachutist lengthens the amount of time that force is experienced, which reduces its effects.

The same principle applies if one were catching a water balloon. In order to keep it from bursting, one needs to catch the balloon in midair, then bring it to a stop slowly by "traveling" with it for a few feet before reducing its momentum down to zero. Once again, there is no way around the fact that one is attempting to bring about a substantial momentum change—a change equal in value to the momentum of the object in movement. Nonetheless, by increasing the time component of impulse, one reduces the effects of force.

In old *Superman* comics, the "Man of Steel" often caught unfortunate people who had fallen, or been pushed, out of tall buildings. The cartoons usually showed him, at a stationary position in midair, catching the person before he or she could hit the ground. In fact, this would not save their lives: the force component of the sudden momentum change involved in being caught would be enough to kill the person. Of course, it is a bit absurd to quibble over scientific accuracy in *Superman,* but in order to make the situation more plausible, the "Man of Steel" should have been shown catching the person, then slowly following through on the trajectory of the fall toward earth.

THE CRACK OF THE BAT: INCREASING IMPULSE. But what if—to once again turn the tables—a strong force is desired? This time, rather than two pool balls striking one another, consider what happens when a batter hits a baseball. Once more, the correlation between momentum change and impulse can create an advantage, if used properly.

As the pitcher hurls the ball toward home plate, it has a certain momentum; indeed, a pitch thrown by a major-league player can send the ball toward the batter at speeds around 100 MPH (160 km/h)—a ball having considerable momentum). In order to hit a line drive or "knock the ball out of the park," the batter must therefore cause a significant change in momentum.

Consider the momentum change in terms of the impulse components. The batter can only apply so much force, but it is possible to magnify impulse greatly by increasing the amount of time over which the force is delivered. This is known in

sports—and it applies as much in tennis or golf as in baseball—as "following through." By increasing the time of impact, the batter has increased impulse and thus, momentum change. Obviously, the mass of the ball has not been altered; the difference, then, is a change in velocity.

How is it possible that in earlier examples, the effects of force were decreased by increasing the time interval, whereas in the baseball illustration, an increase in time interval resulted in a more powerful impact? The answer relates to differences in direction and elasticity. The baseball and the bat are colliding head-on in a relatively elastic situation; by contrast, crumpling cars are inelastic. In the example of a person catching a water balloon, the catcher is moving in the same direction as the balloon, thus reducing momentum change. Even in the case of the paratrooper, the ground is stationary; it does not move toward the parachutist in the way that the baseball moves toward the bat.

WHERE TO LEARN MORE

Beiser, Arthur. *Physics*, 5th ed. Reading, MA: Addison-Wesley, 1991.

Bonnet, Robert L. and Dan Keen. *Science Fair Projects: Physics.* Illustrated by Frances Zweifel. New York: Sterling, 1999.

Fleisher, Paul. *Objects in Motion: Principles of Classical Mechanics.* Minneapolis, MN: Lerner Publications, 2002.

Gardner, Robert. *Experimenting with Science in Sports.* New York: F. Watts, 1993.

"Lesson 1: The Impulse Momentum Change Theorem" (Web site) <http://www.glenbrook.k12.il.us/gbssci/phys/Class/momentum/u4l1a.html> (March 19, 2001).

"Momentum" (Web site). <http://id.mind.net/~zona/mstm/physics/mechanics/momentum/momentum.html> (March 19, 2001).

Physlink.com (Web site). <http://www.physlink.com> (March 7, 2001).

Rutherford, F. James; Gerald Holton; and Fletcher G. Watson. *Project Physics.* New York: Holt, Rinehart, and Winston, 1981.

Schrier, Eric and William F. Allman. *Newton at the Bat: The Science in Sports.* New York: Charles Scribner's Sons, 1984.

Zubrowski, Bernie. *Raceways: Having Fun with Balls and Tracks.* Illustrated by Roy Doty. New York: William Morrow, 1985.

CENTRIPETAL FORCE

CONCEPT

Most people have heard of centripetal and centrifugal force. Though it may be somewhat difficult to keep track of which is which, chances are anyone who has heard of the two concepts remembers that one is the tendency of objects in rotation to move inward, and the other is the tendency of rotating objects to move outward. It may come as a surprise, then, to learn that there is no such thing, strictly speaking, as centrifugal (outward) force. There is only centripetal (inward) force and the inertia that makes objects in rotation under certain situations move outward, for example, a car making a turn, the movement of a roller coaster— even the spinning of a centrifuge.

HOW IT WORKS

Like many other principles in physics, centripetal force ultimately goes back to a few simple precepts relating to the basics of motion. Consider an object in uniform circular motion: an object moves around the center of a circle so that its speed is constant or unchanging.

The formula for speed—or rather, average speed—is distance divided by time; hence, people say, for instance, "miles (or kilometers) per hour." In the case of an object making a circle, distance is equal to the circumference, or distance around, the circle. From geometry, we know that the formula for calculating the circumference of a circle is $2\pi r$, where r is the radius, or the distance from the circumference to the center. The figure π may be rendered as 3.141592..., though in fact, it is an irrational number: the decimal figures continue forever without repetition or pattern.

From the above, it can be discerned that the formula for the average speed of an object moving around a circle is $2\pi r$ divided by time. Furthermore, we can see that there is a proportional relationship between radius and average speed. If the radius of a circle is doubled, but an object at the circle's periphery makes one complete revolution in the same amount of time as before, this means that the average speed has doubled as well. This can be shown by setting up two circles, one with a radius of 2, the other with a radius of 4, and using some arbitrary period of time—say, 2 seconds.

The above conclusion carries with it an interesting implication with regard to speeds at different points along the radius of a circle. Rather than comparing two points moving around the circumferences of two different circles—one twice as big as the other—in the same period of time, these two points could be on the same circle: one at the periphery, and one exactly halfway along the radius. Assuming they both traveled a complete circle in the same period of time, the proportional relationship described earlier would apply. This means, then, that the further out on the circle one goes, the greater the average speed.

VELOCITY = SPEED + DIRECTION

Speed is a scalar, meaning that it has magnitude but no specific direction; by contrast, velocity is a vector—a quantity with both a magnitude (that is, speed) and a direction. For an object in circular motion, the direction of velocity is the same as that in which the object is moving at any given point. Consider the example of the city of Atlanta, Georgia, and Interstate-285, one of several instances in which a city is surrounded by a "loop" highway. Local traffic reporters avoid giv-

TYPICALLY, A CENTRIFUGE CONSISTS OF A BASE; A ROTATING TUBE PERPENDICULAR TO THE BASE; AND VIALS ATTACHED BY MOVABLE CENTRIFUGE ARMS TO THE ROTATING TUBE. THE MOVABLE ARMS ARE HINGED AT THE TOP OF THE ROTATING TUBE, AND THUS CAN MOVE UPWARD AT AN ANGLE APPROACHING 90° TO THE TUBE. WHEN THE TUBE BEGINS TO SPIN, CENTRIPETAL FORCE PULLS THE MATERIAL IN THE VIALS TOWARD THE CENTER. *(Photograph by Charles D. Winters. National Audubon Society Collection/Photo Researchers, Inc. Reproduced by permission.)*

ing mere directional coordinates for spots on that highway (for instance, "southbound on 285"), because the area where traffic moves south depends on whether one is moving clockwise or counterclockwise. Hence, reporters usually say "southbound on the outer loop."

As with cars on I-285, the direction of the velocity vector for an object moving around a circle is a function entirely of its position and the direction of movement—clockwise or counterclockwise—for the circle itself. The direction of the individual velocity vector at any given point may be described as tangential; that is, describing a tangent, or a line that touches the circle at just one point. (By definition, a tangent line cannot intersect the circle.)

It follows, then, that the direction of an object in movement around a circle is changing; hence, its velocity is also changing—and this in turn

means that it is experiencing acceleration. As with the subject of centripetal force and "centrifugal force," most people have a mistaken view of acceleration, believing that it refers only to an increase in speed. In fact, acceleration is a change in velocity, and can thus refer either to a change in speed or direction. Nor must that change be a positive one; in other words, an object undergoing a reduction in speed is also experiencing acceleration.

The acceleration of an object in rotational motion is always toward the center of the circle. This may appear to go against common sense, which should indicate that acceleration moves in the same direction as velocity, but it can, in fact, be proven in a number of ways. One method would be by the addition of vectors, but a "hands-on" demonstration may be more enlightening than an abstract geometrical proof.

It is possible to make a simple accelerometer, a device for measuring acceleration, with a lit candle inside a glass. The candle should be standing at a 90°-angle to the bottom of the glass, attached to it by hot wax as you would affix a burning candle to a plate. When you hold the candle level, the flame points upward; but if you spin the glass in a circle, the flame will point toward the center of that circle—in the direction of acceleration.

MASS × ACCELERATION = FORCE

Since we have shown that acceleration exists for an object spinning around a circle, it is then possible for us to prove that the object experiences some type of force. The proof for this assertion lies in the second law of motion, which defines force as the product of mass and acceleration: hence, where there is acceleration and mass, there must be force. Force is always in the direction of acceleration, and therefore the force is directed toward the center of the circle.

In the above paragraph, we assumed the existence of mass, since all along the discussion has concerned an object spinning around a circle. By definition, an object—that is, an item of matter, rather than an imaginary point—possesses mass. Mass is a measure of inertia, which can be explained by the first law of motion: an object in motion tends to remain in motion, at the same speed and in the same direction (that is, at the same velocity) unless or until some outside force acts on it. This tendency to maintain velocity is inertia. Put another way, it is inertia that causes an object standing still to remain motionless, and

A CLOTHOID LOOP IN A ROLLER COASTER IN HAINES CITY, FLORIDA. AT THE TOP OF A LOOP, YOU FEEL LIGHTER THAN NORMAL; AT THE BOTTOM, YOU FEEL HEAVIER. *(The Purcell Team/Corbis. Reproduced by permission.)*

likewise, it is inertia which dictates that a moving object will "try" to keep moving.

CENTRIPETAL FORCE

Now that we have established the existence of a force in rotational motion, it is possible to give it a name: centripetal force, or the force that causes an object in uniform circular motion to move toward the center of the circular path. This is not a "new" kind of force; it is merely force as applied in circular or rotational motion, and it is absolutely essential. Hence, physicists speak of a "centripetal force requirement": in the absence of centripetal force, an object simply cannot turn. Instead, it will move in a straight line.

The Latin roots of centripetal together mean "seeking the center." What, then, of centrifugal, a word that means "fleeing the center"? It would be correct to say that there is such a thing as centrifugal motion; but centrifugal force is quite a different matter. The difference between centripetal force and a mere centrifugal tendency—a result of inertia rather than of force—can be explained by referring to a familiar example.

REAL-LIFE APPLICATIONS

RIDING IN A CAR

When you are riding in a car and the car accelerates, your body tends to move backward against the seat. Likewise, if the car stops suddenly, your body tends to move forward, in the direction of the dashboard. Note the language here: "tends to move" rather than "is pushed." To say that something is pushed would imply that a force has been applied, yet what is at work here is not a force, but inertia—the tendency of an object in motion to remain in motion, and an object at rest to remain at rest.

A car that is not moving is, by definition, at rest, and so is the rider. Once the car begins moving, thus experiencing a change in velocity, the rider's body still tends to remain in the fixed position. Hence, it is not a force that has pushed the rider backward against the seat; rather, force has pushed the car forward, and the seat moves up to meet the rider's back. When stopping, once again, there is a sudden change in velocity from a certain value down to zero. The rider, meanwhile, is continuing to move forward due to inertia, and thus, his or her body has a tendency to keep moving in the direction of the now-stationary dashboard.

This may seem a bit too simple to anyone who has studied inertia, but because the human mind has such a strong inclination to perceive inertia as a force in itself, it needs to be clarified in the most basic terms. This habit is similar to the experience you have when sitting in a vehicle that is standing still, while another vehicle alongside moves backward. In the first split-second of awareness, your mind tends to interpret the backward motion of the other car as forward motion on the part of the car in which you are sitting—even though your own car is standing still.

Now we will consider the effects of centripetal force, as well as the illusion of centrifugal force. When a car turns to the left, it is undergoing a form of rotation, describing a 90°-angle or one-quarter of a circle. Once again, your body experiences inertia, since it was in motion along with the car at the beginning of the turn, and thus you tend to move forward. The car, at the same time, has largely overcome its own inertia and moved into the leftward turn. Thus the car door itself is moving to the left. As the door meets the right side of your body, you have the sensation of being pushed outward against the door, but in fact what has happened is that the door has moved inward.

The illusion of centrifugal force is so deeply ingrained in the popular imagination that it warrants further discussion below. But while on the subject of riding in an automobile, we need to examine another illustration of centripetal force. It should be noted in this context that for a car to make a turn at all, there must be friction between the tires and the road. Friction is the force that resists motion when the surface of one object comes into contact with the surface of another; yet ironically, while opposing motion, friction also makes relative motion possible.

Suppose, then, that a driver applies the brakes while making a turn. This now adds a force tangential, or at a right angle, to the centripetal force. If this force is greater than the centripetal force—that is, if the car is moving too fast—the vehicle will slide forward rather than making the turn. The results, as anyone who has ever been in this situation will attest, can be disastrous.

The above highlights the significance of the centripetal force requirement: without a sufficient degree of centripetal force, an object simply cannot turn. Curves are usually banked to maximize centripetal force, meaning that the roadway tilts inward in the direction of the curve. This banking causes a change in velocity, and hence, in acceleration, resulting in an additional quantity known as reaction force, which provides the vehicle with the centripetal force necessary for making the turn.

The formula for calculating the angle at which a curve should be banked takes into account the car's speed and the angle of the curve, but does not include the mass of the vehicle itself. As a result, highway departments post signs stating the speed at which vehicles should make the turn, but these signs do not need to include specific statements regarding the weight of given models.

THE CENTRIFUGE

To return to the subject of "centrifugal force"—which, as noted earlier, is really just centrifugal motion—you might ask, "If there is no such thing as centrifugal force, how does a centrifuge work?" Used widely in medicine and a variety of sciences, a centrifuge is a device that separates particles within a liquid. One application, for instance, is to separate red blood cells from plasma.

Typically a centrifuge consists of a base; a rotating tube perpendicular to the base; and two vials attached by movable centrifuge arms to the rotating tube. The movable arms are hinged at the top of the rotating tube, and thus can move upward at an angle approaching 90° to the tube. When the tube begins to spin, centripetal force pulls the material in the vials toward the center.

Materials that are denser have greater inertia, and thus are less responsive to centripetal force. Hence, they seem to be pushed outward, but in fact what has happened is that the less dense material has been pulled inward. This leads to the separation of components, for instance, with plasma on the top and red blood cells on the bottom. Again, the plasma is not as dense, and thus is more easily pulled toward the center of rotation, whereas the red blood cells respond less, and consequently remain on the bottom.

The centrifuge was invented in 1883 by Carl de Laval (1845-1913), a Swedish engineer, who used it to separate cream from milk. During the 1920s, the chemist Theodor Svedberg (1884-1971), who was also Swedish, improved on Laval's work to create the ultracentrifuge, used for separating very small particles of similar weight.

In a typical ultracentrifuge, the vials are no larger than 0.2 in (0.6 cm) in diameter, and these may rotate at speeds of up to 230,000 revolutions per minute. Most centrifuges in use by industry rotate in a range between 1,000 and 15,000 revolutions per minute, but others with scientific applications rotate at a much higher rate, and can produce a force more than 25,000 times as great as that of gravity.

In 1994, researchers at the University of Colorado created a sort of super-centrifuge for simulating stresses applied to dams and other large structures. The instrument has just one centrifuge arm, measuring 19.69 ft (6 m), attached to which is a swinging basket containing a scale model of the structure to be tested. If the model is 1/50 the size of the actual structure, then the centrifuge is set to create a centripetal force 50 times that of gravity.

The Colorado centrifuge has also been used to test the effects of explosions on buildings. Because the combination of forces—centripetal, gravity, and that of the explosion itself—is so great, it takes a very small quantity of explosive to measure the effects of a blast on a model of the building.

Industrial uses of the centrifuge include that for which Laval invented it—separation of cream from milk—as well as the separation of impurities from other substances. Water can be removed from oil or jet fuel with a centrifuge, and likewise, waste-management agencies use it to separate solid materials from waste water prior to purifying the water itself.

Closer to home, a washing machine on spin cycle is a type of centrifuge. As the wet clothes spin, the water in them tends to move outward, separating from the clothes themselves. An even simpler, more down-to-earth centrifuge can be created by tying a fairly heavy weight to a rope and swinging it above one's head: once again, the weight behaves as though it were pushed outward, though in fact, it is only responding to inertia.

ROLLER COASTERS AND CENTRIPETAL FORCE

People ride roller coasters, of course, for the thrill they experience, but that thrill has more to do with centripetal force than with speed. By the late twentieth century, roller coasters capable of speeds above 90 MPH (144 km/h) began to appear in amusement parks around America; but prior to that time, the actual speeds of a roller coaster were not particularly impressive. Seldom, if ever, did they exceed that of a car moving down the highway. On the other hand, the acceleration and centripetal force generated on a roller coaster are high, conveying a sense of weightlessness (and sometimes the opposite of weightlessness) that is memorable indeed.

Few parts of a roller coaster ride are straight and flat—usually just those segments that mark the end of one ride and the beginning of another. The rest of the track is generally composed of dips and hills, banked turns, and in some cases, clothoid loops. The latter refers to a geometric shape known as a clothoid, rather like a teardrop upside-down.

Because of its shape, the clothoid has a much smaller radius at the top than at the bottom—a key factor in the operation of the roller coaster ride through these loops. In days past, roller-coaster designers used perfectly circular loops, which allowed cars to enter them at speeds that were too high, built too much force and resulted in injuries for riders. Eventually, engineers recognized the clothoid as a means of providing a safe, fun ride.

KEY TERMS

ACCELERATION: A change in velocity.

CENTRIFUGAL: A term describing the tendency of objects in uniform circular motion to move away from the center of the circular path. Though the term "centrifugal force" is often used, it is inertia, rather than force, that causes the object to move outward.

CENTRIPETAL FORCE: The force that causes an object in uniform circular motion to move toward the center of the circular path.

INERTIA: The tendency of an object in motion to remain in motion, and of an object at rest to remain at rest.

MASS: A measure of inertia, indicating the resistance of an object to a change in its motion—including a change in velocity.

SCALAR: A quantity that possesses only magnitude, with no specific direction.

Mass, time, and speed are all scalars. A scalar is contrasted with a vector.

SPEED: The rate at which the position of an object changes over a given period of time.

TANGENTIAL: Movement along a tangent, or a line that touches a circle at just one point and does not intersect the circle.

UNIFORM CIRCULAR MOTION: The motion of an object around the center of a circle in such a manner that speed is constant or unchanging.

VECTOR: A quantity that possesses both magnitude and direction. Velocity, acceleration, and weight (which involves the downward acceleration due to gravity) are examples of vectors. It is contrasted with a scalar.

VELOCITY: The speed of an object in a particular direction.

As you move into the clothoid loop, then up, then over, and down, you are constantly changing position. Speed, too, is changing. On the way up the loop, the roller coaster slows due to a decrease in kinetic energy, or the energy that an object possesses by virtue of its movement. At the top of the loop, the roller coaster has gained a great deal of potential energy, or the energy an object possesses by virtue of its position, and its kinetic energy is at zero. But once it starts going down the other side, kinetic energy—and with it speed—increases rapidly once again.

Throughout the ride, you experience two forces, gravity, or weight, and the force (due to motion) of the roller coaster itself, known as normal force. Like kinetic and potential energy—which rise and fall correspondingly with dips and hills—normal force and gravitational force are locked in a sort of "competition" throughout the roller-coaster rider. For the coaster to have its

proper effect, normal force must exceed that of gravity in most places.

The increase in normal force on a roller-coaster ride can be attributed to acceleration and centripetal motion, which cause you to experience something other than gravity. Hence, at the top of a loop, you feel lighter than normal, and at the bottom, heavier. In fact, there has been no real change in your weight: it is, like the idea of "centrifugal force" discussed earlier, a matter of perception.

WHERE TO LEARN MORE

Aylesworth, Thomas G. *Science at the Ball Game.* New York: Walker, 1977.

Beiser, Arthur. *Physics,* 5th ed. Reading, MA: Addison-Wesley, 1991.

Buller, Laura and Ron Taylor. *Forces of Nature.* Illustrations by John Hutchinson and Stan North. New York: Marshall Cavendish, 1990.

"Centrifugal Force—Rotational Motion." *National Aeronautics and Space Administration* (Web site).
<http://observe.ivv.nasa.gov/nasa/space/centrifugal/
centrifugal3.html> (March 5, 2001).

"*Circular and Satellite Motion*" (Web site).
<http://www.glenbrook.k12.il.us/gbssci/phys/Class/
circles/circtoc.html> (March 5, 2001).

Cobb, Vicki. *Why Doesn't the Earth Fall Up? And Other
Not Such Dumb Questions About Motion.* Illustrated
by Ted Enik. New York: Lodestar Books, 1988.

Lefkowitz, R. J. *Push! Pull! Stop! Go! A Book About Forces
and Motion.* Illustrated by June Goldsborough. New
York: Parents' Magazine Press, 1975.

"*Rotational Motion.*" *Physics Department, University of
Guelph* (Web site).
<http://www.physics.uoguelph.ca/tutorials/torque/>
(March 4, 2001).

Schaefer, Lola M. *Circular Movement.* Mankato, MN:
Pebble Books, 2000.

Snedden, Robert. *Forces.* Des Plaines, IL: Heinemann
Library, 1999.

Whyman, Kathryn. *Forces in Action.* New York: Gloucester Press, 1986.

FRICTION

CONCEPT

Friction is the force that resists motion when the surface of one object comes into contact with the surface of another. In a machine, friction reduces the mechanical advantage, or the ratio of output to input: an automobile, for instance, uses one-quarter of its energy on reducing friction. Yet, it is also friction in the tires that allows the car to stay on the road, and friction in the clutch that makes it possible to drive at all. From matches to machines to molecular structures, friction is one of the most significant phenomena in the physical world.

HOW IT WORKS

The definition of friction as "the force that resists motion when the surface of one object comes into contact with the surface of another" does not exactly identify what it is. Rather, the statement describes the manifestation of friction in terms of how other objects respond. A less sophisticated version of such a definition would explain electricity, for instance, as "the force that runs electrical appliances." The reason why friction cannot be more firmly identified is simple: physicists do not fully understand what it is.

Much the same could be said of force, defined by Sir Isaac Newton's (1642-1727) second law of motion as the product of mass multiplied acceleration. The fact is that force is so fundamental that it defies full explanation, except in terms of the elements that compose it, and compared to force, friction is relatively easy to identify. In fact, friction plays a part in the total force that must be opposed in order for movement to take place in many situations. So, too, does grav-

ity—and gravity, unlike force itself, is much easier to explain. Since gravity plays a role in friction, it is worthwhile to review its essentials.

Newton's first law of motion identifies inertia, a tendency of objects in the physical universe that is sometimes mistaken for friction. When an object is in motion or at rest, the first law states, it will remain in that state at a constant velocity (which is zero for an object at rest) unless or until an outside force acts on it. This tendency to remain in a given state of motion is inertia.

Inertia is not a force: on the contrary, a very small quantity of force may accelerate an object, thus overcoming its inertia. Inertia is, however, a component of force, since mass is a measure of inertia. In the case of gravitational force, mass is multiplied by the acceleration due to gravity, which is equal to 32 ft (9.8 m)/sec². People in everyday life are familiar with another term for gravitational force: weight.

Weight, in turn, is an all-important factor in friction, as revealed in the three laws governing the friction between an object at rest and the surface on which it sits. According to the first of these, friction is proportional to the weight of the object. The second law states that friction is not determined by the surface area of the object—that is, the area that touches the surface on which the object rests. In fact, the contact area between object and surface is a dependant variable, a function of weight.

The second law might seem obvious if one were thinking of a relatively elastic object—say, a garbage bag filled with newspapers sitting on the finished concrete floor of a garage. Clearly as more newspapers are added, thus increasing the weight, its surface area would increase as well. But what if one were to compare a large cardboard box (the kind, for instance, in which televisions or computers are shipped) with an ordinary concrete block of the type used in foundations for residential construction? Obviously, the block has more friction against the concrete floor; but at the same time, it is clear that despite its greater weight, the block has less surface area than the box. How can this be?

The answer is that "surface area" is quite literally more than meets the eye. Friction itself occurs at a level invisible to the naked eye, and involves the adhesive forces between molecules on surfaces pushed together by the force of weight. This is similar to the manner in which, when viewed through a high-powered lens, two complementary patches of Velcro™ are revealed as a forest of hooks on the one hand, and a sea of loops on the other.

On a much more intensified level, that of molecular structure, the surfaces of objects appear as mountains and valleys. Nothing, in fact, is smooth when viewed on this scale, and hence, from a molecular perspective, it becomes clear that two objects in contact actually touch one another only in places. An increase of weight, however, begins pushing objects together, causing an increase in the actual—that is, the molecular—area of contact. Hence area of contact is proportional to weight.

Just as the second law regarding friction states that surface area does not determine friction (but rather, weight determines surface area), the third law holds that friction is independent of the speed at which an object is moving along a surface—provided that speed is not zero. The reason for this provision is that an object with no speed (that is, one standing perfectly still) is subject to the most powerful form of friction, static friction.

The latter is the friction that an object at rest must overcome to be set in motion; however, this should not be confused with inertia, which is relatively easy to overcome through the use of force. Inertia, in fact, is far less complicated than static friction, involving only mass rather than weight. Nor is inertia affected by the composition of the materials touching one another.

As stated earlier, friction is proportional to weight, which suggests that another factor is involved. And indeed there is another factor, known as coefficient of friction. The latter, desig-

THE RAISED TREAD ON AN AUTOMOBILE'S TIRES, COUPLED WITH THE ROUGHENED ROAD SURFACE, PROVIDES SUFFICIENT FRICTION FOR THE DRIVER TO BE ABLE TO TURN THE CAR AND TO STOP. (*Photograph by Martyn Goddard/Corbis. Reproduced by permission.*)

nated by the Greek letter mu (μ), is constant for any two types of surface in contact with one another, and provides a means of comparing the friction between them to that between other surfaces. For instance, the coefficient of static friction for wood on wood is 0.5; but for metal on metal with lubrication in between, it is only 0.03. A rubber tire on dry concrete yields the highest coefficient of static friction, 1.0, which is desirable in that particular situation.

Coefficients are much lower for the second type of friction, sliding friction, the frictional resistance experienced by a body in motion. Whereas the earlier figures measured the relative resistance to putting certain objects into motion, the sliding-friction coefficient indicates the relative resistance against those objects once they are moving. To use the same materials mentioned above, the coefficient of sliding friction for wood on wood is 0.3; for two lubricated metals 0.03 (no change); and for a rubber tire on dry concrete 0.7.

Finally, there is a third variety of friction, one in which coefficients are so low as to be neg-

ACTOR JACK BUCHANAN LIT THIS MATCH USING A SIM-PLE DISPLAY OF FRICTION: DRAGGING THE MATCH AGAINST THE BACK OF THE MATCHBOX. *(Photograph by Hulton-Deutsch Collection/Corbis. Reproduced by permission.)*

ligible: rolling friction, or the frictional resistance that a wheeled object experiences when it rolls over a relatively smooth, flat surface. In ideal circumstances, in fact, there would be absolutely no resistance between a wheel and a road. However, there exists no ideal—that is, perfectly rigid—wheel or road; both objects "give" in response to the other, the wheel by flattening somewhat and the road by experiencing indentation.

Up to this point, coefficients of friction have been discussed purely in comparative terms, but in fact, they serve a function in computing frictional force—that is, the force that must be overcome to set an object in motion, or to keep it in motion. Frictional force is equal to the coefficient of friction multiplied by normal force—that is, the perpendicular force that one object or surface exerts on another. On a horizontal plane, normal force is equal to gravity and hence weight. In this equation, the coefficient of friction establishes a limit to frictional force: in order to move an object in a given situation, one must exert a force in excess of the frictional force that keeps it from moving.

REAL-LIFE APPLICATIONS

SELF-MOTIVATION THROUGH FRICTION

Friction, in fact, always opposes movement; why, then, is friction necessary—as indeed it is—for walking, and for keeping a car on the road? The answer relates to the differences between friction and inertia alluded to earlier. In situations of static friction, it is easy to see how a person might confuse friction with inertia, since both serve to keep an object from moving. In situations of sliding or rolling friction, however, it is easier to see the difference between friction and inertia.

Whereas friction is always opposed to movement, inertia is not. When an object is not moving, its inertia does oppose movement—but when the object is in motion, then inertia resists stopping. In the absence of friction or other forces, inertia allows an object to remain in motion forever. Imagine a hockey player hitting a puck across a very, very large rink. Because ice has a much smaller coefficient of friction with regard to the puck than does dirt or asphalt, the puck will travel much further. Still, however, the ice has some friction, and, therefore, the puck will come to a stop at some point.

Now suppose that instead of ice, the surface and objects in contact with it were friction-free, possessing a coefficient of zero. Then what would happen if the player hit the puck? Assuming for the purposes of this thought experiment, that the rink covered the entire surface of Earth, it would travel and travel and travel, ultimately going around the planet. It would never stop, because there would be no friction to stop it, and therefore inertia would have free rein.

The same would be true if one were to firmly push the hockey player with enough force (small in the absence of friction) to set him in motion: he would continue riding around the planet indefinitely, borne by his skates. But what if instead of being set in motion, the hockey player tried to set himself in motion by the action of his skates against the rink's surface?

He would be unable to move even a hair's breadth. The fact is that while static friction opposes the movement of an object from a position of rest to a state of motion, it may—assuming it can be overcome to begin motion at all—be indispensable to that movement. As with the skater in perpetual motion across the rink, the absence of friction means that inertia is "in control;" with friction, however, it is possible to overcome inertia.

FRICTION IN DRIVING A CAR

The same principle applies to a car's tires: if they were perfectly smooth—and, to make matters worse, the road were perfectly smooth as well—the vehicle would keep moving forward when the driver attempted to stop. For this reason, tires are designed with raised tread to maintain a high degree of friction, gripping the road tightly and dispersing water when the roadway is wet.

The force of friction, in fact, pervades the entire operation of a car, and makes it possible for the tires themselves to turn. The turning force, or torque, that the driver exerts on the steering wheel is converted into forces that drive the tires, and these in turn use friction to provide traction. Between steering wheel and tires, of course, are a number of steps, with the engine rotating the crankshaft and transmitting power to the clutch, which applies friction to translate the motion of the crankshaft to the gearbox.

When the driver of a car with a manual transmission presses down on the clutch pedal, this disengages the clutch itself. A clutch is a circular mechanism containing (among other things) a pressure plate, which lifts off the clutch plate. As a result, the flywheel—the instrument that actually transmits force from the crankshaft—is disengaged from the transmission shaft. With the clutch thus disengaged, the driver changes gears, and after the driver releases the clutch pedal, springs return the pressure plate and the clutch plate to their place against the flywheel. The flywheel then turns the transmission shaft.

Controlled friction in the clutch makes this operation possible; likewise the synchromesh within the gearbox uses friction to bring the gearwheels into alignment. This is a complicated process, but at the heart of it is an engagement of gear teeth in which friction forces them to come to the same speed.

Friction is also essential to stopping a car—not just with regard to the tires, but also with respect to the brakes. Whether they are disk brakes or drum brakes, two elements must come together with a force more powerful than the engine's, and friction provides that needed force. In disk brakes, brake pads apply friction to both sides of the spinning disks, and in drum brakes, brake shoes transmit friction to the inside of a spinning drum. This braking force is then transmitted to the tires, which apply friction to the road and thus stop the car.

EFFICIENCY AND FRICTION

The automobile is just one among many examples of a machine that could not operate without friction. The same is true of simple machines such as screws, as well as nails, pliers, bolts, and forceps. At the heart of this relationship is a paradox, however, because friction inevitably reduces the efficiency of machines: a car, as noted earlier, exerts fully one-quarter of its power simply on overcoming the force of friction both within its engine and from air resistance as it travels down the road.

In scientific terms, efficiency or mechanical advantage is measured by the ratio of force output to force input. Clearly, in most situations it is ideal to maximize output and minimize input, and over the years inventors have dreamed of creating a mechanism—a perpetual motion machine—to do just that. In this idealized machine, one would apply a certain amount of energy to set it into operation, and then it would never stop; hence the ratio of output to input would be nearly infinite.

Unfortunately, the perpetual motion machine is a dream every bit as elusive as the mythical Fountain of Youth. At least this is true on Earth, where friction will always cause a system to lose kinetic energy, or the energy of movement. No matter what the design, the machine will eventually lose energy and stop; however, this is not true in outer space, where friction is very small—though it still exists. In space it might truly be possible to set a machine in motion and let inertia do the rest; thus perhaps perpetual motion actually is more than a dream.

It should also be noted that mechanical advantage is not always desirable. A screw is a highly inefficient machine: one puts much more force into screwing it in than the screw will exert once it is in place. Yet this is exactly the purpose of a screw: an "efficient" one, or one that worked its way back out of the place into which it had been screwed, would in fact be of little use.

Once again, it is friction that provides a screw with its strangely efficient form of inefficiency. Nonetheless, friction, in spite of the advantages discussed above, is as undesirable as it is desirable. With friction, there is always something lost; however, there is a physical law that energy does not simply disappear; it just changes form. In the case of friction, the energy that could go to moving the machine is instead translated into sound—or even worse, heat.

WHEN SPARKS FLY

In movement involving friction, molecules vibrate, bringing about a rise in temperature. This can be easily demonstrated by simply rubbing one's hands together quickly, as a person is apt to do when cold: heat increases. For a machine composed of metal parts, this increase in temperature can be disastrous, leading to serious wear and damage. This is why various forms of lubricant are applied to systems subject to friction.

An automobile uses grease and oil, as well as ball bearings, which are tiny uniform balls of metal that imitate the behavior of oil-based substances on a large scale. In a molecule of oil—whether it is a petroleum-related oil or the type of oil that comes from living things—positive and negative electrical charges are distributed throughout the molecule. By contrast, in water the positive charges are at one end of the molecule and the negative at the other. This creates a tight bond as the positive end of one water molecule adheres to the negative end of another. With oil, the relative absence of attraction between molecules means that each is in effect a tiny ball separate from the others. The ball-like molecules "roll" between metal elements, providing the buffer necessary to reduce friction.

Yet for every statement one can make concerning friction, there is always another statement with which to counter it. Earlier it was noted that the wheel, because it reduced friction greatly, provided an enormous technological boost to societies. Yet long before the wheel—hundreds of thousands of years ago—an even more important technological breakthrough occurred when humans made a discovery that depended on maximizing friction: fire, or rather the means of making fire. Unlike the wheel, fire occurs in nature, and can spring from a number of causes, but when human beings harnessed the means of making fire on their own, they had to rely on the heat that comes from friction.

By the early nineteenth century, inventors had developed an easy method of creating fire by using a little stick with a phosphorus tip. This stick, of course, is known as a match. In a strike-anywhere match, the head contains all the chemicals needed to create a spark. To ignite this type of match, one need only create frictional heat by rubbing it against a surface, such as sandpaper, with a high coefficient of friction.

The chemicals necessary for ignition in safety matches, on the other hand, are dispersed between the match head and a treated strip, usually found on the side of the matchbox or matchbook. The chemicals on the tip and those on the striking surface must come into contact for ignition to occur, but once again, there must be friction between the match head and the striking pad. Water reduces friction with its heavy bond, as it does with a car's tires on a rainy day, which explains why matches are useless when wet.

THE OUTER LIMITS OF FRICTION

Clearly friction is a complex subject, and the discoveries of modern physics only promise to add to that complexity. In a February 1999 online article for *Physical Review Focus,* Dana Mackenzie reported that "Engineers hope to make microscopic engines and gears as ordinary in our lives as microscopic circuits are today. But before this dream becomes a reality, they will have to deal with laws of friction that are very different from those that apply to ordinary-sized machines."

The earlier statement that friction is proportional to weight, in fact, applies only in the realm of classical physics. The latter term refers to the studies of physicists up to the end of the nineteenth century, when the concerns were chiefly the workings of large objects whose operations could be discerned by the senses. Modern physics, on the other hand, focuses on atomic and molecular structures, and addresses physical behaviors that could not have been imagined prior to the twentieth century.

According to studies conducted by Alan Burns and others at Sandia National Laboratories in Albuquerque, New Mexico, molecular interactions between objects in very close proximity create a type of friction involving repulsion rather than attraction. This completely upsets the model of friction understood for more than a century, and indicates new frontiers of discovery concerning the workings of friction at a molecular level.

KEY TERMS

ACCELERATION: A change in velocity.

COEFFICIENT OF FRICTION: A figure, constant for a particular pair of surfaces in contact, that can be multiplied by the normal force between them to calculate the frictional force they experience.

FORCE: The product of mass multiplied by acceleration.

FRICTION: The force that resists motion when the surface of one object comes into contact with the surface of another. Varieties including sliding friction, static friction, and rolling friction. The degree of friction between two specific surfaces is proportional to coefficient of friction.

FRICTIONAL FORCE: The force necessary to set an object in motion, or to keep it in motion; equal to normal force multiplied by coefficient of friction.

INERTIA: The tendency of an object in motion to remain in motion, and of an object at rest to remain at rest.

MASS: A measure of inertia, indicating the resistance of an object to a change in its motion—including a change in velocity.

MECHANICAL ADVANTAGE: The ratio of force output to force input in a machine.

NORMAL FORCE: The perpendicular force with which two objects press against one another. On a plane without any incline (which would add acceleration in addition to that of gravity) normal force is the same as weight.

ROLLING FRICTION: The frictional resistance that a circular object experiences when it rolls over a relatively smooth, flat surface. With a coefficient of friction much smaller than that of sliding friction, rolling friction involves by far the least amount of resistance among the three varieties of friction.

SLIDING FRICTION: The frictional resistance experienced by a body in motion. Here the coefficient of friction is greater than that for rolling friction, but less than for static friction.

SPEED: The rate at which the position of an object changes over a given period of time.

STATIC FRICTION: The frictional resistance that a stationary object must overcome before it can go into motion. Its coefficient of friction is greater than that of sliding friction, and thus largest among the three varieties of friction.

VELOCITY: The speed of an object in a particular direction.

WEIGHT: A measure of the gravitational force on an object; the product of mass multiplied by the acceleration due to gravity.

WHERE TO LEARN MORE

Beiser, Arthur. *Physics,* 5th ed. Reading, MA: Addison-Wesley, 1991.

Buller, Laura and Ron Taylor. *Forces of Nature.* Illustrations by John Hutchinson and Stan North. New York: Marshall Cavendish, 1990.

Dixon, Malcolm and Karen Smith. *Forces and Movement.* Mankato, MN: Smart Apple Media, 1998.

"Friction." How Stuff Works (Web site). <http://www.howstuffworks.com/search/index.htm?words=friction> (March 8, 2001).

"Friction and Interactions" (Web site). <http://www.cord.edu/dept/physics/p128/lecture99_12.html> (March 8, 2001).

Levy, Matthys and Richard Panchyk. *Engineering the City: How Infrastructure Works.* Chicago: Chicago Review Press, 2000.

Macaulay, David. *The New Way Things Work.* Boston: Houghton Mifflin, 1998.

Mackenzie, Dana. *"Friction of Molecules." Physical Review Focus* (Web site). <http://focus.aps.org/v3/st9.html (March 8, 2001).

Rutherford, F. James; Gerald Holton; and Fletcher G. Watson. *Project Physics.* New York: Holt, Rinehart, and Winston, 1981.

Skateboard Science (Web site). <http://www.exploratori-um.edu/skateboarding/ (March 8, 2001).

Suplee, Curt. *Everyday Science Explained.* Washington, D.C.: National Geographic Society, 1996.

LAWS OF MOTION

CONCEPT

In all the universe, there are few ideas more fundamental than those expressed in the three laws of motion. Together these explain why it is relatively difficult to start moving, and then to stop moving; how much force is needed to start or stop in a given situation; and how one force relates to another. In their beauty and simplicity, these precepts are as compelling as a poem, and like the best of poetry, they identify something that resonates through all of life. The applications of these three laws are literally endless: from the planets moving through the cosmos to the first seconds of a car crash to the action that takes place when a person walks. Indeed, the laws of motion are such a part of daily life that terms such as inertia, force, and reaction extend into the realm of metaphor, describing emotional processes as much as physical ones.

HOW IT WORKS

The three laws of motion are fundamental to mechanics, or the study of bodies in motion. These laws may be stated in a number of ways, assuming they contain all the components identified by Sir Isaac Newton (1642-1727). It is on his formulation that the following are based:

The Three Laws of Motion

- First law of motion: An object at rest will remain at rest, and an object in motion will remain in motion, at a constant velocity unless or until outside forces act upon it.
- Second law of motion: The net force acting upon an object is a product of its mass multiplied by its acceleration.
- Third law of motion: When one object exerts a force on another, the second object exerts on the first a force equal in magnitude but opposite in direction.

LAWS OF MAN VS. LAWS OF NATURE

These, of course, are not "laws" in the sense that people normally understand that term. Human laws, such as injunctions against stealing or parking in a fire lane, are prescriptive: they state how the world should be. Behind the prescriptive statements of civic law, backing them up and giving them impact, is a mechanism—police, courts, and penalties—for ensuring that citizens obey.

A scientific law operates in exactly the opposite fashion. Here the mechanism for ensuring that nature "obeys" the law comes first, and the "law" itself is merely a descriptive statement concerning evident behavior. With human or civic law, it is clearly possible to disobey: hence, the justice system exists to discourage disobedience. In the case of scientific law, disobedience is clearly impossible—and if it were not, the law would have to be amended.

This is not to say, however, that scientific laws extend beyond their own narrowly defined limits. On Earth, the intrusion of outside forces—most notably friction—prevents objects from behaving perfectly according to the first law of motion. The common-sense definition of friction calls to mind, for instance, the action that a match makes as it is being struck; in its broader scientific meaning, however, friction can be defined as any force that resists relative motion between two bodies in contact.

THE CARGO BAY OF THE SPACE SHUTTLE DISCOVERY, SHOWN JUST AFTER RELEASING A SATELLITE. ONCE RELEASED INTO THE FRICTIONLESS VACUUM AROUND EARTH, THE SATELLITE WILL MOVE INDEFINITELY AROUND EARTH WITHOUT NEED FOR THE MOTIVE POWER OF AN ENGINE. THE PLANET'S GRAVITY KEEPS IT AT A FIXED HEIGHT, AND AT THAT HEIGHT, IT COULD THEORETICALLY CIRCLE EARTH FOREVER. *(Corbis. Reproduced by permission.)*

The operations of physical forces on Earth are continually subject to friction, and this includes not only dry bodies, but liquids, for instance, which are subject to viscosity, or internal friction. Air itself is subject to viscosity, which prevents objects from behaving perfectly in accordance with the first law of motion. Other forces, most notably that of gravity, also come into play to stop objects from moving endlessly once they have been set in motion.

The vacuum of outer space presents scientists with the most perfect natural laboratory for testing the first law of motion: in theory, if they were to send a spacecraft beyond Earth's orbital radius, it would continue travelling indefinitely. But even this craft would likely run into another object, such as a planet, and would then be drawn into its orbit. In such a case, however, it can be said that outside forces have acted upon it, and thus the first law of motion stands.

The orbit of a satellite around Earth illustrates both the truth of the first law, as well as the forces that limit it. To break the force of gravity, a powered spacecraft has to propel the satellite into the exosphere. Yet once it has reached the frictionless vacuum, the satellite will move indefinitely around Earth without need for the motive power of an engine—it will get a "free ride," thanks to the first law of motion. Unlike the hypothetical spacecraft described above, however, it will not go spinning into space, because it is still too close to Earth. The planet's gravity keeps it at a fixed height, and at that height, it could theoretically circle Earth forever.

The first law of motion deserves such particular notice, not simply because it is the first law. Nonetheless, it is first for a reason, because it establishes a framework for describing the behavior of an object in motion. The second law identifies a means of determining the force necessary to move an object, or to stop it from moving, and the third law provides a picture of what happens when two objects exert force on one another.

The first law warrants special attention because of misunderstandings concerning it, which spawned a debate that lasted nearly twenty centuries. Aristotle (384-322 B.C.) was the first scientist to address seriously what is now known as the first law of motion, though in fact, that term would not be coined until about two thousand years after his death. As its title suggests, his *Physics* was a seminal work, a book in which Aristotle attempted to give form to, and thus define the territory of, studies regarding the operation of physical processes. Despite the great philosopher's many achievements, however, *Physics* is a highly flawed work, particularly with regard to what became known as his theory of impetus—that is, the phenomena addressed in the first law of motion.

ARISTOTLE'S MISTAKE

According to Aristotle, a moving object requires a continual application of force to keep it moving: once that force is no longer applied, it ceases to move. You might object that, when a ball is in flight, the force necessary to move it has already been applied: a person has thrown the ball, and it is now on a path that will eventually be stopped by the force of gravity. Aristotle, however, would have maintained that the air itself acts as a force to keep the ball in flight, and that when the ball drops—of course he had no concept of "gravity" as such—it is because the force of the air on the ball is no longer in effect.

These notions might seem patently absurd to the modern mind, but they went virtually unchallenged for a thousand years. Then in the sixth century A.D., the Byzantine philosopher Johannes Philoponus (c. 490-570) wrote a critique of *Physics*. In what sounds very much like a precursor to the first law of motion, Philoponus held that a body will keep moving in the absence of friction or opposition.

He further maintained that velocity is proportional to the positive difference between force and resistance—in other words, that the force propelling an object must be greater than the resistance. As long as force exceeds resistance, Philoponus held, a body will remain in motion. This in fact is true: if you want to push a refrigerator across a carpeted floor, you have to exert enough force not only to push the refrigerator, but also to overcome the friction from the floor itself.

The Arab philosophers Ibn Sina (Avicenna; 980-1037) and Ibn Bâjja (Avempace; fl. c. 1100) defended Philoponus's position, and the French scholar Peter John Olivi (1248-1298) became the first Western thinker to critique Aristotle's statements on impetus. Real progress on the subject, however, did not resume until the time of Jean Buridan (1300-1358), a French physicist who went much further than Philoponus had eight centuries earlier.

In his writing, Buridan offered an amazingly accurate analysis of impetus that prefigured all three laws of motion. It was Buridan's position that one object imparts to another a certain amount of power, in proportion to its velocity and mass, that causes the second object to move a certain distance. This, as will be shown below, was amazingly close to actual fact. He was also correct in stating that the weight of an object may increase or decrease its speed, depending on other circumstances, and that air resistance slows an object in motion.

The true breakthrough in understanding the laws of motion, however, came as the result of work done by three extraordinary men whose lives stretched across nearly 250 years. First came Nicolaus Copernicus (1473-1543), who advanced what was then a heretical notion: that Earth, rather than being the center of the universe, revolved around the Sun along with the other planets. Copernicus made his case purely in terms of astronomy, however, with no direct reference to physics.

GALILEO'S CHALLENGE: THE COPERNICAN MODEL

Galileo Galilei (1564-1642) likewise embraced a heliocentric (Sun-centered) model of the universe—a position the Church forced him to renounce publicly on pain of death. As a result of his censure, Galileo realized that in order to prove the Copernican model, it would be necessary to show why the planets remain in motion as they do. In explaining this, he coined the term inertia to describe the tendency of an object in motion to remain in motion, and an object at rest to remain at rest. Galileo's observations, in fact, formed the foundation for the laws of motion.

In the years that followed Galileo's death, some of the world's greatest scientific minds became involved in the effort to understand the forces that kept the planets in motion around the Sun. Among them were Johannes Kepler (1571-1630), Robert Hooke (1635-1703), and Edmund Halley (1656-1742). As a result of a dispute between Hooke and Sir Christopher Wren (1632-1723) over the subject, Halley brought the question to his esteemed friend Isaac Newton. As it turned out, Newton had long been considering the possibility that certain laws of motion existed, and these he presented in definitive form in his *Principia* (1687).

The impact of the Newton's book, which included his observations on gravity, was nothing short of breathtaking. For the next three centuries, human imagination would be ruled by the Newtonian framework, and only in the twentieth century would the onset of new ideas reveal its limitations. Yet even today, outside the realm of quantum mechanics and relativity theory—in other words, in the world of everyday experience—Newton's laws of motion remain firmly in place.

REAL-LIFE APPLICATIONS

THE FIRST LAW OF MOTION IN A CAR CRASH

It is now appropriate to return to the first law of motion, as formulated by Newton: an object at rest will remain at rest, and an object in motion will remain in motion, at a constant velocity unless or until outside forces act upon it. Examples of this first law in action are literally unlimited.

One of the best illustrations, in fact, involves something completely outside the experience of Newton himself: an automobile. As a car moves down the highway, it has a tendency to remain in motion unless some outside force changes its

velocity. The latter term, though it is commonly understood to be the same as speed, is in fact more specific: velocity can be defined as the speed of an object in a particular direction.

In a car moving forward at a fixed rate of 60 MPH (96 km/h), everything in the car—driver, passengers, objects on the seats or in the trunk—is also moving forward at the same rate. If that car then runs into a brick wall, its motion will be stopped, and quite abruptly. But though its motion has stopped, in the split seconds after the crash it is still responding to inertia: rather than bouncing off the brick wall, it will continue plowing into it.

What, then, of the people and objects in the car? They too will continue to move forward in response to inertia. Though the car has been stopped by an outside force, those inside experience that force indirectly, and in the fragment of time after the car itself has stopped, they continue to move forward—unfortunately, straight into the dashboard or windshield.

It should also be clear from this example exactly why seatbelts, headrests, and airbags in automobiles are vitally important. Attorneys may file lawsuits regarding a client's injuries from airbags, and homespun opponents of the seatbelt may furnish a wealth of anecdotal evidence concerning people who allegedly died in an accident because they were wearing seatbelts; nonetheless, the first law of motion is on the side of these protective devices.

The admittedly gruesome illustration of a car hitting a brick wall assumes that the driver has not applied the brakes—an example of an outside force changing velocity—or has done so too late. In any case, the brakes themselves, if applied too abruptly, can present a hazard, and again, the significant factor here is inertia. Like the brick wall, brakes stop the car, but there is nothing to stop the driver and/or passengers. Nothing, that is, except protective devices: the seat belt to keep the person's body in place, the airbag to cushion its blow, and the headrest to prevent whiplash in rear-end collisions.

Inertia also explains what happens to a car when the driver makes a sharp, sudden turn. Suppose you are is riding in the passenger seat of a car moving straight ahead, when suddenly the driver makes a quick left turn. Though the car's tires turn instantly, everything in the vehicle—its frame, its tires, and its contents—is still respond-

WHEN A VEHICLE HITS A WALL, AS SHOWN HERE IN A CRASH TEST, ITS MOTION WILL BE STOPPED, AND QUITE ABRUPT-LY. BUT THOUGH ITS MOTION HAS STOPPED, IN THE SPLIT SECONDS AFTER THE CRASH IT IS STILL RESPONDING TO INERTIA: RATHER THAN BOUNCING OFF THE BRICK WALL, IT WILL CONTINUE PLOWING INTO IT. *(Photograph by Tim Wright/Corbis. Reproduced by permission.)*

ing to inertia, and therefore "wants" to move forward even as it is turning to the left.

As the car turns, the tires may respond to this shift in direction by squealing: their rubber surfaces were moving forward, and with the sudden turn, the rubber skids across the pavement like a hard eraser on fine paper. The higher the original speed, of course, the greater the likelihood the tires will squeal. At very high speeds, it is possible the car may seem to make the turn "on two wheels"—that is, its two outer tires. It is even possible that the original speed was so high, and the turn so sharp, that the driver loses control of the car.

Here inertia is to blame: the car simply cannot make the change in velocity (which, again, refers both to speed and direction) in time. Even in less severe situations, you are likely to feel that you have been thrown outward against the rider's side door. But as in the car-and-brick-wall illustration used earlier, it is the car itself that first experiences the change in velocity, and thus it responds first. You, the passenger, then, are moving forward even as the car has turned; therefore, rather than being thrown outward, you are simply meeting the leftward-moving door even as you push forward.

FROM PARLOR TRICKS TO SPACE SHIPS

It would be wrong to conclude from the car-related illustrations above that inertia is always harmful. In fact it can help every bit as much as it can potentially harm, a fact shown by two quite different scenarios.

The beneficial quality to the first scenario may be dubious: it is, after all, a mere parlor trick, albeit an entertaining one. In this famous stunt, with which most people are familiar even if they have never seen it, a full table setting is placed on a table with a tablecloth, and a skillful practitioner manages to whisk the cloth out from under the dishes without upsetting so much as a glass. To some this trick seems like true magic, or at least sleight of hand; but under the right conditions, it can be done. (This information, however, carries with it the warning, "Do not try this at home!")

To make the trick work, several things must align. Most importantly, the person doing it has to be skilled and practiced at performing the feat. On a physical level, it is best to minimize the friction between the cloth and settings on the one hand, and the cloth and table on the other. It is also important to maximize the mass (a property that will be discussed below) of the table settings, thus making them resistant to movement. Hence, inertia—which is measured by mass—plays a key role in making the tablecloth trick work.

You might question the value of the tablecloth stunt, but it is not hard to recognize the importance of the inertial navigation system (INS) that guides planes across the sky. Prior to the 1970s, when INS made its appearance, navigation techniques for boats and planes relied on reference to external points: the Sun, the stars, the magnetic North Pole, or even nearby areas of land. This created all sorts of possibilities for error: for instance, navigation by magnet (that is, a compass) became virtually useless in the polar regions of the Arctic and Antarctic.

By contrast, the INS uses no outside points of reference: it navigates purely by sensing the inertial force that results from changes in velocity. Not only does it function as well near the poles as it does at the equator, it is difficult to tamper with an INS, which uses accelerometers in a sealed, shielded container. By contrast, radio signals or radar can be "confused" by signals from the ground—as, for instance, from an enemy unit during wartime.

As the plane moves along, its INS measures movement along all three geometrical axes, and provides a continuous stream of data regarding acceleration, velocity, and displacement. Thanks to this system, it is possible for a pilot leaving California for Japan to enter the coordinates of a half-dozen points along the plane's flight path, and let the INS guide the autopilot the rest of the way.

Yet INS has its limitations, as illustrated by the tragedy that occurred aboard Korean Air Lines (KAL) Flight 007 on September 1, 1983. The plane, which contained 269 people and crew members, departed Anchorage, Alaska, on course for Seoul, South Korea. The route they would fly was an established one called "R-20," and it appears that all the information regarding their flight plan had been entered correctly in the plane's INS.

This information included coordinates for internationally recognized points of reference, actually just spots on the northern Pacific with names such as NABIE, NUKKS, NEEVA, and so on, to NOKKA, thirty minutes east of Japan. Yet, just after passing the fishing village of Bethel, Alaska, on the Bering Sea, the plane started to veer off course, and ultimately wandered into

Soviet airspace over the Kamchatka Peninsula and later Sakhalin Island. There a Soviet Su-15 shot it down, killing all the plane's passengers.

In the aftermath of the Flight 007 shoot-down, the Soviets accused the United States and South Korea of sending a spy plane into their airspace. (Among the passengers was Larry McDonald, a staunchly anti-Communist Congressman from Georgia.) It is more likely, however, that the tragedy of 007 resulted from errors in navigation which probably had something to do with the INS. The fact is that the R-20 flight plan had been designed to keep aircraft well out of Soviet airspace, and at the time KAL 007 passed over Kamchatka, it should have been 200 mi (320 km) to the east—over the Sea of Japan.

Among the problems in navigating a transpacific flight is the curvature of the Earth, combined with the fact that the planet continues to rotate as the aircraft moves. On such long flights, it is impossible to "pretend," as on a short flight, that Earth is flat: coordinates have to be adjusted for the rounded surface of the planet. In addition, the flight plan must take into account that (in the case of a flight from California to Japan), Earth is moving eastward even as the plane moves westward. The INS aboard KAL 007 may simply have failed to correct for these factors, and thus the error compounded as the plane moved further. In any case, INS will eventually be rendered obsolete by another form of navigation technology: the global positioning satellite (GPS) system.

UNDERSTANDING INERTIA

From examples used above, it should be clear that inertia is a more complex topic than you might immediately guess. In fact, inertia as a process is rather straightforward, but confusion regarding its meaning has turned it into a complicated subject.

In everyday terminology, people typically use the word inertia to describe the tendency of a stationary object to remain in place. This is particularly so when the word is used metaphorically: as suggested earlier, the concept of inertia, like numerous other aspects of the laws of motion, is often applied to personal or emotional processes as much as the physical. Hence, you could say, for instance, "He might have changed professions and made more money, but inertia kept him at his old job." Yet you could just as easily say, for example, "He might have taken a vacation, but inertia kept him busy." Because of the misguided way that most people use the term, it is easy to forget that "inertia" equally describes a tendency toward movement or nonmovement: in terms of Newtonian mechanics, it simply does not matter.

The significance of the clause "unless or until outside forces act upon it" in the first law indicates that the object itself must be in equilibrium—that is, the forces acting upon it must be balanced. In order for an object to be in equilibrium, its rate of movement in a given direction must be constant. Since a rate of movement equal to 0 is certainly constant, an object at rest is in equilibrium, and therefore qualifies; but also, any object moving in a constant direction at a constant speed is also in equilibrium.

THE SECOND LAW: FORCE, MASS, ACCELERATION

As noted earlier, the first law of motion deserves special attention because it is the key to unlocking the other two. Having established in the first law the conditions under which an object in motion will change velocity, the second law provides a measure of the force necessary to cause that change.

Understanding the second law requires defining terms that, on the surface at least, seem like a matter of mere common sense. Even inertia requires additional explanation in light of terms related to the second law, because it would be easy to confuse it with momentum.

The measure of inertia is mass, which reflects the resistance of an object to a change in its motion. Weight, on the other hand, measures the gravitational force on an object. (The concept of force itself will require further definition shortly.) Hence a person's mass is the same everywhere in the universe, but their weight would differ from planet to planet.

This can get somewhat confusing when you attempt to convert between English and metric units, because the pound is a unit of weight or force, whereas the kilogram is a unit of mass. In fact it would be more appropriate to set up kilograms against the English unit called the slug (equal to 14.59 kg), or to compare pounds to the metric unit of force, the newton (N), which is equal to the acceleration of one meter per second per second on an object of 1 kg in mass.

Hence, though many tables of weights and measures show that 1 kg is equal to 2.21 lb, this is only true at sea level on Earth. A person with a mass of 100 kg on Earth would have the same mass on the Moon; but whereas he might weigh 221 lb on Earth, he would be considerably lighter on the Moon. In other words, it would be much easier to lift a 221-lb man on the Moon than on Earth, but it would be no easier to push him aside.

To return to the subject of momentum, whereas inertia is measured by mass, momentum is equal to mass multiplied by velocity. Hence momentum, which Newton called "quantity of motion," is in effect inertia multiplied by velocity. Momentum is a subject unto itself; what matters here is the role that mass (and thus inertia) plays in the second law of motion.

According to the second law, the net force acting upon an object is a product of its mass multiplied by its acceleration. The latter is defined as a change in velocity over a given time interval: hence acceleration is usually presented in terms of "feet (or meters) per second per second"—that is, feet or meters per second squared. The acceleration due to gravity is 32 ft (9.8 m) per second per second, meaning that as every second passes, the speed of a falling object is increasing by 32 ft (9.8 m) per second.

The second law, as stated earlier, serves to develop the first law by defining the force necessary to change the velocity of an object. The law was integral to the confirming of the Copernican model, in which planets revolve around the Sun. Because velocity indicates movement in a single (straight) direction, when an object moves in a curve—as the planets do around the Sun—it is by definition changing velocity, or accelerating. The fact that the planets, which clearly possessed mass, underwent acceleration meant that some force must be acting on them: a gravitational pull exerted by the Sun, most massive object in the solar system.

Gravity is in fact one of four types of force at work in the universe. The others are electromagnetic interactions, and "strong" and "weak" nuclear interactions. The other three were unknown to Newton—yet his definition of force is still applicable. Newton's calculation of gravitational force (which, like momentum, is a subject unto itself) made it possible for Halley to determine that the comet he had observed in

1682—the comet that today bears his name—would reappear in 1758, as indeed it has for every 75–76 years since then. Today scientists use the understanding of gravitational force imparted by Newton to determine the exact altitude necessary for a satellite to remain stationary above the same point on Earth's surface.

The second law is so fundamental to the operation of the universe that you seldom notice its application, and it is easiest to illustrate by examples such as those above—of astronomers and physicists applying it to matters far beyond the scope of daily life. Yet the second law also makes it possible, for instance, to calculate the amount of force needed to move an object, and thus people put it into use every day without knowing that they are doing so.

THE THIRD LAW: ACTION AND REACTION

As with the second law, the third law of motion builds on the first two. Having defined the force necessary to overcome inertia, the third law predicts what will happen when one force comes into contact with another force. As the third law states, when one object exerts a force on another, the second object exerts on the first a force equal in magnitude but opposite in direction.

Unlike the second law, this one is much easier to illustrate in daily life. If a book is sitting on a table, that means that the book is exerting a force on the table equal to its mass multiplied by its rate of acceleration. Though it is not moving, the book is subject to the rate of gravitational acceleration, and in fact force and weight (which is defined as mass multiplied by the rate of acceleration due to gravity) are the same. At the same time, the table pushes up on the book with an exactly equal amount of force—just enough to keep it stationary. If the table exerted more force that the book—in other words, if instead of being an ordinary table it were some sort of pneumatic press pushing upward—then the book would fly off the table.

There is no such thing as an unpaired force in the universe. The table rests on the floor just as the book rests on it, and the floor pushes up on the table with a force equal in magnitude to that with which the table presses down on the floor. The same is true for the floor and the supporting beams that hold it up, and for the supporting beams and the foundation of the building, and the building and the ground, and so on.

These pairs of forces exist everywhere. When you walk, you move forward by pushing backward on the ground with a force equal to your mass multiplied by your rate of downward gravitational acceleration. (This force, in other words, is the same as weight.) At the same time, the ground actually pushes back with an equal force. You do not perceive the fact that Earth is pushing you upward, simply because its enormous mass makes this motion negligible—but it does push.

If you were stepping off of a small unmoored boat and onto a dock, however, something quite different would happen. The force of your leap to the dock would exert an equal force against the boat, pushing it further out into the water, and as a result, you would likely end up in the water as well. Again, the reaction is equal and opposite; the problem is that the boat in this illustration is not fixed in place like the ground beneath your feet.

Differences in mass can result in apparently different reactions, though in fact the force is the same. This can be illustrated by imagining a mother and her six-year-old daughter skating on ice, a relatively frictionless surface. Facing one another, they push against each other, and as a result each moves backward. The child, of course, will move backward faster because her mass is less than that of her mother. Because the force they exerted is equal, the daughter's acceleration is greater, and she moves farther.

Ice is not a perfectly frictionless surface, of course: otherwise, skating would be impossible. Likewise friction is absolutely necessary for walking, as you can illustrate by trying to walk on a perfectly slick surface—for instance, a skating rink covered with oil. In this situation, there is still an equally paired set of forces—your body presses down on the surface of the ice with as much force as the ice presses upward—but the lack of friction impedes the physical process of pushing off against the floor.

It will only be possible to overcome inertia by recourse to outside intervention, as for instance if someone who is not on the ice tossed out a rope attached to a pole in the ground. Alternatively, if the person on the ice were carrying a heavy load of rocks, it would be possible to move by throwing the rocks backward. In this situation, you are exerting force on the rock, and this

KEY TERMS

ACCELERATION: A change in velocity over a given time period.

EQUILIBRIUM: A situation in which the forces acting upon an object are in balance.

FRICTION: Any force that resists the motion of body in relation to another with which it is in contact.

INERTIA: The tendency of an object in motion to remain in motion, and of an object at rest to remain at rest.

MASS: A measure of inertia, indicating the resistance of an object to a change in its motion—including a change in velocity. A kilogram is a unit of mass, whereas a pound is a unit of weight. The mass of an object remains the same throughout the universe, whereas its weight is a function of gravity on any given planet.

MECHANICS: The study of bodies in motion.

MOMENTUM: The product of mass multiplied by velocity.

SPEED: The rate at which the position of an object changes over a given period of time.

VELOCITY: The speed of an object in a particular direction.

VISCOSITY: The internal friction in a fluid that makes it resistant to flow.

WEIGHT: A measure of the gravitational force on an object. A pound is a unit of weight, whereas a kilogram is a unit of mass. Weight thus would change from planet to planet, whereas mass remains constant throughout the universe.

---

backward force results in a force propelling the thrower forward.

This final point about friction and movement is an appropriate place to close the discussion on the laws of motion. Where walking or skating are concerned—and in the absence of a bag of rocks or some other outside force—friction is necessary to the action of creating a backward force and therefore moving forward. On the other hand, the absence of friction would make it possible for an object in movement to continue moving indefinitely, in line with the first law of motion. In either case, friction opposes inertia.

The fact is that friction itself is a force. Thus, if you try to slide a block of wood across a floor, friction will stop it. It is important to remember this, lest you fall into the fallacy that bedeviled Aristotle's thinking and thus confused the world for many centuries. The block did not stop moving because the force that pushed it was no longer being applied; it stopped because an opposing force, friction, was greater than the force that was pushing it.

WHERE TO LEARN MORE

Ardley, Neil. *The Science Book of Motion.* San Diego: Harcourt Brace Jovanovich, 1992.

Beiser, Arthur. *Physics,* 5th ed. Reading, MA: Addison-Wesley, 1991.

Chase, Sara B. *Moving to Win: The Physics of Sports.* New York: Messner, 1977.

Fleisher, Paul. *Secrets of the Universe: Discovering the Universal Laws of Science.* Illustrated by Patricia A. Keeler. New York: Atheneum, 1987.

"The Laws of Motion." *How It Flies* (Web site). <http://www.monmouth.com/~jsd/how/htm/motion.html> (February 27, 2001).

Newton, Isaac (translated by Andrew Motte, 1729). *The Principia* (Web site). <http://members.tripod.com/~gravitee/principia.html> (February 27, 2001).

Newton's Laws of Motion (Web site). <http://www.glenbrook.k12.il.us/gbssci/phys/Class/newtlaws/newtloc.html> (February 27, 2001).

"Newton's Laws of Motion." Dryden Flight Research Center, National Aeronautics and Space Administration (NASA) (Web site). <http://www.dfrc.nasa.gov/trc/saic/newton.html> (February 27, 2001).

"Newton's Laws of Motion: Movin' On." Beyond Books (Web site). <http://www.beyondbooks.com/psc91/4.asp> (February 27, 2001).

Roberts, Jeremy. How Do We Know the Laws of Motion? New York: Rosen, 2001.

Suplee, Curt. Everyday Science Explained. Washington, D.C.: National Geographic Society, 1996.

GRAVITY AND GRAVITATION

CONCEPT

Gravity is, quite simply, the force that holds together the universe. People are accustomed to thinking of it purely in terms of the gravitational pull Earth exerts on smaller bodies—a stone, a human being, even the Moon—or perhaps in terms of the Sun's gravitational pull on Earth. In fact, everything exerts a gravitational attraction toward everything else, an attraction commensurate with the two body's relative mass, and inversely related to the distance between them. The earliest awareness of gravity emerged in response to a simple question: why do objects fall when released from any restraining force? The answers, which began taking shape in the sixteenth century, were far from obvious. In modern times, understanding of gravitational force has expanded manyfold: gravity is clearly a law throughout the universe—yet some of the more complicated questions regarding gravitational force are far from settled.

HOW IT WORKS

ARISTOTLE'S MODEL

Greek philosophers of the period from the sixth to the fourth century B.C. grappled with a variety of questions concerning the fundamental nature of physical reality, and the forces that bind that reality into a whole. Among the most advanced thinkers of that period was Democritus (c. 460-370 B.C.), who put forth a hypothesis many thousands of years ahead of its time: that all of matter interacts at the atomic level.

Aristotle (384-322 B.C.), however, rejected the explanation offered by Democritus, an unfortunate circumstance given the fact that the great philosopher exerted an incalculable influence on the development of scientific thought. Aristotle's contributions to the advancement of the sciences were many and varied, yet his influence in physics was at least as harmful as it was beneficial. Furthermore, the fact that intellectual progress began slowing after several fruitful centuries of development in Greece only compounded the error. By the time civilization had reached the Middle Ages (c. 500 A.D.) the Aristotelian model of physical reality had been firmly established, and an entire millennium passed before it was successfully challenged.

Wrong though it was in virtually all particulars, the Aristotelian system offered a comforting symmetry amid the troubled centuries of the early medieval period. It must have been reassuring indeed to believe that the physical universe was as simple as the world of human affairs was complex. According to this neat model, all materials on Earth consisted of four elements: earth, water, air, and fire.

Each element had its natural place. Hence, earth was always the lowest, and in some places, earth was covered by water. Water must then be higher, but clearly air was higher still, since it covered earth and water. Highest of all was fire, whose natural place was in the skies above the air. Reflecting these concentric circles were the orbits of the Sun, the Moon, and the five known planets. Their orbital paths, in the Aristotelian model of the universe—a model developed to a great degree by the astronomer Ptolemy (c. 100-170)—were actually spheres that revolved around Earth with clockwork precision.

On Earth, according to the Aristotelian model, objects tended to fall toward the ground in accordance with the admixtures of differing

BECAUSE OF EARTH'S GRAVITY, THE WOMAN BEING SHOT OUT OF THIS CANNON WILL EVENTUALLY FALL TO THE GROUND RATHER THAN ASCEND INTO OUTER SPACE. *(Underwood & Underwood/Corbis. Reproduced by permission.)*

elements they contained. A rock, for instance, was mostly earth, and hence it sought its own level, the lowest of all four elements. For the same reason, a burning fire rose, seeking the heights that were fire's natural domain. It followed from this that an object falls faster or slower, depending on the relative mixtures of elements in it: or, to use more modern terms, the heavier the object, the faster it falls.

GALILEO TAKES UP THE COPERNICAN CHALLENGE

Over the centuries, a small but significant body of scientists and philosophers—each working independent from the other but building on the ideas of his predecessors—slowly began chipping away at the Aristotelian framework. The pivotal challenge came in the early part of the century, and the thinker who put it forward was not a physicist but an astronomer: Nicolaus Copernicus (1473-1543.)

Based on his study of the planets, Copernicus offered an entirely new model of the universe, one that placed the Sun and not Earth at its center. He was not the first to offer such an idea: half a century after Aristotle's death, Aristarchus (fl. 270 B.C.) had a similar idea, but Ptolemy rejected his heliocentric (Sun-centered) model in favor of the geocentric or Earth-centered one. In subsequent centuries, no less a political authority than the Catholic Church gave its approval to the Ptolemaic system. This system seemed to fit well with a literal interpretation of biblical passages concerning God's relationship with man, and man's relationship to the cosmos; hence, the heliocentric model of Copernicus constituted an offense to morality.

For this reason, Copernicus was hesitant to defend his ideas publicly, yet these concepts found their way into the consciousness of European thinkers, causing a paradigm shift so fundamental that it has been dubbed "the Copernican Revolution." Still, Copernicus offered no explanation as to why the planets behaved as they did: hence, the true leader of the Copernican Revolution was not Copernicus himself but Galileo Galilei (1564-1642.)

Initially, Galileo set out to study and defend the ideas of Copernicus through astronomy, but soon the Church forced him to recant. It is said that after issuing a statement in which he refuted the proposition that Earth moves—a direct attack on the static harmony of the Aristotelian/Ptolemaic model—he protested in private: "*E pur si muove!*" (But it does move!) Placed under house arrest by authorities from Rome, he turned his attention to an effort that, ironically, struck the fatal blow against the old model of the cosmos: a proof of the Copernican system according to the laws of physics.

GRAVITATIONAL ACCELERATION. In the process of defending Copernicus, Galileo actually inaugurated the modern history of physics as a science (as opposed to what it had been during the Middle Ages: a nest of suppositions masquerading as knowledge). Specifically, Galileo set out to test the hypothesis that objects fall as they do, not because of their weight, but as a consequence of gravitational force. If this were so, the acceleration of falling bodies would have to be the same, regardless of weight. Of course, it was clear that a stone fell faster than a feather, but Galileo reasoned that this was a result of factors other than weight, and later investigations confirmed that air resistance and friction, not weight, are responsible for this difference.

On the other hand, if one drops two objects that have similar air resistance but differing weight—say, a large stone and a smaller one—they fall at almost exactly the same rate. To test this directly, however, would have been difficult for Galileo: stones fall so fast that, even if dropped from a great height, they would hit the ground too soon for their rate of fall to be tested with the instruments then available.

Instead, Galileo used the motion of a pendulum, and the behavior of objects rolling or sliding down inclined planes, as his models. On the basis of his observations, he concluded that all bodies are subject to a uniform rate of gravitational acceleration, later calibrated at 32 ft (9.8 m) per second. What this means is that for every 32 ft an object falls, it is accelerating at a rate of 32 ft per second as well; hence, after 2 seconds, it falls at the rate of 64 ft (19.6 m) per second; after 3 seconds, at 96 ft (29.4 m) per second, and so on.

NEWTON DISCOVERS THE PRINCIPLE OF GRAVITY

Building on the work of his distinguished forebear, Sir Isaac Newton (1642-1727)—who, incidentally, was born the same year Galileo died—developed a paradigm for gravitation that, even today, explains the behavior of objects in virtually all situations throughout the universe. Indeed, the Newtonian model reigned until the early

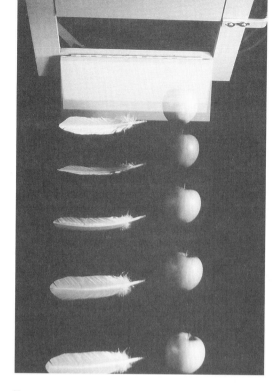

THIS PHOTO SHOWS AN APPLE AND A FEATHER BEING DROPPED IN A VACUUM TUBE. BECAUSE OF THE ABSENCE OF AIR RESISTANCE, THE TWO OBJECTS FALL AT THE SAME RATE. (*Photograph by James A. Sugar/Corbis. Reproduced by permission.*)

twentieth century, when Albert Einstein (1879-1955) challenged it on certain specifics.

Even so, Einstein's relativity did not disprove the Newtonian system as Copernicus and Galileo disproved Aristotle's and Ptolemy's theories; rather, it showed the limitations of Newtonian mechanics for describing the behavior of certain objects and phenomena. However, in the ordinary world of day-to-day experience—the world in which stones drop and heavy objects are hard to lift—the Newtonian system still offers the key to how and why things work as they do. This is particularly the case with regard to gravity and gravitation.

Like Galileo, Newton began in part with the aim of testing hypotheses put forth by an astronomer—in this case Johannes Kepler (1571-1630). In the early years of the seventeenth century, Kepler published his three laws of planetary motion, which together identified the elliptical (oval-shaped) path of the planets around the Sun. Kepler had discovered a mathematical relationship that connected the distances of the planets from the Sun to the period of their revolution around it. Like Galileo with Copernicus, Newton sought to generalize these principles to explain, not only how the planets moved, but also why they did.

Almost everyone has heard the story of Newton and the apple—specifically, that while he was sitting under an apple tree, a falling apple struck him on the head, spurring in him a great intuitive leap that led him to form his theory of gravitation. One contemporary biographer, William Stukely, wrote that he and Newton were sitting in a garden under some apple trees when Newton told him that "...he was just in the same situation, as when formerly, the notion of gravitation came into his mind. It was occasion'd by the fall of an apple, as he sat in a contemplative mood. Why should that apple always descend perpendicularly to the ground, he thought to himself. Why should it not go sideways or upwards, but constantly to the earth's centre?"

The tale of Newton and the apple has become a celebrated myth, rather like that of George Washington and the cherry tree. It is an embellishment of actual events: Newton never said that an apple hit him on the head, just that he was thinking about the way that apples fell. Yet the story has become symbolic of the creative intellectual process that occurs when a thinker makes a vast intuitive leap in a matter of moments. Of course, Newton had spent many years contemplating these ideas, and their development required great effort. What is important is that he brought together the best work of his predecessors, yet transcended all that had gone before—and in the process, forged a model that explained a great deal about how the universe functions.

The result was his *Philosophiae Naturalis Principia Mathematica*, or "Mathematical Principles of Natural Philosophy." Published in 1687, the book—usually referred to simply as the *Principia*—was one of the most influential works ever written. In it, Newton presented his three laws of motion, as well as his law of universal gravitation.

The latter stated that every object in the universe attracts every other one with a force proportional to the masses of each, and inversely proportional to the square of the distance between them. This statement requires some clarification with regard to its particulars, after which it will be reintroduced as a mathematical formula.

MASS AND FORCE. The three laws of motion are a subject unto themselves, covered elsewhere in this volume. However, in order to understand gravitation, it is necessary to understand at least a few rudimentary concepts relating to them. The first law identifies inertia as the tendency of an object in motion to remain in motion, and of an object at rest to remain at rest. Inertia is measured by mass, which—as the second law states—is a component of force.

Specifically, the second law of motion states that force is equal to mass multiplied by acceleration. This means that there is an inverse relationship between mass and acceleration: if force remains constant and one of these factors increases, the other must decrease—a situation that will be discussed in some depth below.

Also, as a result of Newton's second law, it is possible to define weight scientifically. People typically assume that mass and weight are the same, and indeed they are on Earth—or at least, they are close enough to be treated as comparable factors. Thus, tables of weights and measures show that a kilogram, the metric unit of mass, is equal to 2.2 pounds, the latter being the principal unit of weight in the British system.

In fact, this is—if not a case of comparing to apples to oranges—certainly an instance of comparing apples to apple pies. In this instance, the kilogram is the "apple" (a fitting Newtonian metaphor!) and the pound the "apple pie." Just as an apple pie contains apples, but other things as well, the pound as a unit of force contains an additional factor, acceleration, not included in the kilo.

BRITISH VS. SI UNITS. Physicists universally prefer the metric system, which is known in the scientific community as SI (an abbreviation of the French *Système International d'Unités*—that is, "International System of Units"). Not only is SI much more convenient to use, due to the fact that it is based on units of 10; but in discussing gravitation, the unequal relationship between kilograms and pounds makes conversion to British units a tedious and ultimately useless task.

Though Americans prefer the British system to SI, and are much more familiar with pounds than with kilos, the British unit of mass—called the slug—is hardly a household word. By contrast, scientists make regular use of the SI unit of force—named, appropriately enough, the newton. In the metric system, a newton (N) is the amount of force required to accelerate 1 kilogram of mass by 1 meter per second squared (m/s^2) Due to the simplicity of using SI over the British system, certain aspects of the discussion below will be presented purely in terms of SI. Where appropriate, however, conversion to British units will be offered.

CALCULATING GRAVITATIONAL FORCE. The law of universal gravitation can be stated as a formula for calculating the gravitational attraction between two objects of a certain mass, m_1 AND M_2: $F_{grav} = G \cdot (m_1M_2)/R^2$. F_{grav} is gravitational force, and r^2 the square of the distance between m_1 and m_2.

As for G, in Newton's time the value of this number was unknown. Newton was aware simply that it represented a very small quantity: without it, $(m_1m_2)/r^2$ could be quite sizeable for objects of relatively great mass separated by a relatively small distance. When multiplied by this very small number, however, the gravitational attraction would be revealed to be very small as well. Only in 1798, more than a century after Newton's writing, did English physicist Henry Cavendish (1731-1810) calculate the value of G.

As to how Cavendish derived the figure, that is an exceedingly complex subject far beyond the scope of the present discussion. Even to identify G as a number is a challenging task. First of all, it is a unit of force multiplied by squared area, then divided by squared mass: in other words, it is expressed in terms of $(N \cdot m^2)/kg^2$, where N stands for newtons, m for meters, and kg for kilograms. Nor is the coefficient, or numerical value, of G a whole number such as 1. A figure as large as 1, in fact, is astronomically huge compared to G, whose coefficient is $6.67 \cdot 10^{-11}$—in other words, 0.0000000000667.

REAL-LIFE APPLICATIONS

WEIGHT VS. MASS

Before discussing the significance of the gravitational constant, however, at this point it is appropriate to address a few issues that were raised earlier—issues involving mass and weight. In many ways, understanding these properties from the

framework of physics requires setting aside everyday notions.

First of all, why the distinction between weight and mass? People are so accustomed to converting pounds to kilos on Earth that the difference is difficult to comprehend, but if one considers the relation of mass and weight in outer space, the distinction becomes much clearer. Mass is the same throughout the universe, making it a much more fundamental characteristic—and hence, physicists typically speak in terms of mass rather than weight.

Weight, on the other hand, differs according to the gravitational pull of the nearest large body. On Earth, a person weighs a certain amount, but on the Moon, this weight is much less, because the Moon possesses less mass than Earth. Therefore, in accordance with Newton's formula for universal gravitation, it exerts less gravitational pull. By contrast, if one were on Jupiter, it would be almost impossible even to stand up, because the pull of gravity on that planet—with its greater mass—would be vastly greater than on Earth.

It should be noted that mass is not at all a function of size: Jupiter does have a greater mass than Earth, but not because it is bigger. Mass, as noted earlier, is purely a measure of inertia: the more resistant an object is to a change in its velocity, the greater its mass. This in itself yields some results that seem difficult to understand as long as one remains wedded to the concept—true enough on Earth—that weight and mass are identical.

A person might weigh less on the Moon, but it would be just as difficult to move that person from a resting position as it would be to do so on Earth. This is because the person's mass, and hence his or her resistance to inertia, has not changed. Again, this is a mentally challenging concept: is not lifting a person, which implies upward acceleration, not an attempt to counteract their inertia when standing still? Does it not follow that their mass has changed? Understanding the distinction requires a greater clarification of the relationship between mass, gravity, and weight.

F = MA. Newton's second law of motion, stated earlier, shows that force is equal to mass multiplied by acceleration, or in shorthand form, $F = ma$. To reiterate a point already made, if one assumes that force is constant, then mass and acceleration must have an inverse relationship. This can be illustrated by performing a simple experiment.

Suppose one were to apply a certain amount of force to an empty shopping cart. Assuming the floor had just enough friction to allow movement, it would be easy for almost anyone to accelerate the shopping cart. Now assume that the shopping cart were filled with heavy lead balls, so that it weighed, say, 1,102 lb (500 kg). If one applied the same force, it would not move.

What has changed, clearly, is the mass of the shopping cart. Because force remained constant, the rate of acceleration would become very small—in this case, almost infinitesimal. In the first case, with an empty shopping cart, the mass was relatively small, so acceleration was relatively high.

Now to return to the subject of lifting someone on the Moon. It is true that in order to lift that person, one would have to overcome inertia, and, in that sense, it would be as difficult as it is on Earth. But the other component of force, acceleration, has diminished greatly.

Weight is, again, a unit of force, but in calculating weight it is useful to make a slight change to the formula $F = ma$. By definition, the acceleration factor in weight is the downward acceleration due to gravity, usually rendered as g. So one's weight is equal to mg—but on the Moon, g is much smaller than it is on Earth, and hence, the same amount of force yields much greater results.

These facts shed new light on a question that bedeviled physicists at least from the time of Aristotle, until Galileo began clarifying the issue some 2,000 years later: why shouldn't an object of greater mass fall at a different rate than one of smaller mass? There are two answers to that question, one general and one specific. The general answer—that Earth exerts more gravitational pull on an object of greater mass—requires a deeper examination of Newton's gravitational formula. But the more specific answer, relating purely to conditions on Earth, is easily addressed by considering the effect of air resistance.

GRAVITY AND AIR RESISTANCE

One of Galileo's many achievements lay in using an idealized model of reality, one that does not take into account the many complex factors that affect the behavior of objects in the real world.

This permitted physicists to study processes that apparently defy common sense. For instance, in the real world, an apple does drop at a greater rate of speed than does a feather. However, in a vacuum, they will drop at the same rate. Since Galileo's time, it has become commonplace for physicists to discuss specific processes such as gravity with the assumption that all non-pertinent factors (in this case, air resistance or friction) are nonexistent or irrelevant. This greatly simplified the means of testing hypotheses.

Idealization of reality makes it possible to set aside the things people think they know about the real world, where events are complicated due to friction. The latter may be defined as a force that resists motion when the surface of one object comes into contact with the surface of another. If two balls are released in an environment free from friction—one of them simply dropped while the other is rolled down a curved surface or inclined plane—they will reach the bottom at the same time. This seems to go against everything that is known, but that is only because what people "know" is complicated by variables that have nothing to do with gravity.

The same is true for the behavior of falling objects with regard to air resistance. If air resistance were not a factor, one could fire a cannonball over horizontal space and then, when the ball reached the highest point in its trajectory, release another ball from the same height—and again, they would hit the ground at the same time. This is the case, even though the cannonball that was fired from the cannon has to cover a great deal of horizontal space, whereas the dropped ball does not. The fact is that the rate of acceleration due to gravity will be identical for the two balls, and the fact that the ball fired from a cannon also covers a horizontal distance during that same period is irrelevant.

TERMINAL VELOCITY. In the real world, air resistance creates a powerful drag force on falling objects. The faster the rate of fall, the greater the drag force, until the air resistance forces a leveling in the rate of fall. At this point, the object is said to have reached terminal velocity, meaning that its rate of fall will not increase thereafter. Galileo's idealized model, on the other hand, treated objects as though they were falling in a vacuum—space entirely devoid of matter, including air. In such a situation, the rate of acceleration would continue to grow indefinitely.

By means of a graph, one can compare the behavior of an object falling through air with that of an object falling in a vacuum. If the x axis measures time and the y axis downward speed, the rate of an object falling in a vacuum describes a 60°-angle. In other words, the speed of its descent is increasing at a much faster rate than is the rate of time of its descent—as indeed should be the case, in accordance with gravitational acceleration. The behavior of an object falling through air, on the other hand, describes a curve. Up to a point, the object falls at the same rate as it would in a vacuum, but soon velocity begins to increase at a much slower rate than time. Eventually, the curve levels off at the point where the object experiences terminal velocity.

Air resistance and friction have been mentioned separately as though they were two different forces, but in fact air resistance is simply a prominent form of friction. Hence air resistance exerts an upward force to counter the downward force of mass multiplied by gravity—that is, weight. Since g is a constant (32 ft or 9.8 m/sec^2), the greater the weight of the falling object, the longer it takes for air resistance to bring it to terminal velocity.

A feather quickly reaches terminal velocity, whereas it takes much longer for a cannonball to do the same. As a result, a heavier object does take less time to fall, even from a great height, than does a light one—but this is only because of friction, and not because of "elements" seeking their "natural level." Incidentally, if raindrops (which of course fall from a very great height) did not reach terminal velocity, they would cause serious injury by the time they hit the ground.

APPLYING THE GRAVITATIONAL FORMULA

Using Newton's gravitational formula, it is relatively easy to calculate the pull of gravity between two objects. It is also easy to see why the attraction is insignificant unless at least one of the objects has enormous mass. In addition, application of the formula makes it clear why G (the gravitational constant, as opposed to g, the rate of acceleration due to gravity) is such a tiny number.

If two people each have a mass of 45.5 kg (100 lb) and stand 1 m (3.28 ft) apart, m_1m_2 is equal to 2,070 kg (4,555 lb) and r^2 is equal to 1 m^2. Applied to the gravitational formula, this figure is rendered as 2,070 kg^2/1 m^2. This number is then multiplied by gravitational constant, which again is equal to 6.67 • 10^{-11} (N • m^2)/kg^2. The result is a net gravitational force of 0.000000138 N (0.00000003 lb)—about the weight of a single-cell organism!

EARTH, GRAVITY, AND WEIGHT. Though it is certainly interesting to calculate the gravitational force between any two people, computations of gravity are only significant for objects of truly great mass. For instance, there is the Earth, which has a mass of 5.98 • 10^{24} kg—that is, 5.98 *septillion* (1 followed by 24 zeroes) kilograms. And, of course, Earth's mass is relatively minor compared to that of several planets, not to mention the Sun. Yet Earth exerts enough gravitational pull to keep everything on it—living creatures, manmade structures, mountains and other natural features—stable and in place.

One can calculate Earth's gravitational force on any one person—if one wants to take the time to do so using Newton's formula. In fact, it is much simpler than that: gravitational force is equal to weight, or $m • g$. Thus if a woman weighs 100 lb (445 N), this amount is also equal to the gravitational force exerted on her. By dividing 445 N by the acceleration of gravity—9.8 m/sec^2—it is easy to obtain her mass: 45.4 kg.

The use of the *mg* formula for gravitation helps, once again, to explain why heavier objects do not fall faster than lighter ones. The figure for g is a constant, but for the sake of argument, let us assume that it actually becomes larger for objects with a greater mass. This in turn would mean that the gravitational force, or weight, would be bigger than it is—thus creating an irreconcilable logic loop.

Furthermore, one can compare results of two gravitation equations, one measuring the gravitational force between Earth and a large stone, the other measuring the force between Earth and a small stone. (The distance between Earth and each stone is assumed to be the same.) The result will yield a higher quantity for the force exerted on the larger stone—but only because its mass is greater. Clearly, then, the increase of force results only from an increase in mass, not acceleration.

GRAVITY AND CURVED SPACE

As should be clear from Newton's gravitational formula, the force of gravity works both ways: not only does a stone fall toward Earth, but Earth

KEY TERMS

FORCE: The product of mass multiplied by acceleration.

FRICTION: The force that resists motion when the surface of one object comes into contact with the surface of another.

INERTIA: The tendency of an object in motion to remain in motion, and of an object at rest to remain at rest.

INVERSE RELATIONSHIP: A situation involving two variables, in which one of the two increases in direct proportion to the decrease in the other.

LAW OF UNIVERSAL GRAVITATION: A principle, put forth by Sir Isaac Newton (1642-1727), which states that every object in the universe attracts every other one with a force proportional to the masses of

each, and inversely proportional to the square of the distance between them.

MASS: A measure of inertia, indicating the resistance of an object to a change in its motion.

TERMINAL VELOCITY: A term describing the rate of fall for an object experiencing the drag force of air resistance. In a vacuum, the object would continue to accelerate with the force of gravity, but in most real-world situations, air resistance creates a powerful drag force that causes a leveling in the object's rate of fall.

VACUUM: Space entirely devoid of matter, including air.

WEIGHT: A measure of the gravitational force on an object; the product of mass multiplied by the acceleration due to gravity.

actually falls toward it. The mass of Earth is so great compared to that of the stone that the movement of Earth is imperceptible—but it does happen. Furthermore, because Earth is round, when one hurls a projectile at a great distance, Earth curves away from the projectile; but eventually gravity itself forces the projectile to the ground.

However, if one were to fire a rocket at 17,700 MPH (28,500 km/h), at every instant of time the projectile is falling toward Earth with the force of gravity—but the curved Earth would be falling away from it at the same moment as well. Hence, the projectile would remain in constant motion around the planet—that is, it would be in orbit.

The same is true of an artificial satellite's orbit around Earth: even as the satellite falls toward Earth, Earth falls away from it. This same relationship exists between Earth and its great natural satellite, the Moon. Likewise, with the

Sun and its many satellites, including Earth: Earth plunges toward the Sun with every instant of its movement, but at every instant, the Sun falls away.

WHY IS EARTH ROUND? Note that in the above discussion, it was assumed that Earth and the Sun are round. Everyone knows that to be the case, but why? The answer is "Because they have to be"—that is, gravity will not allow them to be otherwise. In fact, the larger the mass of an object, the greater its tendency toward roundness: specifically, the gravitational pull of its interior forces the surface to assume a relatively uniform shape. There is a relatively small vertical differential for Earth's surface: between the lowest point and the highest point is just 12.28 mi (19.6 km)—not a great distance, considering that Earth's radius is about 4,000 mi (6,400 km).

It is true that Earth bulges near the equator, but this is only because it is spinning rapidly on its axis, and thus responding to the centripetal force of its motion, which produces a centrifugal component. If Earth were standing still, it would be much nearer to the shape of a sphere. On the other hand, an object of less mass is more likely to retain a shape that is far less than spherical. This can be shown by reference to the Martian moons Phobos and Deimos, both of which are oblong—and both of which are tiny, in terms of size and mass, compared to Earth's Moon.

Mars itself has a radius half that of Earth, yet its mass is only about 10% of Earth's. In light of what has been said about mass, shape, and gravity, it should not surprising to learn that Mars is also home to the tallest mountain in the solar system. Standing 15 mi (24 km) high, the volcano Olympus Mons is not only much taller than Earth's tallest peak, Mount Everest (29,028 ft [8,848 m]); it is 22% taller than the distance from the top of Mount Everest to the lowest spot on Earth, the Mariana Trench in the Pacific Ocean (-35,797 ft [-10,911 m])

A spherical object behaves with regard to gravitation as though its mass were concentrated near its center. And indeed, 33% of Earth's mass is at is core (as opposed to the crust or mantle), even though the core accounts for only about 20% of the planet's volume. Geologists believe that the composition of Earth's core must be molten iron, which creates the planet's vast electromagnetic field.

THE FRONTIERS OF GRAVITY. The subject of curvature with regard to gravity can be both a threshold or—as it is here—a point of closure. Investigating questions over perceived anomalies in Newton's description of the behavior of large objects in space led Einstein to his General Theory of Relativity, which posited a curved four-dimensional space-time. This led to entirely new notions concerning gravity, mass, and light. But relativity, as well as its relation to gravity, is another subject entirely. Einstein offered a new understanding of gravity, and indeed of physics itself, that has changed the way thinkers both inside and outside the sciences perceive the universe. Here on Earth, however, gravity behaves much as Newton described it more than three centuries ago.

Meanwhile, research in gravity continues to expand, as a visit to the Web site <www.Gravity.org> reveals. Spurred by studies in relativity, a branch of science called relativistic astrophysics has developed as a synthesis of astronomy and physics that incorporates ideas put forth by Einstein and others. The <www.Gravity.org> site presents studies—most of them too abstruse for a reader who is not a professional scientist—across a broad spectrum of disciplines. Among these is bioscience, a realm in which researchers are investigating the biological effects—such as mineral loss and motion sickness—of exposure to low gravity. The results of such studies will ultimately protect the health of the astronauts who participate in future missions to outer space.

WHERE TO LEARN MORE

Ardley, Neil. *The Science Book of Gravity.* San Diego, CA: Harcourt Brace Jovanovich, 1992.

Beiser, Arthur. *Physics,* 5th ed. Reading, MA: Addison-Wesley, 1991.

Bendick, Jeanne. *Motion and Gravity.* New York: F. Watts, 1972.

Dalton, Cindy Devine. *Gravity.* Vero Beach, FL: Rourke, 2001.

David, Leonard. *"Artificial Gravity and Space Travel."* BioScience, March 1992, pp. 155-159.

Exploring Gravity—Curtin University, Australia (Web site). <http://www.curtin.edu.au/curtin/dept/phys-sci/gravity/> (March 18, 2001).

The Gravity Society (Web site). <http://www.gravity.org> (March 18, 2001).

Nardo, Don. *Gravity: The Universal Force.* San Diego, CA: Lucent Books, 1990.

Rutherford, F. James; Gerald Holton; and Fletcher G. Watson. *Project Physics.* New York: Holt, Rinehart, and Winston, 1981.

Stringer, John. *The Science of Gravity.* Austin, TX: Raintree Steck-Vaughn, 2000.

PROJECTILE MOTION

CONCEPT

A projectile is any object that has been thrown, shot, or launched, and ballistics is the study of projectile motion. Examples of projectiles range from a golf ball in flight, to a curve ball thrown by a baseball pitcher to a rocket fired into space. The flight paths of all projectiles are affected by two factors: gravity and, on Earth at least, air resistance.

HOW IT WORKS

The effects of air resistance on the behavior of projectiles can be quite complex. Because effects due to gravity are much simpler and easier to analyze, and since gravity applies in more situations, we will discuss its role in projectile motion first. In most instances on Earth, of course, a projectile will be subject to both forces, but there may be specific cases in which an artificial vacuum has been created, which means it will only be subjected to the force of gravity. Furthermore, in outer space, gravity—whether from Earth or another body—is likely to be a factor, whereas air resistance (unless or until astronomers find another planet with air) will not be.

The acceleration due to gravity is 32 ft (9.8 m)/sec², usually expressed as "per second squared." This means that as every second passes, the speed of a falling object is increasing by 32 ft/sec (9.8 m). Where there is no air resistance, a ball will drop at a velocity of 32 feet per second after one second, 64 ft (19.5 m) per second after two seconds, 96 ft (29.4 m) per second after three seconds, and so on. When an object experiences the ordinary acceleration due to gravity, this figure is rendered in shorthand as g. Actually, the figure of 32 ft (9.8 m) per second squared applies

at sea level, but since the value of g changes little with altitude—it only decreases by 5% at a height of 10 mi (16 km)—it is safe to use this number.

When a plane goes into a high-speed turn, it experiences much higher apparent g. This can be as high as 9 g, which is almost more than the human body can endure. Incidentally, people call these "g-forces," but in fact g is not a measure of force but of a single component, acceleration. On the other hand, since force is the product of mass multiplied by acceleration, and since an aircraft subject to a high g factor clearly experiences a heavy increase in net force, in that sense, the expression "g-force" is not altogether inaccurate.

In a vacuum, where air resistance plays no part, the effects of g are clearly demonstrated. Hence a cannonball and a feather, dropped into a vacuum at the same moment, would fall at exactly the same rate and hit bottom at the same time.

THE CANNONBALL OR THE FEATHER? AIR RESISTANCE VS. MASS

Naturally, air resistance changes the terms of the above equation. As everyone knows, under ordinary conditions, a cannonball falls much faster than a feather, not simply because the feather is lighter than the cannonball, but because the air resists it much better. The speed of descent is a function of air resistance rather than mass, which can be proved with the following experiment. Using two identical pieces of paper—meaning that their mass is exactly the same—wad one up while keeping the other flat. Then drop them. Which one lands first? The wadded piece will fall faster and land first, precisely because it is less air-resistant than the sail-like flat piece.

BECAUSE OF THEIR DESIGN, THE BULLETS IN THIS .357 MAGNUM WILL COME OUT OF THE GUN SPINNING, WHICH GREATLY INCREASES THEIR ACCURACY. *(Photograph by Tim Wright/Corbis. Reproduced by permission.)*

Now to analyze the motion of a projectile in a situation without air resistance. Projectile motion follows the flight path of a parabola, a curve generated by a point moving such that its distance from a fixed point on one axis is equal to its distance from a fixed line on the other axis. In other words, there is a proportional relationship between x and y throughout the trajectory or path of a projectile in motion. Most often this parabola can be visualized as a simple up-and-down curve like the shape of a domed roof. (The Gateway Arch in St. Louis, Missouri, is a steep parabola.)

Instead of referring to the more abstract values of x and y, we will separate projectile motion into horizontal and vertical components. Gravity plays a role only in vertical motion, whereas obviously, horizontal motion is not subject to gravitational force. This means that in the absence of air resistance, the horizontal velocity of a projectile does not change during flight; by contrast, the force of gravity will ultimately reduce its vertical velocity to zero, and this will in turn bring a corresponding drop in its horizontal velocity.

In the case of a cannonball fired at a 45° angle—the angle of maximum efficiency for height and range together—gravity will eventu-

ally force the projectile downward, and once it hits the ground, it can no longer continue on its horizontal trajectory. Not, at least, at the same velocity: if you were to thrust a bowling ball forward, throwing it with both hands from the solar plexus, its horizontal velocity would be reduced greatly once gravity forced it to the floor. Nonetheless, the force on the ball would probably be enough (assuming the friction on the floor was not enormous) to keep the ball moving in a horizontal direction for at least a few more feet.

There are several interesting things about the relationship between gravity and horizontal velocity. Assuming, once again, that air resistance is not a factor, the vertical acceleration of a projectile is *g*. This means that when a cannonball is at the highest point of its trajectory, you could simply drop another cannonball from exactly the same height, and they would land at the same moment. This seems counterintuitive, or opposite to common sense: after all, the cannonball that was fired from the cannon has to cover a great deal of horizontal space, whereas the dropped ball does not. Nonetheless, the rate of acceleration due to gravity will be identical for the two balls, and the fact that the ball fired from a cannon also covers a horizontal distance during that same period is purely incidental.

Gravity, combined with the first law of motion, also makes it possible (in theory at least) for a projectile to keep moving indefinitely. This actually does take place at high altitudes, when a satellite is launched into orbit: Earth's gravitational pull, combined with the absence of air resistance or other friction, ensures that the satellite will remain in constant circular motion around the planet. The same is theoretically possible with a cannonball at very low altitudes: if one could fire a ball at 17,700 MPH (28,500 k/mh), the horizontal velocity would be great enough to put the ball into low orbit around Earth's surface.

The addition of air resistance or airflow to the analysis of projectile motion creates a number of complications, including drag, or the force that opposes the forward motion of an object in airflow. Typically, air resistance can create a drag force proportional to the squared value of a projectile's velocity, and this will cause it to fall far short of its theoretical range.

Shape, as noted in the earlier illustration concerning two pieces of paper, also affects air resistance, as does spin. Due to a principle known as the conservation of angular momentum, an object that is spinning tends to keep spinning; moreover, the orientation of the spin axis (the imaginary "pole" around which the object is spinning) tends to remain constant. Thus spin ensures a more stable flight.

REAL-LIFE APPLICATIONS

BULLETS ON A STRAIGHT SPINNING FLIGHT

One of the first things people think of when they hear the word "ballistics" is the study of gunfire patterns for the purposes of crime-solving. Indeed, this application of ballistics is a significant part of police science, because it allows law-enforcement investigators to determine when, where, and how a firearm was used. In a larger sense, however, the term as applied to firearms refers to efforts toward creating a more effective, predictable, and longer bullet trajectory.

From the advent of firearms in the West during the fourteenth century until about 1500, muskets were hopelessly unreliable. This was because the lead balls they fired had not been fit-

ted to the barrel of the musket. When fired, they bounced erratically off the sides of the barrel, and this made their trajectories unpredictable. Compounding this was the unevenness of the lead balls themselves, and this irregularity of shape could lead to even greater irregularities in trajectory.

Around 1500, however, the first true rifles appeared, and these greatly enhanced the accuracy of firearms. The term rifle comes from the "rifling" of the musket barrels: that is, the barrels themselves were engraved with grooves, a process known as rifling. Furthermore, ammunition-makers worked to improve the production process where the musket balls were concerned, producing lead rounds that were more uniform in shape and size.

Despite these improvements, soldiers over the next three centuries still faced many challenges when firing lead balls from rifled barrels. The lead balls themselves, because they were made of a soft material, tended to become misshapen during the loading process. Furthermore, the gunpowder that propelled the lead balls had a tendency to clog the rifle barrel. Most important of all was the fact that these rifles took time to load—and in a situation of battle, this could cost a man his life.

The first significant change came in the 1840s, when in place of lead balls, armies began using bullets. The difference in shape greatly improved the response of rounds to aerodynamic factors. In 1847, Claude-Etienne Minié, a captain in the French army, developed a bullet made of lead, but with a base that was slightly hollow. Thus when fired, the lead in the round tended to expand, filling the barrel's diameter and gripping the rifling.

As a result, the round came out of the barrel end spinning, and continued to spin throughout its flight. Not only were soldiers able to fire their rifles with much greater accuracy, but thanks to the development of chambers and magazines, they could reload more quickly.

CURVE BALLS, DIMPLED GOLF BALLS, AND OTHER TRICKS WITH SPIN

In the case of a bullet, spin increases accuracy, ensuring that the trajectory will follow an expected path. But sometimes spin can be used in more complex ways, as with a curveball thrown by a baseball pitcher.

The invention of the curveball is credited to Arthur "Candy" Cummings, who as a pitcher for the Brooklyn Excelsiors at the age of 18 in 1867—an era when baseball was still very young—introduced a new throw he had spent several years perfecting. Snapping as he released the ball, he and the spectators (not to mention the startled batter for the opposing team) watched as the pitch arced, then sailed right past the batter for a strike.

The curveball bedeviled baseball players and fans alike for many years thereafter, and many dismissed it as a type of optical illusion. The debate became so heated that in 1941, both *Life* and *Look* magazines ran features using stop-action photography to show that a curveball truly did curve. Even in 1982, a team of researchers from General Motors (GM) and the Massachusetts Institute of Technology (MIT), working at the behest of *Science* magazine, investigated the curveball to determine if it was more than a mere trick.

In fact, the curveball is a trick, but there is nothing fake about it. As the pitcher releases the ball, he snaps his wrist. This puts a spin on the projectile, and air resistance does the rest. As the ball moves toward the plate, its spin moves against the air, which creates an airstream moving against the trajectory of the ball itself. The airstream splits into two lines, one curving over the ball and one curving under, as the ball sails toward home plate.

For the purposes of clarity, assume that you are viewing the throw from a position between third base and home. Thus, the ball is moving from left to right, and therefore the direction of airflow is from right to left. Meanwhile the ball, as it moves into the airflow, is spinning clockwise. This means that the air flowing over the top of the ball is moving in a direction opposite to the spin, whereas that flowing under it is moving in the same direction as the spin.

This creates an interesting situation, thanks to Bernoulli's principle. The latter, formulated by Swiss mathematician and physicist Daniel Bernoulli (1700-1782), holds that where velocity is high, pressure is low—and vice versa. Bernoulli's principle is of the utmost importance to aerodynamics, and likewise plays a significant role in the operation of a curveball. At the top of the

GOLF BALLS ARE DIMPLED BECAUSE THEY TRAVEL MUCH FARTHER THAN NONDIMPLED ONES. *(Photograph by D. Boone/Corbis. Reproduced by permission.)*

ball, its clockwise spin is moving in a direction opposite to the airflow. This produces drag, slowing the ball, increasing pressure, and thus forcing it downward. At the bottom end of the ball, however, the clockwise motion is flowing with the air, thus resulting in higher velocity and lower pressure. As per Bernoulli's principle, this tends to pull the ball downward.

In the 60-ft, 6-in (18.4-m) distance that separates the pitcher's mound from home plate on a regulation major-league baseball field, a curveball can move downward by a foot (0.3048 m) or more. The interesting thing here is that this downward force is almost entirely due to air resistance rather than gravity, though of course gravity eventually brings any pitch to the ground, assuming it has not already been hit, caught, or bounced off a fence.

A curveball represents a case in which spin is used to deceive the batter, but it is just as possible that a pitcher may create havoc at home plate by throwing a ball with little or no spin. This is called a knuckleball, and it is based on the fact that spin in general—though certainly not the deliberate spin of a curveball—tends to ensure a more regular trajectory. Because a knuckleball has no spin, it follows an apparently random path, and thus it can be every bit as tricky for the pitcher as for the batter.

Golf, by contrast, is a sport in which spin is expected: from the moment a golfer hits the ball, it spins backward—and this in turn helps to explain why golf balls are dimpled. Early golf balls, known as featheries, were merely smooth leather pouches containing goose feathers. The smooth surface seemed to produce relatively low drag, and golfers were impressed that a well-hit feathery could travel 150-175 yd (137-160 m).

Then in the late nineteenth century, a professor at St. Andrews University in Scotland realized that a scored or marked ball would travel farther than a smooth one. (The part about St. Andrews may simply be golfing legend, since the course there is regarded as the birthplace of golf in the fifteenth century.) Whatever the case, it is true that a scored ball has a longer trajectory, again as a result of the effect of air resistance on projectile motion.

Airflow produces two varieties of drag on a sphere such as a golf ball: drag due to friction, which is only a small aspect of the total drag, and the much more significant drag that results from the separation of airflow around the ball. As with the curveball discussed earlier, air flows above and below the ball, but the issue here is more complicated than for the curved pitch.

Airflow comes in two basic varieties: laminar, meaning streamlined; or turbulent, indicating an erratic, unpredictable flow. For a jet flying through the air, it is most desirable to create a laminar flow passing over its airfoil, or the curved front surface of the wing. In the case of the golf ball, however, turbulent flow is more desirable.

In laminar flow, the airflow separates quickly, part of it passing over the ball and part passing under. In turbulent flow, however, separation comes later, further back on the ball. Each form of air separation produces a separation region, an area of drag that the ball pulls behind it (so to speak) as it flies through space. But because the separation comes further back on the ball in turbulent flow, the separation region itself is narrower, thus producing less drag.

Clearly, scoring the ball produced turbulent flow, and for a few years in the early twentieth century, manufacturers experimented with designs that included squares, rectangles, and hexagons. In time, they settled on the dimpled design known today. Golf balls made in Britain have 330 dimples, and those in America 336; in either case, the typical drive distance is much, much further than for an unscored ball—180-250 yd (165-229 m).

POWERED PROJECTILES: ROCK-ETS AND MISSILES

The most complex form of projectile widely known in modern life is the rocket or missile. Missiles are unmanned vehicles, most often used in warfare to direct some form of explosive toward an enemy. Rockets, on the other hand, can be manned or unmanned, and may be propulsion vehicles for missiles or for spacecraft. The term rocket can refer either to the engine or to the vehicle it propels.

The first rockets appeared in China during the late medieval period, and were used unsuccessfully by the Chinese against Mongol invaders in the early part of the thirteenth century. Europeans later adopted rocketry for battle, as for instance when French forces under Joan of Arc used crude rockets in an effort to break the siege on Orleans in 1429.

Within a century or so, however, rocketry as a form of military technology became obsolete, though projectile warfare itself remained as effective a method as ever. From the catapults of Roman times to the cannons that appeared in the early Renaissance to the heavy artillery of today, armies have been shooting projectiles against their enemies. The crucial difference between these projectiles and rockets or missiles is that the latter varieties are self-propelled.

Only around the end of World War II did rocketry and missile warfare begin to reappear in new, terrifying forms. Most notable among these was Hitler's V-2 "rocket" (actually a missile), deployed against Great Britain in 1944, but fortunately developed too late to make an impact. The 1950s saw the appearance of nuclear warheads such as the ICBM (intercontinental ballistic missile). These were guided missiles, as opposed to the V-2, which was essentially a huge self-propelled bullet fired toward London.

More effective than the ballistic missile, however, was the cruise missile, which appeared in later decades and which included aerodynamic structures that assisted in guidance and

IN THE CASE OF A ROCKET, LIKE THIS PATRIOT MISSILE BEING LAUNCHED DURING A TEST, PROPULSION COMES BY
EXPELLING FLUID—WHICH IN SCIENTIFIC TERMS CAN MEAN A GAS AS WELL AS A LIQUID—FROM ITS REAR END. MOST
OFTEN THIS FLUID IS A MASS OF HOT GASES PRODUCED BY A CHEMICAL REACTION INSIDE THE ROCKET'S BODY, AND
THIS BACKWARD MOTION CREATES AN EQUAL AND OPPOSITE REACTION FROM THE ATMOSPHERE, PROPELLING THE
ROCKET FORWARD. *(Corbis. Reproduced by permission.)*

maneuvering. In addition to guided or unguided, ballistic or aerodynamic, missiles can be classified in terms of source and target: surface-to-surface, air-to-air, and so on. By the 1970s, the United States had developed an extraordinarily sophisticated surface-to-air missile, the Stinger. Stingers proved a decisive factor in the Afghan-Soviet War (1979-89), when U.S.-supplied Afghan guerrillas used them against Soviet aircraft.

In the period from the late 1940s to the late 1980s, the United States, the Soviet Union, and other smaller nuclear powers stockpiled these warheads, which were most effective precisely because they were never used. Thus, U.S. President Ronald Reagan played an important role in ending the Cold War, because his weapons buildup forced the Soviets to spend money they did not have on building their own arsenal. During the aftermath of the Cold War, America and the newly democratized Russian Federation worked to reduce their nuclear stockpiles. Ironically, this was also the period when sophisticated missiles such as the *Patriot* began gaining widespread use in the Persian Gulf War and later conflicts.

Certain properties unite the many varieties of rocket that have existed across time and

KEY TERMS

ACCELERATION: A change in velocity over a given time period.

AERODYNAMIC: Relating to airflow.

BALLISTICS: The study of projectile motion.

DRAG: The force that opposes the forward motion of an object in airflow. In most cases, its opposite is lift.

FIRST LAW OF MOTION: A principle, formulated by Sir Isaac Newton (1642-1727), which states that an object at rest will remain at rest, and an object in motion will remain in motion, at a constant velocity unless or until outside forces act upon it.

FRICTION: Any force that resists the motion of body in relation to another with which it is in contact.

INERTIA: The tendency of an object in motion to remain in motion, and of an object at rest to remain at rest.

LAMINAR: A term describing a streamlined flow, in which all particles move at the same speed and in the same direction. Its opposite is turbulent flow.

LIFT: An aerodynamic force perpendicular to the direction of the wind. In most cases, its opposite is drag.

MASS: A measure of inertia, indicating the resistance of an object to a change in its motion—including a change in velocity.

PARABOLA: A curve generated by a point moving such that its distance from a fixed point on one axis is equal to its distance from a fixed line on the other axis. As a result, between any two points on the parabola there is a proportional relationship between x and y values.

PROJECTILE: Any object that has been thrown, shot, or launched.

SPECIFIC IMPULSE: A measure of rocket fuel efficiency—specifically, the mass that can be lifted by a particular type of fuel for each pound of fuel consumer (that is, the rocket and its contents) per second of operation time. Figures for specific impulse are rendered in seconds.

SPEED: The rate at which the position of an object changes over a given period of time. Unlike velocity, direction is not a component of speed.

THIRD LAW OF MOTION: A principle, which like the first law of motion was formulated by Sir Isaac Newton. The third law states that when one object exerts a force on another, the second object exerts on the first a force equal in magnitude but opposite in direction.

TRAJECTORY: The path of a projectile in motion, a parabola upward and across space.

TURBULENT: A term describing a highly irregular form of flow, in which a fluid is subject to continual changes in speed and direction. Its opposite is laminar flow.

VELOCITY: The speed of an object in a particular direction.

VISCOSITY: The internal friction in a fluid that makes it resistant to flow.

space—including the relatively harmless fireworks used in Fourth of July and New Year's Eve celebrations around the country. One of the key principles that makes rocket propulsion possible is the third law of motion. Sometimes colloquially put as "For every action, there is an equal and opposite reaction," a more scientifically accurate version of this law would be: "When one object exerts a force on another, the second object exerts on the first a force equal in magnitude but opposite in direction."

In the case of a rocket, propulsion comes by expelling fluid—which in scientific terms can mean a gas as well as a liquid—from its rear. Most often this fluid is a mass of hot gases produced by a chemical reaction inside the rocket's body, and this backward motion creates an equal and opposite reaction from the rocket, propelling it forward.

Before it undergoes a chemical reaction, rocket fuel may be either in solid or liquid form inside the rocket's fuel chamber, though it ends up as a gas when expelled. Both solid and liquid varieties have their advantages and disadvantages in terms of safety, convenience, and efficiency in lifting the craft. Scientists calculate efficiency by a number of standards, among them specific impulse, a measure of the mass that can be lifted by a particular type of fuel for each pound of fuel consumed (that is, the rocket and its contents) per second of operation time. Figures for specific impulse are rendered in seconds.

A spacecraft may be divided into segments or stages, which can be released as specific points

along the flight in part to increase specific impulse. This was the case with the *Saturn 5* rockets that carried astronauts to the Moon in the period 1969-72, but not with the varieties of space shuttle that have flown regular missions since 1981.

The space shuttle is essentially a hybrid of an airplane and rocket, with a physical structure more like that of an aircraft but with rocket power. In fact, the shuttle uses many rockets to maximize efficiency, utilizing no less than 67 rockets—49 of which run on liquid fuel and the rest on solid fuel—at different stages of its flight.

WHERE TO LEARN MORE

"Aerodynamics in Sports Equipment." *K8AIT Principles of Aeronautics—Advanced* (Web site). <http://muttley.ucdavis.edu/Book/Sports/advanced/index.html> (March 2, 2001).

Asimov, Isaac. *Exploring Outer Space: Rockets, Probes, and Satellites.* Revisions and updates by Francis Reddy. Milwaukee, WI: G. Stevens, 1995.

How in the World? Pleasantville, N.Y.: Reader's Digest, 1990.

"*Interesting Properties of Projectile of Motion*" (Web site). <http://www.phy.ntnu.edu.tw/~hwang/projectile3/projectile3.html> (March 2, 2001).

JBM Small Arms Ballistics (Web site). <http://roadrunner.com/~jbm/index_rgt.html> (March 2, 2001).

The Physics of Projectile Motion (Web site). <http://library.thinkquest.org/2779/> (March 2, 2001).

Richardson, Hazel. *How to Build a Rocket.* Illustrated by Scoular Anderson. New York: F. Watts, 2001.

TORQUE

CONCEPT

Torque is the application of force where there is rotational motion. The most obvious example of torque in action is the operation of a crescent wrench loosening a lug nut, and a close second is a playground seesaw. But torque is also crucial to the operation of gyroscopes for navigation, and of various motors, both internal-combustion and electrical.

HOW IT WORKS

Force, which may be defined as anything that causes an object to move or stop moving, is the linchpin of the three laws of motion formulated by Sir Isaac Newton (1642-1727.) The first law states that an object at rest will remain at rest, and an object in motion will remain in motion, unless or until outside forces act upon it. The second law defines force as the product of mass multiplied by acceleration. According to the third law, when one object exerts a force on another, the second object exerts on the first a force equal in magnitude but opposite in direction.

One way to envision the third law is in terms of an active event—for instance, two balls striking one another. As a result of the impact, each flies backward. Given the fact that the force on each is equal, and that force is the product of mass and acceleration (this is usually rendered with the formula $F = ma$), it is possible to make some predictions regarding the properties of mass and acceleration in this interchange. For instance, if the mass of one ball is relatively small compared to that of the other, its acceleration will be correspondingly greater, and it will thus be thrown backward faster.

On the other hand, the third law can be demonstrated when there is no apparent movement, as for instance, when a person is sitting on a chair, and the chair exerts an equal and opposite force upward. In such a situation, when all the forces acting on an object are in balance, that object is said to be in a state of equilibrium.

Physicists often discuss torque within the context of equilibrium, even though an object experiencing net torque is definitely not in equilibrium. In fact, torque provides a convenient means for testing and measuring the degree of rotational or circular acceleration experienced by an object, just as other means can be used to calculate the amount of linear acceleration. In equilibrium, the net sum of all forces acting on an object should be zero; thus in order to meet the standards of equilibrium, the sum of all torques on the object should also be zero.

REAL-LIFE APPLICATIONS

SEESAWS AND WRENCHES

As for what torque is and how it works, it is best discuss it in relationship to actual objects in the physical world. Two in particular are favorites among physicists discussing torque: a seesaw and a wrench turning a lug nut. Both provide an easy means of illustrating the two ingredients of torque, force and moment arm.

In any object experiencing torque, there is a pivot point, which on the seesaw is the balance-point, and which in the wrench-and-lug nut combination is the lug nut itself. This is the area around which all the forces are directed. In each

A SEESAW ROTATES ON AND OFF THE GROUND DUE TO TORQUE IMBALANCE. *(Photograph by Dean Conger/Corbis. Reproduced by permission.)*

case, there is also a place where force is being applied. On the seesaw, it is the seats, each holding a child of differing weight. In the realm of physics, weight is actually a variety of force.

Whereas force is equal to mass multiplied by acceleration, weight is equal to mass multiplied by the acceleration due to gravity. The latter is equal to 32 ft (9.8 m)/sec². This means that for every second that an object experiencing gravitational force continues to fall, its velocity increases at the rate of 32 ft or 9.8 m per second. Thus, the formula for weight is essentially the same as that for force, with a more specific variety of acceleration substituted for the generalized term in the equation for force.

As for moment arm, this is the distance from the pivot point to the vector on which force is being applied. Moment arm is always perpendicular to the direction of force. Consider a wrench operating on a lug nut. The nut, as noted earlier, is the pivot point, and the moment arm is the distance from the lug nut to the place where the person operating the wrench has applied force. The torque that the lug nut experiences is the product of moment arm multiplied by force.

In English units, torque is measured in pound-feet, whereas the metric unit is Newton-meters, or N•m. (One newton is the amount of force that, when applied to 1 kg of mass, will give it an acceleration of 1 m/sec²). Hence if a person were to a grip a wrench 9 in (23 cm) from the pivot point, the moment arm would be 0.75 ft (0.23 m.) If the person then applied 50 lb (11.24 N) of force, the lug nut would be experiencing 37.5 pound-feet (2.59 N•m) of torque.

The greater the amount of torque, the greater the tendency of the object to be put into rotation. In the case of a seesaw, its overall design, in particular the fact that it sits on the ground, means that its board can never undergo anything close to 360° rotation; nonetheless, the board does rotate within relatively narrow parameters. The effects of torque can be illustrated by imagining the clockwise rotational behavior of a seesaw viewed from the side, with a child sitting on the left and a teenager on the right.

Suppose the child weighs 50 lb (11.24 N) and sits 3 ft (0.91 m) from the pivot point, giving her side of the seesaw a torque of 150 pound-feet (10.28 N•m). On the other side, her teenage sister weighs 100 lb (22.48 N) and sits 6 ft (1.82 m) from the center, creating a torque of 600 pound-feet (40.91 N•m). As a result of the torque imbalance, the side holding the teenager will rotate clockwise, toward the ground, causing the child's side to also rotate clockwise—off the ground.

TORQUE, ALONG WITH ANGULAR MOMENTUM, IS THE LEADING FACTOR DICTATING THE MOTION OF A GYROSCOPE. HERE, A WOMAN RIDES INSIDE A GIANT GYROSCOPE AT AN AMUSEMENT PARK. *(Photograph by Richard Cummins/Corbis. Reproduced by permission.)*

In order for the two to balance one another perfectly, the torque on each side has to be adjusted. One way would be by changing weight, but a more likely remedy is a change in position, and therefore, of moment arm. Since the teenager weighs exactly twice as much as the child, the moment arm on the child's side must be exactly twice as long as that on the teenager's.

Hence, a remedy would be for the two to switch positions with regard to the pivot point. The child would then move out an additional 3 ft (.91 m), to a distance of 6 ft (1.83 m) from the pivot, and the teenager would cut her distance from the pivot point in half, to just 3 ft (.91 m). In fact, however, any solution that gave the child a moment arm twice as long as that of the teenager would work: hence, if the teenager sat 1 ft (.3 m) from the pivot point, the child should be at 2 ft (.61 m) in order to maintain the balance, and so on.

On the other hand, there are many situations in which you may be unable to increase force, but can increase moment arm. Suppose you were trying to disengage a particularly stubborn lug nut, and after applying all your force, it still would not come loose. The solution would be to increase moment arm, either by grasping the wrench further from the pivot point, or by using a longer wrench.

For the same reason, on a door, the knob is placed as far as possible from the hinges. Here the hinge is the pivot point, and the door itself is the moment arm. In some situations of torque, however, moment arm may extend over "empty space," and for this reason, the handle of a wrench is not exactly the same as its moment arm. If one applies force on the wrench at a 90°-angle to the handle, then indeed handle and moment arm are identical; however, if that force were at a 45° angle, then the moment arm would be outside the handle, because moment arm and force are always perpendicular. And if one were to pull the wrench away from the lug nut, then there would be 0° difference between the direction of force and the pivot point—meaning that moment arm (and hence torque) would also be equal to zero.

Gyroscopes

A gyroscope consists of a wheel-like disk, called a flywheel, mounted on an axle, which in turn is mounted on a larger ring perpendicular to the plane of the wheel itself. An outer circle on the same plane as the flywheel provides structural stability, and indeed, the gyroscope may include several such concentric rings. Its focal point, however, is the flywheel and the axle. One end of the axle is typically attached to some outside object, while the other end is left free to float.

Once the flywheel is set spinning, gravity has a tendency to pull the unattached end of the axle downward, rotating it on an axis perpendicular to that of the flywheel. This should cause the gyroscope to fall over, but instead it begins to spin a third axis, a horizontal axis perpendicular both to the plane of the flywheel and to the direction of gravity. Thus, it is spinning on three axes, and as a result becomes very stable—that is, very resistant toward outside attempts to upset its balance.

This in turn makes the gyroscope a valued instrument for navigation: due to its high degree of gyroscopic inertia, it resists changes in orientation, and thus can guide a ship toward its destination. Gyroscopes, rather than magnets, are often the key element in a compass. A magnet will point to magnetic north, some distance from "true north" (that is, the North Pole.) But with a gyroscope whose axle has been aligned with true north before the flywheel is set spinning, it is possible to possess a much more accurate directional indicator. For this reason, gyroscopes are used on airplanes—particularly those flying over the poles—as well as submarines and even the Space Shuttle.

Torque, along with angular momentum, is the leading factor dictating the motion of a gyroscope. Think of angular momentum as the momentum (mass multiplied by velocity) that a turning object acquires. Due to a principle known as the conservation of angular momentum, a spinning object has a tendency to reach a constant level of angular momentum, and in order to do this, the sum of the external torques acting on the system must be reduced to zero. Thus angular momentum "wants" or "needs" to cancel out torque.

The "right-hand rule" can help you to understand the torque in a system such as the gyroscope. If you extend your right hand, palm downward, your fingers are analogous to the moment arm. Now if you curl your fingers downward, toward the ground, then your fingertips point in the direction of g—that is, gravitational force. At that point, your thumb (involuntarily, due to the bone structure of the hand) points in the direction of the torque vector.

When the gyroscope starts to spin, the vectors of angular momentum and torque are at odds with one another. Were this situation to persist, it would destabilize the gyroscope; instead, however, the two come into alignment. Using the right-hand rule, the torque vector on a gyroscope is horizontal in direction, and the vector of angular momentum eventually aligns with

KEY TERMS

ACCELERATION: A change in velocity over a given time period.

EQUILIBRIUM: A situation in which the forces acting upon an object are in balance.

FORCE: The product of mass multiplied by acceleration.

INERTIA: The tendency of an object in motion to remain in motion, and of an object at rest to remain at rest.

MASS: A measure of inertia, indicating the resistance of an object to a change in its motion—including a change in velocity.

MOMENT ARM: For an object experiencing torque, moment arm is the distance from the pivot or balance point to the vector on which force is being applied.

Moment arm is always perpendicular to the direction of force.

SPEED: The rate at which the position of an object changes over a given period of time.

TORQUE: The product of moment arm multiplied by force.

VECTOR: A quantity that possesses both magnitude and direction. By contrast, a scalar quantity is one that possesses only magnitude, with no specific direction.

VELOCITY: The speed of an object in a particular direction.

WEIGHT: A measure of the gravitational force on an object; the product of mass multiplied by the acceleration due to gravity.

it. To achieve this, the gyroscope experiences what is known as gyroscopic precession, pivoting along its support post in an effort to bring angular momentum into alignment with torque. Once this happens, there is no net torque on the system, and the conservation of angular momentum is in effect.

TORQUE IN COMPLEX MACHINES

Torque is a factor in several complex machines such as the electric motor that—with variations—runs most household appliances. It is especially important to the operation of automobiles, playing a significant role in the engine and transmission.

An automobile engine produces energy, which the pistons or rotor convert into torque for transmission to the wheels. Though torque is greatest at high speeds, the amount of torque needed to operate a car does not always vary proportionately with speed. At moderate speeds and on level roads, the engine does not need to provide a great deal of torque. But when the car is starting, or climbing a steep hill, it is important

that the engine supply enough torque to keep the car running; otherwise it will stall. To allocate torque and speed appropriately, the engine may decrease or increase the number of revolutions per minute to which the rotors are subjected.

Torque comes from the engine, but it has to be supplied to the transmission. In an automatic transmission, there are two principal components: the automatic gearbox and the torque converter. It is the job of the torque converter to transmit power from the flywheel of the engine to the gearbox, and it has to do so as smoothly as possible. The torque converter consists of three elements: an impeller, which is turned by the engine flywheel; a reactor that passes this motion on to a turbine; and the turbine itself, which turns the input shaft on the automatic gearbox. An infusion of oil to the converter assists the impeller and turbine in synchronizing movement, and this alignment of elements in the torque converter creates a smooth relationship between engine and gearbox. This also leads to an increase in the car's overall torque—that is, its turning force.

Torque is also important in the operation of electric motors, found in everything from vacuum cleaners and dishwashers to computer printers and videocassette recorders to subway systems and water-pumping stations. Torque in the context of electricity involves reference to a number of concepts beyond the scope of this discussion: current, conduction, magnetic field, and other topics relevant to electromagnetic force.

WHERE TO LEARN MORE

Beiser, Arthur. *Physics,* 5th ed. Reading, MA: Addison-Wesley, 1991.

Macaulay, David. *The New Way Things Work.* Boston: Houghton Mifflin, 1998.

"Rotational Motion." Physics Department, University of Guelph (Web site). <http://www.physics.uoguelph.ca/tutorials/torque/> (March 4, 2001).

"Rotational Motion—Torque." Lee College (Web site). <http://www.lee.edu/mathscience/physics/physics/Courses/LabManual/2b/2b.html> (March 4, 2001).

Schweiger, Peggy E. *"Torque"* (Web site). <http://www.cyberclassrooms.net/~pschweiger/rot-mot.html> (March 4, 2001).

"Torque and Rotational Motion" (Web site). <http://online.cctt.org/curriculumguide/units/torque.asp> (March 4, 2001).

FLUID MECHANICS

FLUID MECHANICS

AERODYNAMICS

BERNOULLI'S PRINCIPLE

BUOYANCY

FLUID MECHANICS

CONCEPT

The term "fluid" in everyday language typically refers only to liquids, but in the realm of physics, fluid describes any gas or liquid that conforms to the shape of its container. Fluid mechanics is the study of gases and liquids at rest and in motion. This area of physics is divided into fluid statics, the study of the behavior of stationary fluids, and fluid dynamics, the study of the behavior of moving, or flowing, fluids. Fluid dynamics is further divided into hydrodynamics, or the study of water flow, and aerodynamics, the study of airflow. Applications of fluid mechanics include a variety of machines, ranging from the waterwheel to the airplane. In addition, the study of fluids provides an understanding of a number of everyday phenomena, such as why an open window and door together create a draft in a room.

HOW IT WORKS

THE CONTRAST BETWEEN FLUIDS AND SOLIDS

To understand fluids, it is best to begin by contrasting their behavior with that of solids. Whereas solids possess a definite volume and a definite shape, these physical characteristics are not so clearly defined for fluids. Liquids, though they possess a definite volume, have no definite shape—a factor noted above as one of the defining characteristics of fluids. As for gases, they have neither a definite shape nor a definite volume.

One of several factors that distinguishes fluids from solids is their response to compression, or the application of pressure in such a way as to reduce the size or volume of an object. A solid is highly noncompressible, meaning that it resists compression, and if compressed with a sufficient force, its mechanical properties alter significantly. For example, if one places a drinking glass in a vise, it will resist a small amount of pressure, but a slight increase will cause the glass to break.

Fluids vary with regard to compressibility, depending on whether the fluid in question is a liquid or a gas. Most gases tend to be highly compressible—though air, at low speeds at least, is not among them. Thus, gases such as propane fuel can be placed under high pressure. Liquids tend to be noncompressible: unlike a gas, a liquid can be compressed significantly, yet its response to compression is quite different from that of a solid—a fact illustrated below in the discussion of hydraulic presses.

One way to describe a fluid is "anything that flows"—a behavior explained in large part by the interaction of molecules in fluids. If the surface of a solid is disturbed, it will resist, and if the force of the disturbance is sufficiently strong, it will deform—as for instance, when a steel plate begins to bend under pressure. This deformation will be permanent if the force is powerful enough, as was the case in the above example of the glass in a vise. By contrast, when the surface of a liquid is disturbed, it tends to flow.

MOLECULAR BEHAVIOR OF FLUIDS AND SOLIDS. At the molecular level, particles of solids tend to be definite in their arrangement and close to one another. In the case of liquids, molecules are close in proximity, though not as much so as solid molecules, and the arrangement is random. Thus, with a glass of water, the molecules of glass (which at

IN A WIDE, UNCONSTRICTED REGION, A RIVER FLOWS SLOWLY. HOWEVER, IF ITS FLOW IS NARROWED BY CANYON WALLS, AS WITH WYOMING'S BIGHORN RIVER, THEN IT SPEEDS UP DRAMATICALLY. *(Photograph by Kevin R. Morris/Corbis. Reproduced by permission.)*

relatively low temperatures is a solid) in the container are fixed in place while the molecules of water contained by the glass are not. If one portion of the glass were moved to another place on the glass, this would change its structure. On the other hand, no significant alteration occurs in the character of the water if one portion of it is moved to another place within the entire volume of water in the glass.

As for gas molecules, these are both random in arrangement and far removed in proximity. Whereas solid particles are slow-moving and have a strong attraction to one another, liquid molecules move at moderate speeds and exert a moderate attraction on each other. Gas molecules are extremely fast-moving and exert little or no attraction.

Thus, if a solid is released from a container pointed downward, so that the force of gravity moves it, it will fall as one piece. Upon hitting a floor or other surface, it will either rebound, come to a stop, or deform permanently. A liquid, on the other hand, will disperse in response to impact, its force determining the area over which the total volume of liquid is distributed. But for a gas, assuming it is lighter than air, the downward pull of gravity is not even required to disperse it:

once the top on a container of gas is released, the molecules begin to float outward.

FLUIDS UNDER PRESSURE

As suggested earlier, the response of fluids to pressure is one of the most significant aspects of fluid behavior and plays an important role within both the statics and dynamics subdisciplines of fluid mechanics. A number of interesting principles describe the response to pressure, on the part of both fluids at rest inside a container, and fluids which are in a state of flow.

Within the realm of hydrostatics, among the most important of all statements describing the behavior of fluids is Pascal's principle. This law is named after Blaise Pascal (1623-1662), a French mathematician and physicist who discovered that the external pressure applied on a fluid is transmitted uniformly throughout its entire body. The understanding offered by Pascal's principle later became the basis for one of the most important machines ever developed, the hydraulic press.

HYDROSTATIC PRESSURE AND BUOYANCY. Some nineteen centuries before Pascal, the Greek mathematician, physicist, and inventor Archimedes (c. 287-212 B.C.) discovered a precept of fluid statics that had implications at least as great as those of Pascal's principle. This was Archimedes's principle, which explains the buoyancy of an object immersed in fluid. According to Archimedes's principle, the buoyant force exerted on the object is equal to the weight of the fluid it displaces.

Buoyancy explains both how a ship floats on water, and how a balloon floats in the air. The pressures of water at the bottom of the ocean, and of air at the surface of Earth, are both examples of hydrostatic pressure—the pressure that exists at any place in a body of fluid due to the weight of the fluid above. In the case of air pressure, air is pulled downward by the force of Earth's gravitation, and air along the planet's surface has greater pressure due to the weight of the air above it. At great heights above Earth's surface, however, the gravitational force is diminished, and thus the air pressure is much smaller.

Water, too, is pulled downward by gravity, and as with air, the fluid at the bottom of the ocean has much greater pressure due to the weight of the fluid above it. Of course, water is much heavier than air, and therefore, water at even a moderate depth in the ocean has enormous pressure. This pressure, in turn, creates a buoyant force that pushes upward.

If an object immersed in fluid—a balloon in the air, or a ship on the ocean—weighs less that the fluid it displaces, it will float. If it weighs more, it will sink or fall. The balloon itself may be "heavier than air," but it is not as heavy as the air it has displaced. Similarly, an aircraft carrier contains a vast weight in steel and other material, yet it floats, because its weight is not as great as that of the displaced water.

BERNOULLI'S PRINCIPLE. Archimedes and Pascal contributed greatly to what became known as fluid statics, but the father of fluid mechanics, as a larger realm of study, was the Swiss mathematician and physicist Daniel Bernoulli (1700-1782). While conducting experiments with liquids, Bernoulli observed that when the diameter of a pipe is reduced, the water flows faster. This suggested to him that some force must be acting upon the water, a force that he reasoned must arise from differences in pressure.

Specifically, the slower-moving fluid in the wider area of pipe had a greater pressure than the portion of the fluid moving through the narrower part of the pipe. As a result, he concluded that

pressure and velocity are inversely related—in other words, as one increases, the other decreases. Hence, he formulated Bernoulli's principle, which states that for all changes in movement, the sum of static and dynamic pressure in a fluid remains the same.

A fluid at rest exerts pressure—what Bernoulli called "static pressure"—on its container. As the fluid begins to move, however, a portion of the static pressure—proportional to the speed of the fluid—is converted to what Bernoulli called dynamic pressure, or the pressure of movement. In a cylindrical pipe, static pressure is exerted perpendicular to the surface of the container, whereas dynamic pressure is parallel to it.

According to Bernoulli's principle, the greater the velocity of flow in a fluid, the greater the dynamic pressure and the less the static pressure. In other words, slower-moving fluid exerts greater pressure than faster-moving fluid. The discovery of this principle ultimately made possible the development of the airplane.

REAL-LIFE APPLICATIONS

BERNOULLI'S PRINCIPLE IN ACTION

As fluid moves from a wider pipe to a narrower one, the volume of the fluid that moves a given distance in a given time period does not change. But since the width of the narrower pipe is smaller, the fluid must move faster (that is, with greater dynamic pressure) in order to move the same amount of fluid the same distance in the same amount of time. Observe the behavior of a river: in a wide, unconstricted region, it flows slowly, but if its flow is narrowed by canyon walls, it speeds up dramatically.

Bernoulli's principle ultimately became the basis for the airfoil, the design of an airplane's wing when seen from the end. An airfoil is shaped like an asymmetrical teardrop laid on its side, with the "fat" end toward the airflow. As air hits the front of the airfoil, the airstream divides, part of it passing over the wing and part passing under. The upper surface of the airfoil is curved, however, whereas the lower surface is much straighter.

As a result, the air flowing over the top has a greater distance to cover than the air flowing under the wing. Since fluids have a tendency to compensate for all objects with which they come into contact, the air at the top will flow faster to meet the other portion of the airstream, the air flowing past the bottom of the wing, when both reach the rear end of the airfoil. Faster airflow, as demonstrated by Bernoulli, indicates lower pressure, meaning that the pressure on the bottom of the wing keeps the airplane aloft.

CREATING A DRAFT. Among the most famous applications of Bernoulli's principle is its use in aerodynamics, and this is discussed in the context of aerodynamics itself elsewhere in this book. Likewise, a number of other applications of Bernoulli's principle are examined in an essay devoted to that topic. Bernoulli's principle, for instance, explains why a shower curtain tends to billow inward when the water is turned on; in addition, it shows why an open window and door together create a draft.

Suppose one is in a hotel room where the heat is on too high, and there is no way to adjust the thermostat. Outside, however, the air is cold, and thus, by opening a window, one can presumably cool down the room. But if one opens the window without opening the front door of the room, there will be little temperature change. The only way to cool off will be by standing next to the window: elsewhere in the room, the air will be every bit as stuffy as before. But if the door leading to the hotel hallway is opened, a nice cool breeze will blow through the room. Why?

With the door closed, the room constitutes an area of relatively high pressure compared to the pressure of the air outside the window. Because air is a fluid, it will tend to flow into the room, but once the pressure inside reaches a certain point, it will prevent additional air from entering. The tendency of fluids is to move from high-pressure to low-pressure areas, not the other way around. As soon as the door is opened, the relatively high-pressure air of the room flows into the relatively low-pressure area of the hallway. As a result, the air pressure in the room is reduced, and the air from outside can now enter. Soon a wind will begin to blow through the room.

A WIND TUNNEL. The above scenario of wind flowing through a room describes a rudimentary wind tunnel. A wind tunnel is a chamber built for the purpose of examining the characteristics of airflow in contact with solid objects, such as aircraft and automobiles. The wind tunnel represents a safe and judicious use of the properties of fluid mechanics. Its purpose is to test the interaction of airflow and solids in relative motion: in other words, either the aircraft has to be moving against the airflow, as it does in flight, or the airflow can be moving against a stationary aircraft. The first of these choices, of course, poses a number of dangers; on the other hand, there is little danger in exposing a stationary craft to winds at speeds simulating that of the aircraft in flight.

The first wind tunnel was built in England in 1871, and years later, aircraft pioneers Orville (1871-1948) and Wilbur (1867-1912) Wright used a wind tunnel to improve their planes. By the late 1930s, the U.S. National Advisory Committee for Aeronautics (NACA) was building wind tunnels capable of creating speeds equal to 300 MPH (480 km/h); but wind tunnels built after World War II made these look primitive. With the development of jet-powered flight, it became necessary to build wind tunnels capable of simulating winds at the speed of sound—760 MPH (340 m/s). By the 1950s, wind tunnels were being used to simulate hypersonic speeds—that is, speeds of Mach 5 (five times the speed of sound) and above. Researchers today use helium to create wind blasts at speeds up to Mach 50.

FLUID MECHANICS FOR PERFORMING WORK

HYDRAULIC PRESSES. Though applications of Bernoulli's principle are among the most dramatic examples of fluid mechanics in operation, the everyday world is filled with instances of other ideas at work. Pascal's principle, for instance, can be seen in the operation of any number of machines that represent variations on the idea of a hydraulic press. Among these is the hydraulic jack used to raise a car off the floor of an auto mechanic's shop.

Beneath the floor of the shop is a chamber containing a quantity of fluid, and at either end of the chamber are two large cylinders side by side. Each cylinder holds a piston, and valves control flow between the two cylinders through the channel of fluid that connects them. In accordance with Pascal's principle, when one applies force by pressing down the piston in one cylinder (the input cylinder), this yields a uniform pressure that causes output in the second cylinder, pushing up a piston that raises the car.

Another example of a hydraulic press is the hydraulic ram, which can be found in machines ranging from bulldozers to the hydraulic lifts used by firefighters and utility workers to reach heights. In a hydraulic ram, however, the characteristics of the input and output cylinders are reversed from those of a car jack. For the car jack, the input cylinder is long and narrow, while the output cylinder is wide and short. This is because the purpose of a car jack is to raise a heavy object through a relatively short vertical range of movement—just high enough so that the mechanic can stand comfortably underneath the car.

In the hydraulic ram, the input or master cylinder is short and squat, while the output or slave cylinder is tall and narrow. This is because the hydraulic ram, in contrast to the car jack, carries a much lighter cargo (usually just one person) through a much greater vertical range—for instance, to the top of a tree or building.

PUMPS. A pump is a device made for moving fluid, and it does so by utilizing a pressure difference, causing the fluid to move from an area of higher pressure to one of lower pressure. Its operation is based on aspects both of Pascal's and Bernoulli's principles—though, of course, humans were using pumps thousands of years before either man was born.

A siphon hose used to draw gas from a car's fuel tank is a very simple pump. Sucking on one end of the hose creates an area of low pressure compared to the relatively high-pressure area of the gas tank. Eventually, the gasoline will come out of the low-pressure end of the hose.

The piston pump, slightly more complex, consists of a vertical cylinder along which a piston rises and falls. Near the bottom of the cylinder are two valves, an inlet valve through which fluid flows into the cylinder, and an outlet valve through which fluid flows out. As the piston moves upward, the inlet valve opens and allows fluid to enter the cylinder. On the downstroke, the inlet valve closes while the outlet valve opens, and the pressure provided by the piston forces the fluid through the outlet valve.

One of the most obvious applications of the piston pump is in the engine of an automobile. In this case, of course, the fluid being pumped is gasoline, which pushes the pistons up and down

PUMPS FOR DRAWING USABLE WATER FROM THE GROUND ARE UNDOUBTEDLY THE OLDEST PUMPS KNOWN. *(Photograph by Richard Cummins/Corbis. Reproduced by permission.)*

by providing a series of controlled explosions created by the spark plug's ignition of the gas. In another variety of piston pump—the kind used to inflate a basketball or a bicycle tire—air is the fluid being pumped. Then there is a pump for water. Pumps for drawing usable water from the ground are undoubtedly the oldest known variety, but there are also pumps designed to remove water from areas where it is undesirable; for example, a bilge pump, for removing water from a boat, or the sump pump used to pump flood water out of a basement.

FLUID POWER. For several thousand years, humans have used fluids—in particular water—to power a number of devices. One of the great engineering achievements of ancient times was the development of the waterwheel, which included a series of buckets along the rim that made it possible to raise water from the river below and disperse it to other points. By about 70 B.C., Roman engineers recognized that they could use the power of water itself to turn wheels and grind grain. Thus, the waterwheel became one of the first mechanisms in which an inanimate

KEY TERMS

AERODYNAMICS: An area of fluid dynamics devoted to studying the properties and characteristics of airflow.

ARCHIMEDES'S PRINCIPLE: A rule of physics stating that the buoyant force of an object immersed in fluid is equal to the weight of the fluid displaced by the object. It is named after the Greek mathematician, physicist, and inventor, Archimedes (c. 287-212 B.C.), who first identified it.

BERNOULLI'S PRINCIPLE: A proposition, credited to Swiss mathematician and physicist Daniel Bernoulli (1700-1782), which maintains that slower-moving fluid exerts greater pressure than faster-moving fluid.

BUOYANCY: The tendency of an object immersed in a fluid to float. This can be explained by Archimedes's principle.

COMPRESSION: To reduce in size or volume by applying pressure.

FLUID: Any substance, whether gas or liquid, that conforms to the shape of its container.

FLUID DYNAMICS: An area of fluid mechanics devoted to studying of the behavior of moving, or flowing, fluids. Fluid dynamics is further divided into hydrodynamics and aerodynamics.

FLUID MECHANICS: The study of the behavior of gases and liquids at rest and in motion. The major divisions of fluid mechanics are fluid statics and fluid dynamics.

FLUID STATICS: An area of fluid mechanics devoted to studying the behavior of stationary fluids.

HYDRODYNAMICS: An area of fluid dynamics devoted to studying the properties and characteristics of water flow.

HYDROSTATIC PRESSURE: The pressure that exists at any place in a body of fluid due to the weight of the fluid above.

PASCAL'S PRINCIPLE: A statement, formulated by French mathematician and physicist Blaise Pascal (1623-1662), which holds that the external pressure applied on a fluid is transmitted uniformly throughout the entire body of that fluid.

PRESSURE: The ratio of force to surface area, when force is applied in a direction perpendicular to that surface.

TURBINE: A machine that converts the kinetic energy (the energy of movement) in fluids to useable mechanical energy by passing the stream of fluid through a series of fixed and moving fans or blades.

WIND TUNNEL: A chamber built for the purpose of examining the characteristics of airflow in relative motion against solid objects such as aircraft and automobiles.

source (as opposed to the effort of humans or animals) created power.

The water clock, too, was another ingenious use of water developed by the ancients. It did not use water for power; rather, it relied on gravity—a concept only dimly understood by ancient peo-

ples—to move water from one chamber of the clock to another, thus, marking a specific interval of time. The earliest clocks were sundials, which were effective for measuring time, provided the Sun was shining, but which were less useful for measuring periods shorter than an hour. Hence, the development of the hourglass, which used sand, a solid that in larger quantities exhibits the behavior of a fluid. Then, in about 270 B.C., Ctesibius of Alexandria (fl. c. 270-250 B.C.) used gearwheel technology to devise a constant-flow water clock called a "clepsydra." Use of water clocks prevailed for more than a thousand years, until the advent of the first mechanical clocks.

During the medieval period, fluids provided power to windmills and water mills, and at the dawn of the Industrial Age, engineers began applying fluid principles to a number of sophisticated machines. Among these was the turbine, a machine that converts the kinetic energy (the energy of movement) in fluids to useable mechanical energy by passing the stream of fluid through a series of fixed and moving fans or blades. A common house fan is an example of a turbine in reverse: the fan adds energy to the passing fluid (air), whereas a turbine extracts energy from fluids such as air and water.

The turbine was developed in the mid-eighteenth century, and later it was applied to the extraction of power from hydroelectric dams, the first of which was constructed in 1894. Today, hydroelectric dams provide electric power to millions of homes around the world. Among the most dramatic examples of fluid mechanics in action, hydroelectric dams are vast in size and equally impressive in the power they can generate using a completely renewable resource: water.

A hydroelectric dam forms a huge steel-and-concrete curtain that holds back millions of tons of water from a river or other body. The water nearest the top—the "head" of the dam—has enormous potential energy, or the energy that an object possesses by virtue of its position. Hydroelectric power is created by allowing controlled streams of this water to flow downward, gathering kinetic energy that is then transferred to powering turbines, which in turn generate electric power.

WHERE TO LEARN MORE

Aerodynamics for Students (Web site). <http://www.ae.su.oz.au/aero/contents.html> (April 8, 2001).

Beiser, Arthur. *Physics*, 5th ed. Reading, MA: Addison-Wesley, 1991.

Chahrour, Janet. *Flash! Bang! Pop! Fizz!: Exciting Science for Curious Minds*. Illustrated by Ann Humphrey Williams. Hauppauge, N.Y.: Barron's, 2000.

"Educational Fluid Mechanics Sites." *Virginia Institute of Technology* (Web site). <http://www.eng.vt.edu/fluids/links/edulinks.htm> (April 8, 2001).

Fleisher, Paul. *Liquids and Gases: Principles of Fluid Mechanics*. Minneapolis, MN: Lerner Publications, 2002.

Institute of Fluid Mechanics (Web site). <http://www.ts.go.dlr.de> (April 8, 2001).

K8AIT Principles of Aeronautics Advanced Text (web site). <http://wings.ucdavis.edu/Book/advanced.html> (February 19, 2001).

Macaulay, David. *The New Way Things Work*. Boston: Houghton Mifflin, 1998.

Sobey, Edwin J. C. *Wacky Water Fun with Science: Science You Can Float, Sink, Squirt, and Sail*. Illustrated by Bill Burg. New York: McGraw-Hill, 2000.

Wood, Robert W. *Mechanics Fundamentals*. Illustrated by Bill Wright. Philadelphia: Chelsea House, 1997.

AERODYNAMICS

CONCEPT

Though the term "aerodynamics" is most commonly associated with airplanes and the overall science of flight, in fact, its application is much broader. Simply put, aerodynamics is the study of airflow and its principles, and applied aerodynamics is the science of improving manmade objects such as airplanes and automobiles in light of those principles. Aside from the obvious application to these heavy forms of transportation, aerodynamic concepts are also reflected in the simplest of manmade flying objects—and in the natural model for all studies of flight, a bird's wings.

HOW IT WORKS

All physical objects on Earth are subject to gravity, but gravity is not the only force that tends to keep them pressed to the ground. The air itself, though it is invisible, operates in such a way as to prevent lift, much as a stone dropped into the water will eventually fall to the bottom. In fact, air behaves much like water, though the downward force is not as great due to the fact that air's pressure is much less than that of water. Yet both are media through which bodies travel, and air and water have much more in common with one another than either does with a vacuum.

Liquids such as water and gasses such as air are both subject to the principles of fluid dynamics, a set of laws that govern the motion of liquids and vapors when they come in contact with solid surfaces. In fact, there are few significant differences—for the purposes of the present discussion—between water and air with regard to their behavior in contact with solid surfaces.

When a person gets into a bathtub, the water level rises uniformly in response to the fact that a solid object is taking up space. Similarly, air currents blow over the wings of a flying aircraft in such a way that they meet again more or less simultaneously at the trailing edge of the wing. In both cases, the medium adjusts for the intrusion of a solid object. Hence within the parameters of fluid dynamics, scientists typically use the term "fluid" uniformly, even when describing the movement of air.

The study of fluid dynamics in general, and of air flow in particular, brings with it an entire vocabulary. One of the first concepts of importance is viscosity, the internal friction in a fluid that makes it resistant to flow and resistant to objects flowing through it. As one might suspect, viscosity is a far greater factor with water than with air, the viscosity of which is less than two percent that of water. Nonetheless, near a solid surface—for example, the wing of an airplane—viscosity becomes a factor because air tends to stick to that surface.

Also significant are the related aspects of density and compressibility. At speeds below 220 MPH (354 km/h), the compressibility of air is not a significant factor in aerodynamic design. However, as air flow approaches the speed of sound—660 MPH (1,622 km/h)—compressibility becomes a significant factor. Likewise temperature increases greatly when airflow is supersonic, or faster than the speed of sound.

All objects in the air are subject to two types of airflow, laminar and turbulent. Laminar flow is smooth and regular, always moving at the same speed and in the same direction. This type of airflow is also known as streamlined flow, and under these conditions every particle of fluid that passes a particular point follows a path identical to all particles that passed that point earlier. This may be illustrated by imagining a stream flowing around a twig.

By contrast, in turbulent flow the air is subject to continual changes in speed and direction—as for instance when a stream flows over shoals of rocks. Whereas the mathematical model of laminar airflow is rather straightforward, conditions are much more complex in turbulent flow, which typically occurs in the presence either of obstacles or of high speeds.

Absent the presence of viscosity, and thus in conditions of perfect laminar flow, an object behaves according to Bernoulli's principle, sometimes known as Bernoulli's equation. Named after the Swiss mathematician and physicist Daniel Bernoulli (1700-1782), this proposition goes to the heart of that which makes an airplane fly.

While conducting experiments concerning the conservation of energy in liquids, Bernoulli observed that when the diameter of a pipe is reduced, the water flows faster. This suggested to him that some force must be acting upon the water, a force that he reasoned must arise from differences in pressure. Specifically, the slower-moving fluid had a greater pressure than the portion of the fluid moving through the narrower part of the pipe. As a result, he concluded that pressure and velocity are inversely related.

Bernoulli's principle states that for all changes in movement, the sum of static and dynamic pressure in a fluid remain the same. A fluid at rest exerts static pressure, which is the same as what people commonly mean when they say "pressure," as in "water pressure." As the fluid begins to move, however, a portion of the static pressure—proportional to the speed of the fluid—is converted to what scientists call dynamic pressure, or the pressure of movement. The greater the speed, the greater the dynamic pressure and the less the static pressure. Bernoulli's findings would prove crucial to the design of aircraft in the twentieth century, as engineers learned how to use currents of faster and slower air for keeping an airplane aloft.

Very close to the surface of an object experiencing airflow, however, the presence of viscosity plays havoc with the neat proportions of the Bernoulli's principle. Here the air sticks to the object's surface, slowing the flow of nearby air and creating a "boundary layer" of slow-moving

A TYPICAL PAPER AIRPLANE HAS LOW ASPECT RATIO WINGS, A TERM THAT REFERS TO THE SIZE OF THE WINGSPAN COMPARED TO THE CHORD. IN SUBSONIC FLIGHT, HIGHER ASPECT RATIOS ARE USUALLY PREFERRED. *(Photograph by Bruce Burkhardt/Corbis. Reproduced by permission.)*

air. At the beginning of the flow—for instance, at the leading edge of an airplane's wing—this boundary layer describes a laminar flow; but the width of the layer increases as the air moves along the surface, and at some point it becomes turbulent.

These and a number of other factors contribute to the coefficients of drag and lift. Simply put, drag is the force that opposes the forward motion of an object in airflow, whereas lift is a force perpendicular to the direction of the wind, which keeps the object aloft. Clearly these concepts can be readily applied to the operation of an airplane, but they also apply in the case of an automobile, as will be shown later.

REAL-LIFE APPLICATIONS

How a Bird Flies—and Why a Human Being Cannot

Birds are exquisitely designed (or adapted) for flight, and not simply because of the obvious fact

BIRDS LIKE THESE FAIRY TERNS ARE SUPREME EXAMPLES OF AERODYNAMIC PRINCIPLES, FROM THEIR LOW BODY WEIGHT AND LARGE STERNUM AND PECTORALIS MUSCLES TO THEIR LIGHTWEIGHT FEATHERS. *(Corbis. Reproduced by permission.)*

that they have wings. Thanks to light, hollow bones, their body weight is relatively low, giving them the advantage in overcoming gravity and remaining aloft. Furthermore, a bird's sternum or breast bone, as well as its pectoralis muscles (those around the chest) are enormous in proportion to its body size, thus helping it to achieve the thrust necessary for flight. And finally, the bird's lightweight feathers help to provide optimal lift and minimal drag.

A bird's wing is curved along the top, a crucial aspect of its construction. As air passes over the leading edge of the wing, it divides, and because of the curve, the air on top must travel a greater distance before meeting the air that flowed across the bottom. The tendency of airflow, as noted earlier, is to correct for the presence of solid objects. Therefore, in the absence of outside factors such as viscosity, the air on top "tries" to travel over the wing in the same amount of time that it takes the air below to travel under the wing. As shown by Bernoulli, the fast-moving air above the wing exerts less pressure than the slow-moving air below it; hence there is a difference in pressure between the air below and the air above, and this keeps the wing aloft.

When a bird beats its wings, its downstrokes propel it, and as it rises above the ground, the force of aerodynamic lift helps push its wings upward in preparation for the next downstroke. However, to reduce aerodynamic drag during the upstroke, the bird folds its wings, thus decreasing its wingspan. Another trick that birds execute instinctively is the moving of their wings forward and backward in order to provide balance. They also "know" how to flap their wings in a direction almost parallel to the ground when they need to fly slowly or hover.

Witnessing the astonishing aerodynamic feats of birds, humans sought the elusive goal of flight from the earliest of times. This was symbolized by the Greek myth of Icarus and Daedalus, who escaped from a prison in Crete by constructing a set of bird-like wings and flying away. In the world of physical reality, however, the goal would turn out to be unattainable as long as humans attempted to achieve flight by imitating birds.

As noted earlier, a bird's physiology is quite different from that of a human being. There is simply no way that a human can fly by flapping his arms—nor will there ever be a man strong enough to do so, no matter how apparently well-designed his mechanical wings are. Indeed, to be capable of flying like a bird, a man would have to have a chest so enormous in proportion to his body that he would be hideous in appearance.

Not realizing this, humans for centuries attempted to fly like birds—with disastrous results. An English monk named Eilmer (b. 980) attempted to fly off the tower of Malmesbury Abbey with a set of wings attached to his arms and feet. Apparently Eilmer panicked after gliding some 600 ft (about 200 m) and suddenly plummeted to earth, breaking both of his legs. At least he lived; more tragic was the case of Abul Qasim Ibn Firnas (d. 873), an inventor from Cordoba in Arab Spain who devised and demonstrated a glider. Much of Cordoba's population came out to see him demonstrate his flying machine, but after covering just a short distance, the craft fell to earth. Severely wounded, Ibn Firnas died shortly afterward.

The first real progress in the development of flying machines came when designers stopped trying to imitate birds and instead used the principle of buoyancy. Hence in 1783, the French brothers Jacques-Etienne and Joseph-Michel Montgolfier constructed the first practical balloon.

Balloons and their twentieth-century descendant, the dirigible, had a number of obvious drawbacks, however. Without a motor, a balloon could not be guided, and even with a motor, dirigibles proved highly dangerous. At that stage, most dirigibles used hydrogen, a gas that is cheap and plentiful, but extremely flammable. After the *Hindenburg* exploded in 1937, the age of passenger travel aboard airships was over.

However, the German military continued to use dirigibles for observation purposes, as did the United States forces in World War II. Today airships, the most famous example being the Goodyear Blimp, are used not only for observation but for advertising. Scientists working in rain forests, for instance, use dirigibles to glide above the forest canopy; as for the Goodyear Blimp, it provides television networks with "eye in the sky" views of large sporting events.

The first man to make a serious attempt at creating a heavier-than-air flying machine (as opposed to a balloon, which uses gases that are lighter than air) was Sir George Cayley (1773-1857), who in 1853 constructed a glider. It is interesting to note that in creating this, the fore-runner of the modern airplane, Cayley went back to an old model: the bird. After studying the physics of birds' flight for many years, he equipped his glider with an extremely wide wingspan, used the lightest possible materials in its construction, and designed it with exceptionally smooth surfaces to reduce drag.

The only thing that in principle differentiated Cayley's craft from a modern airplane was its lack of an engine. In those days, the only possible source of power was a steam engine, which would have added far too much weight to his aircraft. However, the development of the internal-combustion engine in the nineteenth century overcame that obstacle, and in 1903 Orville and Wilbur Wright achieved the dream of flight that had intrigued and eluded human beings for centuries.

AIRPLANES: GETTING ALOFT, STAYING ALOFT, AND REMAINING STABLE

Once engineers and pilots took to the air, they encountered a number of factors that affect flight. In getting aloft and staying aloft, an aircraft is subject to weight, lift, drag, and thrust.

As noted earlier, the design of an airplane wing takes advantage of Bernoulli's principle to give it lift. Seen from the end, the wing has the shape of a long teardrop lying on its side, with the large end forward, in the direction of airflow, and the narrow tip pointing toward the rear. (Unlike a teardrop, however, an airplane's wing is asymmetrical, and the bottom side is flat.) This cross-section is known as an airfoil, and the greater curvature of its upper surface in comparison to the lower side is referred to as the airplane's camber. The front end of the airfoil is also curved, and the chord line is an imaginary straight line connecting the spot where the air hits the front—known as the stagnation point—to the rear, or trailing edge, of the wing.

Again in accordance with Bernoulli's principle, the shape of the airflow facilitates the spread of laminar flow around it. The slower-moving currents beneath the airfoil exert greater pressure than the faster currents above it, giving lift to the aircraft.

Another parameter influencing the lift coefficient (that is, the degree to which the aircraft experiences lift) is the size of the wing: the longer the wing, the greater the total force exerted beneath it, and the greater the ratio of this pressure to that of the air above. The size of a modern aircraft's wing is actually somewhat variable, due to the presence of flaps at the trailing edge.

With regard to the flaps, however, it should be noted that they have different properties at different stages of flight: in takeoff, they provide lift, but in stable flight they increase drag, and for that reason the pilot retracts them. In preparing for landing, as the aircraft slows and descends, the extended flaps then provide stability and assist in the decrease of speed.

Speed, too, encourages lift: the faster the craft, the faster the air moves over the wing. The pilot affects this by increasing or decreasing the power of the engine, thus regulating the speed with which the plane's propellers turn. Another highly significant component of lift is the airfoil's angle of attack—the orientation of the airfoil with regard to the air flow, or the angle that the chord line forms with the direction of the air stream.

Up to a point, increasing the angle of attack provides the aircraft with extra lift because it moves the stagnation point from the leading edge down along the lower surface; this increases the low-pressure area of the upper surface. However, if the pilot increases the angle of attack too much, this affects the boundary layer of slow-moving air, causing the aircraft to go into a stall.

Together the engine provides the propellers with power, and this gives the aircraft thrust, or propulsive force. In fact, the propeller blades constitute miniature wings, pivoted at the center and powered by the engine to provide rotational motion. As with the wings of the aircraft, the blades have a convex forward surface and a narrow trailing edge. Also like the aircraft wings, their angle of attack (or pitch) is adjusted at different points for differing effects. In stable flight, the pilot increases the angle of attack for the propeller blades sharply as against airflow, whereas at takeoff and landing the pitch is dramatically reduced. During landing, in fact, the pilot actually reverses the direction of the propeller blades, turning them into a brake on the aircraft's forward motion—and producing that lurching sensation that a passenger experiences as the aircraft slows after touching down.

By this point there have been several examples regarding the use of the same technique alternately to provide lift or—when slowing or preparing to land—drag. This apparent inconsistency results from the fact that the characteristics of air flow change drastically from situation to situation, and in fact, air never behaves as perfectly as it does in a textbook illustration of Bernoulli's principle.

Not only is the aircraft subject to air viscosity—the air's own friction with itself—it also experiences friction drag, which results from the fact that no solid can move through a fluid without experiencing a retarding force. An even greater drag factor, accounting for one-third of that which an aircraft experiences, is induced drag. The latter results because air does not flow in perfect laminar streams over the airfoil; rather, it forms turbulent eddies and currents that act against the forward movement of the plane.

In the air, an aircraft experiences forces that tend to destabilize flight in each of three dimensions. Pitch is the tendency to rotate forward or backward; yaw, the tendency to rotate on a horizontal plane; and roll, the tendency to rotate vertically on the axis of its fuselage. Obviously, each of these is a terrifying prospect, but fortunately, pilots have a solution for each. To prevent pitching, they adjust the angle of attack of the horizontal tail at the rear of the craft. The vertical rear tail plays a part in preventing yawing, and to prevent rolling, the pilot raises the tips of the main wings so that the craft assumes a V-shape when seen from the front or back.

The above factors of lift, drag, thrust, and weight, as well as the three types of possible destabilization, affect all forms of heavier-than-air flying machines. But since the 1944 advent of jet engines, which travel much faster than piston-driven engines, planes have flown faster and faster, and today some craft such as the *Concorde* are capable of supersonic flight. In these situations, air compressibility becomes a significant issue.

Sound is transmitted by the successive compression and expansion of air. But when a plane is traveling at above Mach 1.2—the Mach number indicates the speed of an aircraft in relation to the speed of sound—there is a significant discrepancy between the speed at which sound is traveling away from the craft, and the speed at which the craft is moving away from the sound. Eventually the compressed sound waves build up, resulting in a shock wave.

Down on the ground, the shock wave manifests as a "sonic boom"; meanwhile, for the aircraft, it can cause sudden changes in pressure, density, and temperature, as well as an increase in drag and a loss of stability. To counteract this effect, designers of supersonic and hypersonic (Mach 5 and above) aircraft are altering wing design, using a much narrower airfoil and swept-back wings.

One of the pioneers in this area is Richard Whitcomb of the National Aeronautics and Space Administration (NASA). Whitcomb has designed a supercritical airfoil for a proposed hypersonic plane, which would ascend into outer space in the course of a two-hour flight—all the time needed for it to travel from Washington, D.C., to Tokyo, Japan. Before the craft can become operational, however, researchers will have to figure out ways to control temperatures and keep the plane from bursting into flame as it reenters the atmosphere.

Much of the research for improving the aerodynamic qualities of such aircraft takes place in wind tunnels. First developed in 1871, these use powerful fans to create strong air currents, and over the years the top speed in wind tunnels has been increased to accommodate testing on supersonic and hypersonic aircraft. Researchers today use helium to create wind blasts at speeds up to Mach 50.

THROWN AND FLOWN: THE AERO-DYNAMICS OF SMALL OBJECTS

Long before engineers began to dream of sending planes into space for transoceanic flight—about 14,000 years ago, in fact—many of the features that make an airplane fly were already present in the boomerang. It might seem backward to move from a hypersonic jet to a boomerang, but in fact, it is easier to appreciate the aerodynamics of small objects, including the kite and even the paper airplane, once one comprehends the larger picture.

There is a certain delicious irony in the fact that the first manmade object to take flight was constructed by people who never advanced beyond the Stone Age until the nineteenth century, when the Europeans arrived in Australia. As the ethnobotanist Jared Diamond showed in his groundbreaking work *Guns, Germs, and Steel: The Fates of Human Societies* (1997), this was not because the Aborigines of Australia were less

intelligent than Europeans. In fact, as Diamond showed, an individual would actually have to be smarter to figure out how to survive on the limited range of plants and animals available in Australia prior to the introduction of Eurasian flora and fauna. Hence the wonder of the boomerang, one of the most ingenious inventions ever fashioned by humans in a "primitive" state.

Thousands of years before Bernoulli, the boomerang's designers created an airfoil consistent with Bernoulli's principle. The air below exerts more pressure than the air above, and this, combined with the factors of gyroscopic stability and gyroscopic precession, gives the boomerang flight.

Gyroscopic stability can be illustrated by spinning a top: the action of spinning itself keeps the top stable. Gyroscopic precession is a much more complex process: simply put, the leading wing of the boomerang—the forward or upward edge as it spins through the air—creates more lift than the other wing. At this point it should be noted that, contrary to the popular image, a boomerang travels on a plane perpendicular to that of the ground, not parallel. Hence any thrower who knows what he or she is doing tosses the boomerang not with a side-arm throw, but overhand.

And of course a boomerang does not just sail through the air; a skilled thrower can make it come back as if by magic. This is because the force of the increased lift that it experiences in flight, combined with gyroscopic precession, turns it around. As noted earlier, in different situations the same force that creates lift can create drag, and as the boomerang spins downward the increasing drag slows it. Certainly it takes great skill for a thrower to make a boomerang come back, and for this reason, participants in boomerang competitions often attach devices such as flaps to increase drag on the return cycle.

Another very early example of an aerodynamically sophisticated humanmade device—though it is quite recent compared to the boomerang—is the kite, which first appeared in China in about 1000 B.C. The kite's design borrows from avian anatomy, particularly the bird's light, hollow bones. Hence a kite, in its simplest form, consists of two crossed strips of very light wood such as balsa, with a lightweight fabric stretched over them.

Kites can come in a variety of shapes, though for many years the well-known diamond shape has been the most popular, in part because its aerodynamic qualities make it easiest for the novice kite-flyer to handle. Like birds and boomerangs, kites can "fly" because of the physical laws embodied in Bernoulli's principle: at the best possible angle of attack, the kite experiences a maximal ratio of pressure from the slower-moving air below as against the faster-moving air above.

For centuries, when the kite represented the only way to put a humanmade object many hundreds of feet into the air, scientists and engineers used them for a variety of experiments. Of course, the most famous example of this was Benjamin Franklin's 1752 experiment with electricity. More significant to the future of aerodynamics were investigations made half a century later by Cayley, who recognized that the kite, rather than the balloon, was an appropriate model for the type of heavier-than-air flight he intended.

In later years, engineers built larger kites capable of lifting men into the air, but the advent of the airplane rendered kites obsolete for this purpose. However, in the 1950s an American engineer named Francis Rogallo invented the flexible kite, which in turn spawned the delta wing kite used by hang gliders. During the 1960s, Domina Jolbert created the parafoil, an even more efficient device, which took nonmechanized human flight perhaps as far as it can go.

Akin to the kite, glider, and hang glider is that creation of childhood fancy, the paper airplane. In its most basic form—and paper airplane enthusiasts are capable of fairly complex designs—a paper airplane is little more than a set of wings. There are a number or reasons for this, not least the fact that in most cases, a person flying a paper airplane is not as concerned about pitch, yaw, and roll as a pilot flying with several hundred passengers on board would be.

However, when fashioning a paper airplane it is possible to add a number of design features, for instance by folding flaps upward at the tail. These become the equivalent of the elevator, a control surface along the horizontal edge of a real aircraft's tail, which the pilot rotates upward to provide stability. But as noted by Ken Blackburn, author of several books on paper airplanes, it is not necessarily the case that an airplane must have a tail; indeed, some of the most sophisticated craft in the sky today—including the fearsome B-2 "Stealth" bomber—do not have tails.

A typical paper airplane has low aspect ratio wings, a term that refers to the size of the wingspan compared to the chord line. In subsonic flight, higher aspect ratios are usually preferred, and this is certainly the case with most "real" gliders; hence their wings are longer, and their chord lines shorter. But there are several reasons why this is not the case with a paper airplane.

First of all, as Blackburn noted wryly on his Web site, "Paper is a lousy building material. There is a reason why real airplanes are not made of paper." He stated the other factors governing paper airplanes' low aspect ratio in similarly whimsical terms. First, "Low aspect ratio wings are easier to fold...."; second, "Paper airplane gliding performance is not usually very important...."; and third, "Low-aspect ratio wings look faster, especially if they are swept back."

The reason why low-aspect ratio wings look faster, Blackburn suggested, is that people see them on jet fighters and the *Concorde*, and assume that a relatively narrow wing span with a long chord line yields the fastest speeds. And indeed they do—but only at supersonic speeds. Below the speed of sound, high-aspect ratio wings are best for preventing drag. Furthermore, as Blackburn went on to note, low-aspect ratio wings help the paper airplane to withstand the relatively high launch speeds necessary to send them into longer glides.

In fact, a paper airplane is not subject to anything like the sort of design constraints affecting a real craft. All real planes look somewhat similar, because the established combinations, ratios, and dimensions of wings, tails, and fuselage work best. Certainly there is a difference in basic appearance between subsonic and supersonic aircraft—but again, all supersonic jets have more or less the same low-aspect, swept wing. "With paper airplanes," Blackburn wrote, "it's easy to make airplanes that don't look like real airplanes" since "The mission of a paper airplane is [simply] to provide a good time for the pilot."

AERODYNAMICS ON THE GROUND

The preceding discussions of aerodynamics in action have concerned the behavior of objects off the ground. But aerodynamics is also a factor in wheeled transport on Earth's surface, whether by bicycle, automobile, or some other variation.

On a bicycle, the rider accounts for 65-80% of the drag, and therefore his or her position with regard to airflow is highly important. Thus, from as early as the 1890s, designers of racing bikes have favored drop handlebars, as well as a seat and frame that allow a crouched position. Since the 1980s, bicycle designers have worked to eliminate all possible extra lines and barriers to airflow, including the crossbar and chainstays.

A typical bicycle's wheel contains 32 or 36 cylindrical spokes, and these can affect aerodynamics adversely. As the wheel rotates, the airflow behind the spoke separates, creating turbulence and hence drag. For this reason, some of the most advanced bicycles today use either aerodynamic rims, which reduce the length of the spokes, three-spoke aerodynamic wheels, or even solid wheels.

The rider's gear can also serve to impede or enhance his velocity, and thus modern racing helmets have a streamlined shape—rather like that of an airfoil. The best riders, such as those who compete in the Olympics or the Tour de France, have bikes custom-designed to fit their own body shape.

One interesting aspect of aerodynamics where it concerns bicycle racing is the phenomenon of "drafting." Riders at the front of a pack, like riders pedaling alone, consume 30-40% more energy than do riders in the middle of a pack. The latter are benefiting from the efforts of bicyclists in front of them, who put up most of the wind resistance. The same is true for bicyclists who ride behind automobiles or motorcycles.

The use of machine-powered pace vehicles to help in achieving extraordinary speeds is far from new. Drafting off of a railroad car with specially designed aerodynamic shields, a rider in 1896 was able to exceed 60 MPH (96 km/h), a then unheard-of speed. Today the record is just under 167 MPH (267 km/h). Clearly one must be a highly skilled, powerful rider to approach anything like this speed; but design factors also come into play, and not just in the case of the pace vehicle. Just as supersonic jets are quite different from ordinary planes, super high-speed bicycles are not like the average bike; they are designed in such a way that they must be moving faster than 60 MPH before the rider can even pedal.

A PROFESSIONAL BICYCLE RACER'S STREAMLINED HELMET AND CROUCHED POSITION HELP TO IMPROVE AIRFLOW, THUS INCREASING SPEED. *(Photograph by Ronnen Eshel/Corbis. Reproduced by permission.)*

With regard to automobiles, as noted earlier, aerodynamics has a strong impact on body design. For this reason, cars over the years have become steadily more streamlined and aerodynamic in appearance, a factor that designers balance with aesthetic appeal. Today's Chrysler PT Cruiser, which debuted in 2000, may share outward features with 1930s and 1940s cars, but the PT Cruiser's design is much more sound aerodynamically—not least because a modern vehicle can travel much, much faster than the cars driven by previous generations.

Nowhere does the connection between aerodynamics and automobiles become more crucial than in the sport of auto racing. For race-car drivers, drag is always a factor to be avoided and counteracted by means ranging from drafting to altering the body design to reduce the airflow under the vehicle. However, as strange as it may seem, a car—like an airplane—is also subject to lift.

It was noted earlier that in some cases lift can be undesirable in an airplane (for instance, when trying to land), but it is virtually always undesirable in an automobile. The greater the speed, the greater the lift force, which increases

KEY TERMS

AERODYNAMICS: The study of air flow and its principles. Applied aerodynamics is the science of improving manmade objects in light of those principles.

AIRFOIL: The design of an airplane's wing when seen from the end, a shape intended to maximize the aircraft's response to airflow.

ANGLE OF ATTACK: The orientation of the airfoil with regard to the airflow, or the angle that the chord line forms with the direction of the air stream.

BERNOULLI'S PRINCIPLE: A proposition, credited to Swiss mathematician and physicist Daniel Bernoulli (1700-1782), which maintains that slower-moving fluid exerts greater pressure than faster-moving fluid.

CAMBER: The enhanced curvature on the upper surface of an airfoil.

CHORD LINE: The distance, along an imaginary straight line, from the stagnation point of an airfoil to the rear, or trailing edge.

DRAG: The force that opposes the forward motion of an object in airflow.

LAMINAR: A term describing a streamlined flow, in which all particles move at

the same speed and in the same direction. Its opposite is turbulent flow.

LIFT: An aerodynamic force perpendicular to the direction of the wind. For an aircraft, lift is the force that raises it off the ground and keeps it aloft.

PITCH: The tendency of an aircraft in flight to rotate forward or backward; see also yaw and roll.

ROLL: The tendency of an aircraft in flight to rotate vertically on the axis of its fuselage; see also pitch and yaw.

STAGNATION POINT: The spot where airflow hits the leading edge of an airfoil.

SUPERSONIC: Faster than Mach 1, or the speed of sound—660 MPH (1,622 km/h). Speeds above Mach 5 are referred to as hypersonic.

TURBULENT: A term describing a highly irregular form of flow, in which a fluid is subject to continual changes in speed and direction. Its opposite is laminar flow.

VISCOSITY: The internal friction in a fluid that makes it resistant to flow.

YAW: The tendency of an aircraft in flight to rotate on a horizontal plane; see also Pitch and Roll.

the threat of instability. For this reason, builders of race cars design their vehicles for negative lift: hence a typical family car has a lift coefficient of about 0.03, whereas a race car is likely to have a coefficient of -3.00.

Among the design features most often used to reduce drag while achieving negative lift is a rear-deck spoiler. The latter has an airfoil shape, but its purpose is different: to raise the rear stagnation point and direct air flow so that it does

not wrap around the vehicle's rear end. Instead, the spoiler creates a downward force to stabilize the rear, and it may help to decrease drag by reducing the separation of airflow (and hence the creation of turbulence) at the rear window.

Similar in concept to a spoiler, though somewhat different in purpose, is the aerodynamically curved shield that sits atop the cab of most modern eighteen-wheel transport trucks. The purpose of the shield becomes apparent when the truck is moving at high speeds: wind resistance becomes strong, and if the wind were to hit the truck's trailer head-on, it would be as though the air were pounding a brick wall. Instead, the shield scoops air upward, toward the rear of the truck. At the rear may be another panel, patented by two young engineers in 1994, that creates a drag-reducing vortex between panel and truck.

WHERE TO LEARN MORE

Cockpit Physics (Department of Physics, United States Air Force Academy web site.). <http://www.usafa.af.mil/dfp/cockpit-phys/> (February 19, 2001).

K8AIT Principles of Aeronautics Advanced Text. (web site). <http://wings.ucdavis.edu/Book/advanced.html> (February 19, 2001).

Macaulay, David. *The New Way Things Work.* Boston: Houghton Mifflin, 1998.

Blackburn, Ken. *Paper Airplane Aerodynamics.* (web site). <http://www.geocities.com/CapeCanaveral/1817/paero.html> (February 19, 2001).

Schrier, Eric and William F. Allman. *Newton at the Bat: The Science in Sports.* New York: Charles Scribner's Sons, 1984.

Smith, H. C. *The Illustrated Guide to Aerodynamics.* Blue Ridge Summit, PA: Tab Books, 1992.

Stever, H. Guyford, James J. Haggerty, and the Editors of Time-Life Books. *Flight.* New York: Time-Life Books, 1965.

Suplee, Curt. *Everyday Science Explained.* Washington, D.C.: National Geographic Society, 1996.

BERNOULLI'S PRINCIPLE

CONCEPT

Bernoulli's principle, sometimes known as Bernoulli's equation, holds that for fluids in an ideal state, pressure and density are inversely related: in other words, a slow-moving fluid exerts more pressure than a fast-moving fluid. Since "fluid" in this context applies equally to liquids and gases, the principle has as many applications with regard to airflow as to the flow of liquids. One of the most dramatic everyday examples of Bernoulli's principle can be found in the airplane, which stays aloft due to pressure differences on the surface of its wing; but the truth of the principle is also illustrated in something as mundane as a shower curtain that billows inward.

HOW IT WORKS

The Swiss mathematician and physicist Daniel Bernoulli (1700-1782) discovered the principle that bears his name while conducting experiments concerning an even more fundamental concept: the conservation of energy. This is a law of physics that holds that a system isolated from all outside factors maintains the same total amount of energy, though energy transformations from one form to another take place.

For instance, if you were standing at the top of a building holding a baseball over the side, the ball would have a certain quantity of potential energy—the energy that an object possesses by virtue of its position. Once the ball is dropped, it immediately begins losing potential energy and gaining kinetic energy—the energy that an object possesses by virtue of its motion. Since the total energy must remain constant, potential and kinetic energy have an inverse relationship: as the value of one variable decreases, that of the other increases in exact proportion.

The ball cannot keep falling forever, losing potential energy and gaining kinetic energy. In fact, it can never gain an amount of kinetic energy greater than the potential energy it possessed in the first place. At the moment before the ball hits the ground, its kinetic energy is equal to the potential energy it possessed at the top of the building. Correspondingly, its potential energy is zero—the same amount of kinetic energy it possessed before it was dropped.

Then, as the ball hits the ground, the energy is dispersed. Most of it goes into the ground, and depending on the rigidity of the ball and the ground, this energy may cause the ball to bounce. Some of the energy may appear in the form of sound, produced as the ball hits bottom, and some will manifest as heat. The total energy, however, will not be lost: it will simply have changed form.

Bernoulli was one of the first scientists to propose what is known as the kinetic theory of gases: that gas, like all matter, is composed of tiny molecules in constant motion. In the 1730s, he conducted experiments in the conservation of energy using liquids, observing how water flows through pipes of varying diameter. In a segment of pipe with a relatively large diameter, he observed, water flowed slowly, but as it entered a segment of smaller diameter, its speed increased.

It was clear that some force had to be acting on the water to increase its speed. Earlier, Robert Boyle (1627-1691) had demonstrated that pressure and volume have an inverse relationship, and Bernoulli seems to have applied Boyle's findings to the present situation. Clearly the volume of water flowing through the narrower pipe at any given moment was less than that flowing through the wider one. This suggested, according to Boyle's law, that the pressure in the wider pipe must be greater.

As fluid moves from a wider pipe to a narrower one, the volume of that fluid that moves a given distance in a given time period does not change. But since the width of the narrower pipe is smaller, the fluid must move faster in order to achieve that result. One way to illustrate this is to observe the behavior of a river: in a wide, unconstricted region, it flows slowly, but if its flow is narrowed by canyon walls (for instance), then it speeds up dramatically.

The above is a result of the fact that water is a fluid, and having the characteristics of a fluid, it adjusts its shape to fit that of its container or other solid objects it encounters on its path. Since the volume passing through a given length of pipe during a given period of time will be the same, there must be a decrease in pressure. Hence Bernoulli's conclusion: the slower the rate of flow, the higher the pressure, and the faster the rate of flow, the lower the pressure.

Bernoulli published the results of his work in *Hydrodynamica* (1738), but did not present his ideas or their implications clearly. Later, his friend the German mathematician Leonhard Euler (1707-1783) generalized his findings in the statement known today as Bernoulli's principle.

THE VENTURI TUBE

Also significant was the work of the Italian physicist Giovanni Venturi (1746-1822), who is credited with developing the Venturi tube, an instrument for measuring the drop in pressure that takes place as the velocity of a fluid increases. It consists of a glass tube with an inward-sloping area in the middle, and manometers, devices for measuring pressure, at three places: the entrance, the point of constriction, and the exit. The Venturi meter provided a consistent means of demonstrating Bernoulli's principle.

Like many propositions in physics, Bernoulli's principle describes an ideal situation in the absence of other forces. One such force is

viscosity, the internal friction in a fluid that makes it resistant to flow. In 1904, the German physicist Ludwig Prandtl (1875-1953) was conducting experiments in liquid flow, the first effort in well over a century to advance the findings of Bernoulli and others. Observing the flow of liquid in a tube, Prandtl found that a tiny portion of the liquid adheres to the surface of the tube in the form of a thin film, and does not continue to move. This he called the viscous boundary layer.

Like Bernoulli's principle itself, Prandtl's findings would play a significant part in aerodynamics, or the study of airflow and its principles. They are also significant in hydrodynamics, or the study of water flow and its principles, a discipline Bernoulli founded.

LAMINAR VS. TURBULENT FLOW

Air and water are both examples of fluids, substances which—whether gas or liquid—conform to the shape of their container. The flow patterns of all fluids may be described in terms either of laminar flow, or of its opposite, turbulent flow.

Laminar flow is smooth and regular, always moving at the same speed and in the same direction. Also known as streamlined flow, it is characterized by a situation in which every particle of fluid that passes a particular point follows a path identical to all particles that passed that point earlier. A good illustration of laminar flow is what occurs when a stream flows around a twig.

By contrast, in turbulent flow, the fluid is subject to continual changes in speed and direction—as, for instance, when a stream flows over shoals of rocks. Whereas the mathematical model of laminar flow is rather straightforward, conditions are much more complex in turbulent flow, which typically occurs in the presence of obstacles or high speeds.

Turbulent flow makes it more difficult for two streams of air, separated after hitting a barrier, to rejoin on the other side of the barrier; yet that is their natural tendency. In fact, if a single air current hits an airfoil—the design of an airplane's wing when seen from the end, a streamlined shape intended to maximize the aircraft's response to airflow—the air that flows over the top will "try" to reach the back end of the airfoil at the same time as the air that flows over the

A KITE'S DESIGN, PARTICULARLY ITS USE OF LIGHTWEIGHT FABRIC STRETCHED OVER TWO CROSSED STRIPS OF VERY
LIGHT WOOD, MAKES IT WELL-SUITED FOR FLIGHT, BUT WHAT KEEPS IT IN THE AIR IS A DIFFERENCE IN AIR PRESSURE.
AT THE BEST POSSIBLE ANGLE OF ATTACK, THE KITE EXPERIENCES AN IDEAL RATIO OF PRESSURE FROM THE SLOW-
ER-MOVING AIR BELOW VERSUS THE FASTER-MOVING AIR ABOVE, AND THIS GIVES IT LIFT. *(Roger Ressmeyer/Corbis. Repro-
duced by permission.)*

bottom. In order to do so, it will need to speed
up—and this, as will be shown below, is the basis
for what makes an airplane fly.

When viscosity is absent, conditions of perfect laminar flow exist: an object behaves in complete alignment with Bernoulli's principle. Of

THE BOOMERANG, DEVELOPED BY AUSTRALIA'S ABORIGINAL PEOPLE, FLIES THROUGH THE AIR ON A PLANE PER-
PENDICULAR TO THE GROUND, RATHER THAN PARALLEL. AS IT FLIES, THE BOOMERANG BECOMES BOTH A GYROSCOPE
AND AN AIRFOIL, AND THIS DUAL ROLE GIVES IT AERODYNAMIC LIFT. *(Bettmann/Corbis. Reproduced by permission.)*

course, though ideal conditions seldom occur in
the real world, Bernoulli's principle provides a
guide for the behavior of planes in flight, as well
as a host of everyday things.

REAL-LIFE APPLICATIONS

FLYING MACHINES

For thousands of years, human beings vainly
sought to fly "like a bird," not realizing that this is
literally impossible, due to differences in phys-
iognomy between birds and *homo sapiens*. No
man has ever been born (or ever will be) who
possesses enough strength in his chest that he
could flap a set of attached wings and lift his
body off the ground. Yet the bird's physical structure proved highly useful to designers of practical flying machines.

A bird's wing is curved along the top, so that
when air passes over the wing and divides, the
curve forces the air on top to travel a greater distance than the air on the bottom. The tendency
of airflow, as noted earlier, is to correct for the
presence of solid objects and to return to its original pattern as quickly as possible. Hence, when

the air hits the front of the wing, the rate of flow
at the top increases to compensate for the greater
distance it has to travel than the air below the
wing. And as shown by Bernoulli, fast-moving
fluid exerts less pressure than slow-moving fluid;
therefore, there is a difference in pressure
between the air below and the air above, and this
keeps the wing aloft.

Only in 1853 did Sir George Cayley (1773-
1857) incorporate the avian airfoil to create history's first workable (though engine-less) flying
machine, a glider. Much, much older than Cayley's glider, however, was the first manmade flying machine built "according to Bernoulli's principle"—only it first appeared in about 12,000
B.C., and the people who created it had had little
contact with the outside world until the late eighteenth century A.D. This was the boomerang, one
of the most ingenious devices ever created by a
stone-age society—in this case, the Aborigines of
Australia.

Contrary to the popular image, a
boomerang flies through the air on a plane perpendicular to the ground, rather than parallel.
Hence, any thrower who properly knows how
tosses the boomerang not with a side-arm throw,
but overhand. As it flies, the boomerang becomes

both a gyroscope and an airfoil, and this dual role gives it aerodynamic lift.

Like the gyroscope, the boomerang imitates a top; spinning keeps it stable. It spins through the air, its leading wing (the forward or upward wing) creating more lift than the other wing. As an airfoil, the boomerang is designed so that the air below exerts more pressure than the air above, which keeps it airborne.

Another very early example of a flying machine using Bernoulli's principles is the kite, which first appeared in China in about 1000 B.C. The kite's design, particularly its use of lightweight fabric stretched over two crossed strips of very light wood, makes it well-suited for flight, but what keeps it in the air is a difference in air pressure. At the best possible angle of attack, the kite experiences an ideal ratio of pressure from the slower-moving air below versus the faster-moving air above, and this gives it lift.

Later Cayley studied the operation of the kite, and recognized that it—rather than the balloon, which at first seemed the most promising apparatus for flight—was an appropriate model for the type of heavier-than-air flying machine he intended to build. Due to the lack of a motor, however, Cayley's prototypical airplane could never be more than a glider: a steam engine, then state-of-the-art technology, would have been much too heavy.

Hence, it was only with the invention of the internal-combustion engine that the modern airplane came into being. On December 17, 1903, at Kitty Hawk, North Carolina, Orville (1871-1948) and Wilbur (1867-1912) Wright tested a craft that used a 25-horsepower engine they had developed at their bicycle shop in Ohio. By maximizing the ratio of power to weight, the engine helped them overcome the obstacles that had dogged recent attempts at flight, and by the time the day was over, they had achieved a dream that had eluded men for more than four millennia.

Within fifty years, airplanes would increasingly obtain their power from jet rather than internal-combustion engines. But the principle that gave them flight, and the principle that kept them aloft once they were airborne, reflected back to Bernoulli's findings of more than 160 years before their time. This is the concept of the airfoil.

As noted earlier, an airfoil has a streamlined design. Its shape is rather like that of an elongat-

ed, asymmetrical teardrop lying on its side, with the large end toward the direction of airflow, and the narrow tip pointing toward the rear. The greater curvature of its upper surface in comparison to the lower side is referred to as the airplane's camber. The front end of the airfoil is also curved, and the chord line is an imaginary straight line connecting the spot where the air hits the front—known as the stagnation point—to the rear, or trailing edge, of the wing.

Again, in accordance with Bernoulli's principle, the shape of the airflow facilitates the spread of laminar flow around it. The slower-moving currents beneath the airfoil exert greater pressure than the faster currents above it, giving lift to the aircraft. Of course, the aircraft has to be moving at speeds sufficient to gain momentum for its leap from the ground into the air, and here again, Bernoulli's principle plays a part.

Thrust comes from the engines, which run the propellers—whose blades in turn are designed as miniature airfoils to maximize their power by harnessing airflow. Like the aircraft wings, the blades' angle of attack—the angle at which airflow hits it. In stable flight, the pilot greatly increases the angle of attack (also called pitched), whereas at takeoff and landing, the pitch is dramatically reduced.

DRAWING FLUIDS UPWARD: ATOMIZERS AND CHIMNEYS

A number of everyday objects use Bernoulli's principle to draw fluids upward, and though in terms of their purposes, they might seem very different—for instance, a perfume atomizer vs. a chimney—they are closely related in their application of pressure differences. In fact, the idea behind an atomizer for a perfume spray bottle can also be found in certain garden-hose attachments, such as those used to provide a high-pressure car wash.

The air inside the perfume bottle is moving relatively slowly; therefore, according to Bernoulli's principle, its pressure is relatively high, and it exerts a strong downward force on the perfume itself. In an atomizer there is a narrow tube running from near the bottom of the bottle to the top. At the top of the perfume bottle, it opens inside another tube, this one perpendicular to the first tube. At one end of the horizontal tube is a simple squeeze-pump which causes air to flow quickly through it. As a result, the pressure toward the top of the bottle is reduced, and the perfume flows upward along the vertical tube, drawn from the area of higher pressure at the bottom. Once it is in the upper tube, the squeeze-pump helps to eject it from the spray nozzle.

A carburetor works on a similar principle, though in that case the lower pressure at the top draws air rather than liquid. Likewise a chimney draws air upward, and this explains why a windy day outside makes for a better fire inside. With wind blowing over the top of the chimney, the air pressure at the top is reduced, and tends to draw higher-pressure air from down below.

The upward pull of air according to the Bernoulli principle can also be illustrated by what is sometimes called the "Hoover bugle"—a name perhaps dating from the Great Depression, when anything cheap or contrived bore the appellation "Hoover" as a reflection of popular dissatisfaction with President Herbert Hoover. In any case, the Hoover bugle is simply a long corrugated tube that, when swung overhead, produces musical notes.

You can create a Hoover bugle using any sort of corrugated tube, such as vacuum-cleaner hose or swimming-pool drain hose, about 1.8 in (4 cm) in diameter and 6 ft (1.8 m) in length. To operate it, you should simply hold the tube in both hands, with extra length in the leading hand—that is, the right hand, for most people. This is the hand with which to swing the tube over your head, first slowly and then faster, observing the changes in tone that occur as you change the pace.

The vacuum hose of a Hoover tube can also be returned to a version of its original purpose in an illustration of Bernoulli's principle. If a piece of paper is torn into pieces and placed on a table, with one end of the tube just above the paper and the other end spinning in the air, the paper tends to rise. It is drawn upward as though by a vacuum cleaner—but in fact, what makes it happen is the pressure difference created by the movement of air.

In both cases, reduced pressure draws air from the slow-moving region at the bottom of the tube. In the case of the Hoover bugle, the corrugations produce oscillations of a certain frequency. Slower speeds result in slower oscillations and hence lower frequency, which produces a lower tone. At higher speeds, the opposite is

true. There is little variation in tones on a Hoover bugle: increasing the velocity results in a frequency twice that of the original, but it is difficult to create enough speed to generate a third tone.

SPIN, CURVE, AND PULL: THE COUNTERINTUITIVE PRINCIPLE

There are several other interesting illustrations—sometimes fun and in one case potentially tragic—of Bernoulli's principle. For instance, there is the reason why a shower curtain billows inward once the shower is turned on. It would seem logical at first that the pressure created by the water would push the curtain outward, securing it to the side of the bathtub.

Instead, of course, the fast-moving air generated by the flow of water from the shower creates a center of lower pressure, and this causes the curtain to move away from the slower-moving air outside. This is just one example of the ways in which Bernoulli's principle creates results that, on first glance at least, seem counterintuitive—that is, the opposite of what common sense would dictate.

Another fascinating illustration involves placing two empty soft drink cans parallel to one another on a table, with a couple of inches or a few centimeters between them. At that point, the air on all sides has the same slow speed. If you were to blow directly between the cans, however, this would create an area of low pressure between them. As a result, the cans push together. For ships in a harbor, this can be a frightening prospect: hence, if two crafts are parallel to one another and a strong wind blows between them, there is a possibility that they may behave like the cans.

Then there is one of the most illusory uses of Bernoulli's principle, that infamous baseball pitcher's trick called the curve ball. As the ball moves through the air toward the plate, its velocity creates an air stream moving against the trajectory of the ball itself. Imagine it as two lines, one curving over the ball and one curving under, as the ball moves in the opposite direction.

In an ordinary throw, the effects of the airflow would not be particularly intriguing, but in this case, the pitcher has deliberately placed a "spin" on the ball by the manner in which he has thrown it. How pitchers actually produce spin is a complex subject unto itself, involving grip,

KEY TERMS

AERODYNAMICS: The study of airflow and its principles. Applied aerodynamics is the science of improving manmade objects in light of those principles.

AIRFOIL: The design of an airplane's wing when seen from the end, a shape intended to maximize the aircraft's response to airflow.

ANGLE OF ATTACK: The orientation of the airfoil with regard to the airflow, or the angle that the chord line forms with the direction of the air stream.

BERNOULLI'S PRINCIPLE: A proposition, credited to Swiss mathematician and physicist Daniel Bernoulli (1700-1782), which maintains that slower-moving fluid exerts greater pressure than faster-moving fluid.

CAMBER: The enhanced curvature on the upper surface of an airfoil.

CHORD LINE: The distance, along an imaginary straight line, from the stagnation point of an airfoil to the rear, or trailing edge.

CONSERVATION OF ENERGY: A law of physics which holds that within a system isolated from all other outside factors, the total amount of energy remains the same, though transformations of energy from one form to another take place.

FLUID: Any substance, whether gas or liquid, that conforms to the shape of its container.

HYDRODYNAMICS: The study of water flow and its principles.

INVERSE RELATIONSHIP: A situation involving two variables, in which one of the two increases in direct proportion to the decrease in the other.

KINETIC ENERGY: The energy that an object possesses by virtue of its motion.

LAMINAR: A term describing a streamlined flow, in which all particles move at the same speed and in the same direction. Its opposite is turbulent flow.

LIFT: An aerodynamic force perpendicular to the direction of the wind. For an aircraft, lift is the force that raises it off the ground and keeps it aloft.

MANOMETERS: Devices for measuring pressure in conjunction with a Venturi tube.

POTENTIAL ENERGY: The energy that an object possesses by virtue of its position. **STAGNATION POINT:** The spot where airflow hits the leading edge of an airfoil.

TURBULENT: A term describing a highly irregular form of flow, in which a fluid is subject to continual changes in speed and direction. Its opposite is laminar flow.

VENTURI TUBE: An instrument, consisting of a glass tube with an inward-sloping area in the middle, for measuring the drop in pressure that takes place as the velocity of a fluid increases.

VISCOSITY: The internal friction in a fluid that makes it resistant to flow.

wrist movement, and other factors, and in any case, the fact of the spin is more important than the way in which it was achieved.

If the direction of airflow is from right to left, the ball, as it moves into the airflow, is spinning clockwise. This means that the air flowing over the ball is moving in a direction opposite to the spin, whereas that flowing under it is moving in the same direction. The opposite forces produce a drag on the top of the ball, and this cuts down on the velocity at the top compared to that at the bottom of the ball, where spin and airflow are moving in the same direction.

Thus the air pressure is higher at the top of the ball, and as per Bernoulli's principle, this tends to pull the ball downward. The curve ball—of which there are numerous variations, such as the fade and the slider—creates an unpredictable situation for the batter, who sees the ball leave the pitcher's hand at one altitude, but finds to his dismay that it has dropped dramatically by the time it crosses the plate.

A final illustration of Bernoulli's often counterintuitive principle neatly sums up its effects on the behavior of objects. To perform the experiment, you need only an index card and a flat surface. The index card should be folded at the ends so that when the card is parallel to the surface, the ends are perpendicular to it. These folds should be placed about half an inch (about one centimeter) from the ends.

At this point, it would be handy to have an unsuspecting person—someone who has not studied Bernoulli's principle—on the scene, and challenge him or her to raise the card by blowing under it. Nothing could seem easier, of course: by blowing under the card, any person would naturally assume, the air will lift it. But of course this is completely wrong according to Bernoulli's principle. Blowing under the card, as illustrated, will create an area of high velocity and low pressure. This will do nothing to lift the card: in fact, it only pushes the card more firmly down on the table.

WHERE TO LEARN MORE

Beiser, Arthur. *Physics,* 5th ed. Reading, MA: Addison-Wesley, 1991.

"Bernoulli's Principle: Explanations and Demos." (Web site). <http://207.10.97.102/physicszone/lesson/02forces/bernoull/bernoul l.html> (February 22, 2001).

Cockpit Physics (Department of Physics, United States Air Force Academy. Web site.). <http://www.usafa.af.mil/dfp/cockpit-phys/> (February 19, 2001).

K8AIT Principles of Aeronautics Advanced Text. (Web site). <http://wings.ucdavis.edu/Book/advanced.html> (February 19, 2001).

Schrier, Eric and William F. Allman. *Newton at the Bat: The Science in Sports.* New York: Charles Scribner's Sons, 1984.

Smith, H. C. *The Illustrated Guide to Aerodynamics.* Blue Ridge Summit, PA: Tab Books, 1992.

Stever, H. Guyford, James J. Haggerty, and the Editors of Time-Life Books. *Flight.* New York: Time-Life Books, 1965.

BUOYANCY

CONCEPT

The principle of buoyancy holds that the buoyant or lifting force of an object submerged in a fluid is equal to the weight of the fluid it has displaced. The concept is also known as Archimedes's principle, after the Greek mathematician, physicist, and inventor Archimedes (c. 287-212 B.C.), who discovered it. Applications of Archimedes's principle can be seen across a wide vertical spectrum: from objects deep beneath the oceans to those floating on its surface, and from the surface to the upper limits of the stratosphere and beyond.

HOW IT WORKS

ARCHIMEDES DISCOVERS BUOYANCY

There is a famous story that Sir Isaac Newton (1642-1727) discovered the principle of gravity when an apple fell on his head. The tale, an exaggerated version of real events, has become so much a part of popular culture that it has been parodied in television commercials. Almost equally well known is the legend of how Archimedes discovered the concept of buoyancy.

A native of Syracuse, a Greek colony in Sicily, Archimedes was related to one of that city's kings, Hiero II (308?-216 B.C.). After studying in Alexandria, Egypt, he returned to his hometown, where he spent the remainder of his life. At some point, the royal court hired (or compelled) him to set about determining the weight of the gold in the king's crown. Archimedes was in his bath pondering this challenge when suddenly it occurred to him that the buoyant force of a submerged object is equal to the weight of the fluid displaced by it.

He was so excited, the legend goes, that he jumped out of his bath and ran naked through the streets of Syracuse shouting "Eureka!" (I have found it). Archimedes had recognized a principle of enormous value—as will be shown—to shipbuilders in his time, and indeed to shipbuilders of the present.

Concerning the history of science, it was a particularly significant discovery; few useful and enduring principles of physics date to the period before Galileo Galilei (1564-1642.) Even among those few ancient physicists and inventors who contributed work of lasting value—Archimedes, Hero of Alexandria (c. 65-125 A.D.), and a few others—there was a tendency to miss the larger implications of their work. For example, Hero, who discovered steam power, considered it useful only as a toy, and as a result, this enormously significant discovery was ignored for seventeen centuries.

In the case of Archimedes and buoyancy, however, the practical implications of the discovery were more obvious. Whereas steam power must indeed have seemed like a fanciful notion to the ancients, there was nothing farfetched about oceangoing vessels. Shipbuilders had long been confronted with the problem of how to keep a vessel afloat by controlling the size of its load on the one hand, and on the other hand, its tendency to bob above the water. Here, Archimedes offered an answer.

BUOYANCY AND WEIGHT

Why does an object seem to weigh less underwater than above the surface? How is it that a ship made of steel, which is obviously heavier than water, can float? How can we determine whether a balloon will ascend in the air, or a submarine will descend in the water? These and other questions are addressed by the principle of buoyancy, which can be explained in terms of properties—most notably, gravity—unknown to Archimedes.

To understand the factors at work, it is useful to begin with a thought experiment. Imagine a certain quantity of fluid submerged within a larger body of the same fluid. Note that the terms "liquid" or "water" have not been used: not only is "fluid" a much more general term, but also, in general physical terms and for the purposes of the present discussion, there is no significant difference between gases and liquids. Both conform to the shape of the container in which they are placed, and thus both are fluids.

To return to the thought experiment, what has been posited is in effect a "bag" of fluid—that is, a "bag" made out of fluid and containing fluid no different from the substance outside the "bag." This "bag" is subjected to a number of forces. First of all, there is its weight, which tends to pull it to the bottom of the container. There is also the pressure of the fluid all around it, which varies with depth: the deeper within the container, the greater the pressure.

Pressure is simply the exertion of force over a two-dimensional area. Thus it is as though the fluid is composed of a huge number of two-dimensional "sheets" of fluid, each on top of the other, like pages in a newspaper. The deeper into the larger body of fluid one goes, the greater the pressure; yet it is precisely this increased force at the bottom of the fluid that tends to push the "bag" upward, against the force of gravity.

Now consider the weight of this "bag." Weight is a force—the product of mass multiplied by acceleration—that is, the downward acceleration due to Earth's gravitational pull. For an object suspended in fluid, it is useful to substitute another term for mass. Mass is equal to volume, or the amount of three-dimensional space occupied by an object, multiplied by density. Since density is equal to mass divided by volume, this means that volume multiplied by density is the same as mass.

We have established that the weight of the fluid "bag" is Vdg, where V is volume, d is density, and g is the acceleration due to gravity. Now imagine that the "bag" has been replaced by a solid object of exactly the same size. The solid object will experience exactly the same degree of pressure as the imaginary "bag" did—and hence, it will also experience the same buoyant force pushing it up from the bottom. This means that buoyant force is equal to the weight—Vdg—of displaced fluid.

Buoyancy is always a double-edged proposition. If the buoyant force on an object is greater than the weight of that object—in other words, if the object weighs less than the amount of water it has displaced—it will float. But if the buoyant force is less than the object's weight, the object will sink. Buoyant force is not the same as net force: if the object weighs more than the water it displaces, the force of its weight cancels out and in fact "overrules" that of the buoyant force.

At the heart of the issue is density. Often, the density of an object in relation to water is referred to as its specific gravity: most metals, which are heavier than water, are said to have a high specific gravity. Conversely, petroleum-based products typically float on the surface of water, because their specific gravity is low. Note the close relationship between density and weight where buoyancy is concerned: in fact, the most buoyant objects are those with a relatively high volume and a relatively low density.

This can be shown mathematically by means of the formula noted earlier, whereby density is equal to mass divided by volume. If $Vd = V(m/V)$, an increase in density can only mean an increase in mass. Since weight is the product of mass multiplied by g (which is assumed to be a constant figure), then an increase in density means an increase in mass and hence, an increase in weight—not a good thing if one wants an object to float.

REAL-LIFE APPLICATIONS

STAYING AFLOAT

In the early 1800s, a young Mississippi River flatboat operator submitted a patent application describing a device for "buoying vessels over shoals." The invention proposed to prevent a problem he had often witnessed on the river—boats grounded on sandbars—by equipping the boats with adjustable buoyant air chambers. The young man even whittled a model of his invention, but he was not destined for fame as an inventor; instead, Abraham Lincoln (1809-1865) was famous for much else. In fact Lincoln had a sound idea with his proposal to use buoyant force in protecting boats from running aground.

Buoyancy on the surface of water has a number of easily noticeable effects in the real world. (Having established the definition of fluid, from this point onward, the fluids discussed will be primarily those most commonly experienced: water and air.) It is due to buoyancy that fish, human swimmers, icebergs, and ships stay afloat. Fish offer an interesting application of volume change as a means of altering buoyancy: a fish has an internal swim bladder, which is filled with gas. When it needs to rise or descend, it changes the volume in its swim bladder, which then changes its density. The examples of swimmers and icebergs directly illustrate the principle of density—on the part of the water in the first instance, and on the part of the object itself in the second.

To a swimmer, the difference between swimming in fresh water and salt water shows that buoyant force depends as much on the density of the fluid as on the volume displaced. Fresh water has a density of 62.4 lb/ft³ (9,925 N/m³), whereas that of salt water is 64 lb/ft³ (10,167 N/m³). For this reason, salt water provides more buoyant force than fresh water; in Israel's Dead Sea, the saltiest body of water on Earth, bathers experience an enormous amount of buoyant force.

Water is an unusual substance in a number of regards, not least its behavior as it freezes. Close to the freezing point, water thickens up, but once it turns to ice, it becomes less dense. This is why ice cubes and icebergs float. However, their low density in comparison to the water around them means that only part of an iceberg stays atop the surface. The submerged percentage of an iceberg is the same as the ratio of the density of ice to that of water: 89%.

SHIPS AT SEA

Because water itself is relatively dense, a high-volume, low-density object is likely to displace a quantity of water more dense—and heavier—than the object itself. By contrast, a steel ball dropped into the water will sink straight to the bottom, because it is a low-volume, high-density object that outweighs the water it displaced.

This brings back the earlier question: how can a ship made out of steel, with a density of 487 lb/ft³ (77,363 N/m³), float on a salt-water ocean with an average density of only about one-eighth that amount? The answer lies in the design of the ship's hull. If the ship were flat like a raft, or if all the steel in it were compressed into a ball, it would indeed sink. Instead, however, the hollow hull displaces a volume of water heavier than the ship's own weight: once again, volume has been maximized, and density minimized.

For a ship to be seaworthy, it must maintain a delicate balance between buoyancy and stability. A vessel that is too light—that is, too much volume and too little density—will bob on the top of the water. Therefore, it needs to carry a certain amount of cargo, and if not cargo, then water or some other form of ballast. Ballast is a heavy substance that increases the weight of an object experiencing buoyancy, and thereby improves its stability.

Ideally, the ship's center of gravity should be vertically aligned with its center of buoyancy. The center of gravity is the geometric center of the ship's weight—the point at which weight above is equal to weight below, weight fore is equal to weight aft, and starboard (right-side) weight is equal to weight on the port (left) side. The center of buoyancy is the geometric center of its submerged volume, and in a stable ship, it is some distance directly below center of gravity.

Displacement, or the weight of the fluid that is moved out of position when an object is immersed, gives some idea of a ship's stability. If a ship set down in the ocean causes 1,000 tons (8.896 · 10⁶ N) of water to be displaced, it is said to possess a displacement of 1,000 tons. Obviously, a high degree of displacement is desirable. The principle of displacement helps to explain how an aircraft carrier can remain afloat, even though it weighs many thousands of tons.

DOWN TO THE DEPTHS

A submarine uses ballast as a means of descending and ascending underwater: when the submarine captain orders the crew to take the craft down, the craft is allowed to take water into its ballast tanks. If, on the other hand, the command is given to rise toward the surface, a valve will be opened to release compressed air into the tanks. The air pushes out the water, and causes the craft to ascend.

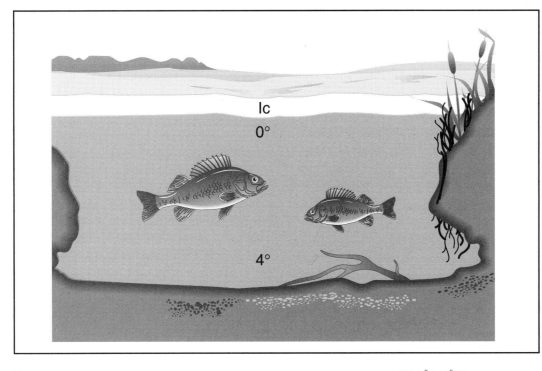

THE MOLECULAR STRUCTURE OF WATER BEGINS TO EXPAND ONCE IT COOLS BEYOND 39.4°F (4°C) AND CONTINUES TO EXPAND UNTIL IT BECOMES ICE. FOR THIS REASON, ICE IS LESS DENSE THAN WATER, FLOATS ON THE SURFACE, AND RETARDS FURTHER COOLING OF DEEPER WATER, WHICH ACCOUNTS FOR THE SURVIVAL OF FRESHWATER PLANT AND ANIMAL LIFE THROUGH THE WINTER. FOR THEIR PART, FISH CHANGE THE VOLUME OF THEIR INTERNAL SWIM BLADDER IN ORDER TO ALTER THEIR BUOYANCY.

A submarine is an underwater ship; its streamlined shape is designed to ease its movement. On the other hand, there are certain kinds of underwater vessels, known as submersibles, that are designed to sink—in order to observe or collect data from the ocean floor. Originally, the idea of a submersible was closely linked to that of diving itself. An early submersible was the diving bell, a device created by the noted English astronomer Edmund Halley (1656-1742.)

Though his diving bell made it possible for Halley to set up a company in which hired divers salvaged wrecks, it did not permit divers to go beyond relatively shallow depths. First of all, the diving bell received air from the surface: in Halley's time, no technology existed for taking an oxygen supply below. Nor did it provide substantial protection from the effects of increased pressure at great depths.

PERILS OF THE DEEP. The most immediate of those effects is, of course, the tendency of an object experiencing such pressure to simply implode like a tin can in a vise. Furthermore, the human body experiences several severe reactions to great depth: under water, nitrogen gas accumulates in a diver's bodily tissues, pro-

ducing two different—but equally frightening—effects.

Nitrogen is an inert gas under normal conditions, yet in the high pressure of the ocean depths it turns into a powerful narcotic, causing nitrogen narcosis—often known by the poetic-sounding name "rapture of the deep." Under the influence of this deadly euphoria, divers begin to think themselves invincible, and their altered judgment can put them into potentially fatal situations.

Nitrogen narcosis can occur at depths as shallow as 60 ft (18.29 m), and it can be overcome simply by returning to the surface. However, one should not return to the surface too quickly, particularly after having gone down to a significant depth for a substantial period of time. In such an instance, on returning to the surface nitrogen gas will bubble within the body, producing decompression sickness—known colloquially as "the bends." This condition may manifest as itching and other skin problems, joint pain, choking, blindness, seizures, unconsciousness, and even permanent neurological defects such as paraplegia.

French physiologist Paul Bert (1833-1886) first identified the bends in 1878, and in 1907, John Scott Haldane (1860-1936) developed a method for counteracting decompression sickness. He calculated a set of decompression tables that advised limits for the amount of time at given depths. He recommended what he called stage decompression, which means that the ascending diver stops every few feet during ascension and waits for a few minutes at each level, allowing the body tissues time to adjust to the new pressure. Modern divers use a decompression chamber, a sealed container that simulates the stages of decompression.

BATHYSPHERE, SCUBA, AND BATHYSCAPHE.

In 1930, the American naturalist William Beebe (1877-1962) and American engineer Otis Barton created the bathysphere. This was the first submersible that provided the divers inside with adequate protection from external pressure. Made of steel and spherical in shape, the bathysphere had thick quartz windows and was capable of maintaining ordinary atmosphere pressure even when lowered by a cable to relatively great depths. In 1934, a bathysphere descended to what was then an extremely impressive depth: 3,028 ft (923 m). However, the bathysphere was difficult to operate and maneuver, and in time it was be replaced by a more workable vessel, the bathyscaphe.

Before the bathyscaphe appeared, however, in 1943, two Frenchmen created a means for divers to descend without the need for any sort of external chamber. Certainly a diver with this new apparatus could not go to anywhere near the same depths as those approached by the bathysphere; nonetheless, the new aqualung made it possible to spend an extended time under the surface without need for air. It was now theoretically feasible for a diver to go below without any need for help or supplies from above, because he carried his entire oxygen supply on his back. The name of one of inventors, Emile Gagnan, is hardly a household word; but that of the other—Jacques Cousteau (1910-1997)—certainly is. So, too, is the name of their invention: the *self-contained underwater breathing apparatus*, better known as scuba.

The most important feature of the scuba gear was the demand regulator, which made it possible for the divers to breathe air at the same pressure as their underwater surroundings. This in turn facilitated breathing in a more normal, comfortable manner. Another important feature of a modern diver's equipment is a buoyancy compensation device. Like a ship atop the water, a diver wants to have only so much buoyancy—not so much that it causes him to surface.

As for the bathyscaphe—a term whose two Greek roots mean "deep" and "boat"—it made its debut five years after scuba gear. Built by the Swiss physicist and adventurer Auguste Piccard (1884-1962), the bathyscaphe consisted of two compartments: a heavy steel crew cabin that was resistant to sea pressure, and above it, a larger, light container called a float. The float was filled with gasoline, which in this case was not used as fuel, but to provide extra buoyancy, because of the gasoline's low specific gravity.

When descending, the occupants of the bathyscaphe—there could only be two, since the pressurized chamber was just 79 in (2.01 m) in diameter—released part of the gasoline to decrease buoyancy. They also carried iron ballast pellets on board, and these they released when preparing to ascend. Thanks to battery-driven screw propellers, the bathyscaphe was much more maneuverable than the bathysphere had ever been; furthermore, it was designed to reach depths that Beebe and Barton could hardly have conceived.

REACHING NEW DEPTHS.

It took several years of unsuccessful dives, but in 1953 a bathyscaphe set the first of many depth records. This first craft was the *Trieste,* manned by Piccard and his son Jacques, which descended 10,335 ft (3,150 m) below the Mediterranean, off Capri, Italy. A year later, in the Atlantic Ocean off Dakar, French West Africa (now Senegal), French divers Georges Houot and Pierre-Henri Willm reached 13,287 ft (4,063 m) in the *FNRS 3.*

Then in 1960, Jacques Piccard and United States Navy Lieutenant Don Walsh set a record that still stands: 35,797 ft (10,911 m)—23% greater than the height of Mt. Everest, the world's tallest peak. This they did in the *Trieste* some 250 mi (402 km) southeast of Guam at the Mariana Trench, the deepest spot in the Pacific Ocean and indeed the deepest spot on Earth. Piccard and Walsh went all the way to the bottom, a descent that took them 4 hours, 48 minutes. Coming up took 3 hours, 17 minutes.

Thirty-five years later, in 1995, the Japanese craft *Kaiko* also made the Mariana descent and confirmed the measurements of Piccard and Walsh. But the achievement of the *Kaiko* was not nearly as impressive of that of the *Trieste's* two-man crew: the *Kaiko,* in fact, had no crew. By the 1990s, sophisticated remote-sensing technology had made it possible to send down unmanned ocean expeditions, and it became less necessary to expose human beings to the incredible risks encountered by the Piccards, Walsh, and others.

FILMING TITANIC. An example of such an unmanned vessel is the one featured in the opening minutes of the Academy Award-winning motion picture *Titanic* (1997). The vessel itself, whose sinking in 1912 claimed more than 1,000 lives, rests at such a great depth in the North Atlantic that it is impractical either to raise it, or to send manned expeditions to explore the interior of the wreck. The best solution, then, is a remotely operated vessel of the kind also used for purposes such as mapping the ocean floor, exploring for petroleum and other deposits, and gathering underwater plate technology data.

The craft used in the film, which has "arms" for grasping objects, is of a variety specially designed for recovering items from shipwrecks. For the scenes that showed what was supposed to be the *Titanic* as an active vessel, director James Cameron used a 90% scale model that depicted the ship's starboard side—the side hit by the iceberg. Therefore, when showing its port side, as when it was leaving the Southampton, England, dock on April 15, 1912, all shots had to be reversed: the actual signs on the dock were in reverse lettering in order to appear correct when seen in the final version. But for scenes of the wrecked vessel lying at the bottom of the ocean, Cameron used the real *Titanic.*

To do this, he had to use a submersible; but he did not want to shoot only from inside the submersible, as had been done in the 1992 IMAX film *Titanica.* Therefore, his brother Mike Cameron, in cooperation with Panavision, built a special camera that could withstand 400 atm ($3.923 \cdot 10^7$ Pa)—that is, 400 times the air pressure at sea level. The camera was attached to the outside of the submersible, which for these external shots was manned by Russian submarine operators.

Because the special camera only held twelve minutes' worth of film, it was necessary to make a total of twelve dives. On the last two, a remotely operated submersible entered the wreck, which

THE DIVERS PICTURED HERE HAVE ASCENDED FROM A SUNKEN SHIP AND HAVE STOPPED AT THE 10-FT (3-M) DECOMPRESSION LEVEL TO AVOID GETTING DECOMPRESSION SICKNESS, BETTER KNOWN AS THE "BENDS." *(Photograph, copyright Jonathan Blair/Corbis. Reproduced by permission.)*

would have been too dangerous for the humans in the manned craft. Cameron had intended the remotely operated submersible as a mere prop, but in the end its view inside the ruined *Titanic* added one of the most poignant touches in the entire film. To these he later added scenes involving objects specific to the film's plot, such as the safe. These he shot in a controlled underwater environment designed to look like the interior of the *Titanic.*

INTO THE SKIES

In the earlier description of Piccard's bathyscaphe design, it was noted that the craft consisted of two compartments: a heavy steel crew cabin resistant to sea pressure, and above it a larger, light container called a float. If this sounds rather like the structure of a hot-air balloon, there is no accident in that.

In 1931, nearly two decades before the bathyscaphe made its debut, Piccard and another Swiss scientist, Paul Kipfer, set a record of a different kind with a balloon. Instead of going lower than anyone ever had, as Piccard and his son Jacques did in 1953—and as Jacques and Walsh did in an even greater way in 1960—Piccard and Kipfer went higher than ever, ascending to 55,563 ft (16,940 m). This made them the first two men to penetrate the stratosphere, which is the next atmospheric layer above the troposphere, a layer approximately 10 mi (16.1 km) high that covers the surface of Earth.

Piccard, without a doubt, experienced the greatest terrestrial altitude range of any human being over a lifetime: almost 12.5 mi (20.1 km) from his highest high to his lowest low, 84% of it above sea level and the rest below. His career, then, was a tribute to the power of buoyant force—and to the power of overcoming buoyant force for the purpose of descending to the ocean depths. Indeed, the same can be said of the Piccard family as a whole: not only did Jacques set the world's depth record, but years later, Jacques's son Bertrand took to the skies for another record-setting balloon flight.

In 1999, Bertrand Piccard and British balloon instructor Brian Wilson became the first men to circumnavigate the globe in a balloon, the *Breitling Orbiter 3*. The craft extended 180 ft (54.86) from the top of the envelope—the part of the balloon holding buoyant gases—to the bottom of the gondola, the part holding riders. The pressurized cabin had one bunk in which one pilot could sleep while the other flew, and up front was a computerized control panel which allowed the pilot to operate the burners, switch propane tanks, and release empty ones. It took Piccard and Wilson just 20 days to circle the Earth—a far cry from the first days of ballooning two centuries earlier.

THE FIRST BALLOONS. The Piccard family, though Swiss, are francophone; that is, they come from the French-speaking part of Switzerland. This is interesting, because the history of human encounters with buoyancy—below the ocean and even more so in the air—has been heavily dominated by French names. In fact, it was the French brothers, Joseph-Michel (1740-1810) and Jacques-Etienne (1745-1799) Montgolfier, who launched the first balloon in 1783. These two became to balloon flight what two other brothers, the Americans Orville and Wilbur Wright, became with regard to the invention that superseded the balloon twelve decades later: the airplane.

On that first flight, the Montgolfiers sent up a model 30 ft (9.15 m) in diameter, made of linen-lined paper. It reached a height of 6,000 ft (1,828 m), and stayed in the air for 10 minutes before coming back down. Later that year, the Montgolfiers sent up the first balloon flight with living creatures—a sheep, a rooster, and a duck—and still later in 1783, Jean-François Pilatre de Rozier (1756-1785) became the first human being to ascend in a balloon.

Rozier only went up 84 ft (26 m), which was the length of the rope that tethered him to the ground. As the makers and users of balloons learned how to use ballast properly, however, flight times were extended, and balloon flight became ever more practical. In fact, the world's first military use of flight dates not to the twentieth century but to the eighteenth—1794, specifically, when France created a balloon corps.

HOW A BALLOON FLOATS. There are only three gases practical for lifting a balloon: hydrogen, helium, and hot air. Each is much less than dense than ordinary air, and this gives them their buoyancy. In fact, hydrogen is the lightest gas known, and because it is cheap to produce, it would be ideal—except for the fact that it is extremely flammable. After the 1937 crash of the airship *Hindenburg*, the era of hydrogen use for lighter-than-air transport effectively ended.

Helium, on the other hand, is perfectly safe and only slightly less buoyant than hydrogen. This makes it ideal for balloons of the sort that children enjoy at parties; but helium is expensive, and therefore impractical for large balloons. Hence, hot air—specifically, air heated to a temperature of about 570°F (299°C), is the only truly viable option.

Charles's law, one of the laws regarding the behavior of gases, states that heating a gas will increase its volume. Gas molecules, unlike their liquid or solid counterparts, are highly non-attractive—that is, they tend to spread toward relatively great distances from one another. There is already a great deal of empty space between gas molecules, and the increase in volume only increases the amount of empty space. Hence, density is lowered, and the balloon floats.

AIRSHIPS. Around the same time the Montgolfier brothers launched their first balloons, another French designer, Jean-Baptiste-Marie Meusnier, began experimenting with a

ONCE CONSIDERED OBSOLETE, BLIMPS ARE ENJOYING A RENAISSANCE AMONG SCIENTISTS AND GOVERNMENT AGEN-
CIES. THE BLIMP PICTURED HERE, THE AEROSTAT BLIMP, IS EQUIPPED WITH RADAR FOR DRUG ENFORCEMENT AND
INSTRUMENTS FOR WEATHER OBSERVATION. *(Corbis. Reproduced by permission.)*

more streamlined, maneuverable model. Early balloons, after all, could only be maneuvered along one axis, up and down: when it came to moving sideways or forward and backward, they were largely at the mercy of the elements.

It was more than a century before Meusnier's idea—the prototype for an airship—became a reality. In 1898, Alberto Santos-Dumont of Brazil combined a balloon with a propeller powered by an internal-combustion instrument, creating a machine that improved on the balloon, much as the bathyscaphe later improved on the bathysphere. Santos-Dumont's airship was non-rigid, like a balloon. It also used hydrogen, which is apt to contract during descent and collapse the envelope. To counter this problem, Santos-Dumont created the ballonet, an internal airbag designed to provide buoyancy and stabilize flight.

One of the greatest figures in the history of lighter-than-air flight—a man whose name, along with blimp and dirigible, became a synonym for the airship—was Count Ferdinand von Zeppelin (1838-1917). It was he who created a lightweight structure of aluminum girders and rings that made it possible for an airship to remain rigid under varying atmospheric condi-

tions. Yet Zeppelin's earliest launches, in the decade that followed 1898, were fraught with a number of problems—not least of which were disasters caused by the flammability of hydrogen.

Zeppelin was finally successful in launching airships for public transport in 1911, and the quarter-century that followed marked the golden age of airship travel. Not that all was "golden" about this age: in World War I, Germany used airships as bombers, launching the first London blitz in May 1915. By the time Nazi Germany initiated the more famous World War II London blitz 25 years later, ground-based anti-aircraft technology would have made quick work of any zeppelin; but by then, airplanes had long since replaced airships.

During the 1920s, though, airships such as the *Graf Zeppelin* competed with airplanes as a mode of civilian transport. It is a hallmark of the perceived safety of airships over airplanes at the time that in 1928, the *Graf Zeppelin* made its first transatlantic flight carrying a load of passengers. Just a year earlier, Charles Lindbergh had made the first-ever solo, nonstop transatlantic flight in an airplane. Today this would be the equivalent of someone flying to the Moon, or perhaps even Mars, and there was no question of carrying pas-

KEY TERMS

ARCHIMEDES'S PRINCIPLE: A rule of physics which holds that the buoyant force of an object immersed in fluid is equal to the weight of the fluid displaced by the object. It is named after the Greek mathematician, physicist, and inventor Archimedes (c. 287-212 B.C.), who first identified it.

BALLAST: A heavy substance that, by increasing the weight of an object experiencing buoyancy, improves its stability.

BUOYANCY: The tendency of an object immersed in a fluid to float. This can be explained by Archimedes's principle.

DENSITY: Mass divided by volume.

DISPLACEMENT: A measure of the weight of the fluid that has had to be moved out of position so that an object can be immersed. If a ship set down in the ocean causes 1,000 tons of water to be displaced, it is said to possess a displacement of 1,000 tons.

FLUID: Any substance, whether gas or liquid, that conforms to the shape of its container.

FORCE: The product of mass multiplied by acceleration.

MASS: A measure of inertia, indicating the resistance of an object to a change in its motion. For an object immerse in fluid, mass is equal to volume multiplied by density.

PRESSURE: The exertion of force over a two-dimensional area; hence the formula for pressure is force divided by area. The British system of measures typically reckons pressure in pounds per square inch. In metric terms, this is measured in terms of newtons (N) per square meter, a figure known as a pascal (Pa.)

SPECIFIC GRAVITY: The density of an object or substance relative to the density of water; or more generally, the ratio between the densities of two objects or substances.

VOLUME: The amount of three-dimensional space occupied by an object. Volume is usually measured in cubic units.

WEIGHT: A force equal to mass multiplied by the acceleration due to gravity (32 ft/9.8 m/sec^2). For an object immersed in fluid, weight is the same as volume multiplied by density multiplied by gravitational acceleration.

---

sengers. Furthermore, Lindbergh was celebrated as a hero for the rest of his life, whereas the passengers aboard the *Graf Zeppelin* earned no more distinction for bravery than would pleasure-seekers aboard a cruise.

THE LIMITATIONS OF LIGHTER-THAN-AIR TRANSPORT. For a few years, airships constituted the luxury liners of the skies; but the *Hindenburg* crash signaled the end of relatively widespread airship transport. In any case, by the time of the 1937 *Hindenburg* crash, lighter-than-air transport was no longer the leading contender in the realm of flight technology.

Ironically enough, by 1937 the airplane had long since proved itself more viable—even though it was actually heavier than air. The principles that make an airplane fly have little to do with buoyancy as such, and involve differences in pressure rather than differences in density. Yet the replacement of lighter-than-air craft on the cutting edge of flight did not mean that balloons and airships were relegated to the museum; instead, their purposes changed.

The airship enjoyed a brief resurgence of interest during World War II, though purely as a surveillance craft for the United States military. In the period after the war, the U.S. Navy hired the Goodyear Tire and Rubber Company to produce airships, and as a result of this relationship Goodyear created the most visible airship since the *Graf Zeppelin* and the *Hindenburg:* the Goodyear Blimp.

BLIMPS AND BALLOONS: ON THE CUTTING EDGE? The blimp, known to viewers of countless sporting events, is much better-suited than a plane or helicopter to providing TV cameras with an aerial view of a stadium—and advertisers with a prominent billboard. Military forces and science communities have also found airships useful for unexpected purposes. Their virtual invisibility with regard to radar has reinvigorated interest in blimps on the part of the U.S. Department of Defense, which has discussed plans to use airships as radar platforms in a larger Strategic Air Initiative. In addition, French scientists have used airships for studying rain forest treetops or canopies.

Balloons have played a role in aiding space exploration, which is emblematic of the relationship between lighter-than-air transport and more advanced means of flight. In 1961, Malcolm D. Ross and Victor A. Prother of the U.S. Navy set the balloon altitude record with a height of 113,740 ft (34,668 m.) The technology that enabled their survival at more than 21 mi (33.8 km) in the air was later used in creating life-support systems for astronauts.

Balloon astronomy provides some of the clearest images of the cosmos: telescopes mounted on huge, unmanned balloons at elevations as high as 120,000 ft (35,000 m)—far above the dust and smoke of Earth—offer high-resolution images. Balloons have even been used on other planets: for 46 hours in 1985, two balloons launched by the unmanned Soviet expedition to Venus collected data from the atmosphere of that planet.

American scientists have also considered a combination of a large hot-air balloon and a smaller helium-filled balloon for gathering data on the surface and atmosphere of Mars during expeditions to that planet. As the air balloon is heated by the Sun's warmth during the day, it would ascend to collect information on the atmosphere. (In fact the "air" heated would be from the atmosphere of Mars, which is composed primarily of carbon dioxide.) Then at night when Mars cools, the air balloon would lose its buoyancy and descend, but the helium balloon would keep it upright while it collected data from the ground.

WHERE TO LEARN MORE

"*Buoyancy*" (Web site). <http://www.aquaholic.com/gasses/laws.htm> (March 12, 2001).

"*Buoyancy*" (Web site). <http://www.uncwil.edu/nurc/aquarius/lessons/buoyancy.htm> (March 12, 2001).

"*Buoyancy Basics*" *Nova/PBS* (Web site). <http://www.pbs.org/wgbh/nova/lasalle/buoybasics.html> (March 12, 2001).

Challoner, Jack. *Floating and Sinking.* Austin, TX: Raintree Steck-Vaughn, 1997.

Cobb, Allan B. *Super Science Projects About Oceans.* New York: Rosen, 2000.

Gibson, Gary. *Making Things Float and Sink.* Illustrated by Tony Kenyon. Brookfield, CT: Copper Beeck Brooks, 1995.

Taylor, Barbara. *Liquid and Buoyancy.* New York: Warwick Press, 1990.

STATICS

STATICS AND EQUILIBRIUM

PRESSURE

ELASTICITY

STATICS AND EQUILIBRIUM

CONCEPT

Statics, as its name suggests, is the study of bodies at rest. Those bodies may be acted upon by a variety of forces, but as long as the lines of force meet at a common point and their vector sum is equal to zero, the body itself is said to be in a state of equilibrium. Among the topics of significance in the realm of statics is center of gravity, which is relatively easy to calculate for simple bodies, but much more of a challenge where aircraft or ships are concerned. Statics is also applied in analysis of stress on materials—from a picture frame to a skyscraper.

HOW IT WORKS

Equilibrium and Vectors

Essential to calculations in statics is the use of vectors, or quantities that have both magnitude and direction. By contrast, a scalar has only magnitude. If one says that a certain piece of property has an area of one acre, there is no directional component. Nor is there a directional component involved in the act of moving the distance of 1 mi (1.6 km), since no statement has been made as to the direction of that mile. On the other hand, if someone or something experiences a displacement, or change in position, of 1 mi to the northeast, then what was a scalar description has been placed in the language of vectors.

Not only are mass and speed (as opposed to velocity) considered scalars; so too is time. This might seem odd at first glance, but—on Earth at least, and outside any special circumstances posed by quantum mechanics—time can only move forward. Hence, direction is not a factor. By contrast, force, equal to mass multiplied by acceleration, is a vector. So too is weight, a specific type of force equal to mass multiplied by the acceleration due to gravity (32 ft or [9.8 m] / sec^2). Force may be in any direction, but the direction of weight is always downward along a vertical plane.

VECTOR SUMS. Adding scalars is simple, since it involves mere arithmetic. The addition of vectors is more challenging, and usually requires drawing a diagram, for instance, if trying to obtain a vector sum for the velocity of a car that has maintained a uniform speed, but has changed direction several times.

One would begin by representing each vector as an arrow on a graph, with the tail of each vector at the head of the previous one. It would then be possible to draw a vector from the tail of the first to the head of the last. This is the sum of the vectors, known as a resultant, which measures the net change.

Suppose, for instance, that a car travels north 5 mi (8 km), east 2 mi (3.2 km), north 3 mi (4.8 km), east 3 mi, and finally south 3 mi. One must calculate its net displacement—in other words, not the sum of all the miles it has traveled, but the distance and direction between its starting point and its end point. First, one draws the vectors on a piece of graph paper, using a logical system that treats the y axis as the north-south plane, and the x axis as the east-west plane. Each vector should be in the form of an arrow pointing in the appropriate direction.

Having drawn all the vectors, the only remaining one is between the point where the car's journey ends and the starting point—that is, the resultant. The number of sides to the resulting shape is always one more than the number of vectors being added; the final side is the resultant.

In this particular case, the answer is fairly easy. Because the car traveled north 5 mi and ultimately moved east by 5 mi, returning to a position of 5 mi north, the segment from the resultant forms the hypotenuse of an equilateral (that is, all sides equal) right triangle. By applying the Pythagorean theorem, which states that the square of the length of the hypotenuse is equal to the sum of the squares of the other two sides, one quickly arrives at a figure of 7.07 m (11.4 km) in a northeasterly direction. This is the car's net displacement.

CALCULATING FORCE AND TENSION IN EQUILIBRIUM

Using vector sums, it is possible to make a number of calculations for objects in equilibrium, but these calculations are somewhat more challenging than those in the car illustration. One form of equilibrium calculation involves finding tension, or the force exerted by a supporting object on an object in equilibrium—a force that is always equal to the amount of weight supported. (Another way of saying this is that if the tension on the supporting object is equal to the weight it supports, then the supported object is in equilibrium.)

In calculations for tension, it is best to treat the supporting object—whether it be a rope, picture hook, horizontal strut or some other item—as though it were weightless. One should begin by drawing a free-body diagram, a sketch showing all the forces acting on the supported object. It is not necessary to show any forces (other than weight) that the object itself exerts, since those do not contribute to its equilibrium.

RESOLVING X AND Y COMPONENTS. As with the distance vector graph discussed above, next one must equate these forces to the x and y axes. The distance graph example involved only segments already parallel to x and y, but suppose—using the numbers already discussed—the graph had called for the car to move in a perfect 45°-angle to the northeast along a distance of 7.07 mi. It would then have been easy to resolve this distance into an x component (5 mi east) and a y component (5 mi north)—which are equal to the other two sides of the equilateral triangle.

This resolution of x and y components is more challenging for calculations involving equilibrium, but once one understands the principle involved, it is easy to apply. For example, imagine a box suspended by two ropes, neither of which is at a 90°-angle to the box. Instead, each rope is at an acute angle, rather like two segments of a chain holding up a sign.

The x component will always be the product of tension (that is, weight) multiplied by the cosine of the angle. In a right triangle, one angle is always equal to 90°, and thus by definition, the other two angles are acute, or less than 90°. The angle of either rope is acute, and in fact, the rope itself may be considered the hypotenuse of an imaginary triangle. The base of the triangle is the x axis, and the angle between the base and the hypotenuse is the one under consideration.

Hence, we have the use of the cosine, which is the ratio between the adjacent leg (the base) of the triangle and the hypotenuse. Regardless of the size of the triangle, this figure is a constant for any particular angle. Likewise, to calculate the y component of the angle, one uses the sine, or the ratio between the opposite side and the hypotenuse. Keep in mind, once again, that the adjacent leg for the angle is by definition the same as the x axis, just as the opposite leg is the same as the y axis. The cosine (abbreviated cos), then, gives the x component of the angle, as the sine (abbreviated sin) does the y component.

REAL-LIFE APPLICATIONS

EQUILIBRIUM AND CENTER OF GRAVITY IN REAL OBJECTS

Before applying the concept of vector sums to matters involving equilibrium, it is first necessary to clarify the nature of equilibrium itself—what it is and what it is not. Earlier it was stated that an object is in equilibrium if the vector sum of the forces acting on it are equal to zero—as long as those forces meet at a common point.

This is an important stipulation, because it is possible to have lines of force that cancel one another out, but nonetheless cause an object to move. If a force of a certain magnitude is applied to the right side of an object, and a line of force of equal magnitude meets it exactly from the left, then the object is in equilibrium. But if the line of force from the right is applied to the top of the object, and the line of force from the left to the bottom, then they do not meet at a common point, and the object is not in equilibrium. Instead, it is experiencing torque, which will cause it to rotate.

VARIETIES OF EQUILIBRIUM. There are two basic conditions of equilibrium. The term "translational equilibrium" describes an object that experiences no linear (straight-line) acceleration; on the other hand, an object experiencing no rotational acceleration (a component of torque) is said to be in rotational equilibrium.

Typically, an object at rest in a stable situation experiences both linear and rotational equilibrium. But equilibrium itself is not necessarily stable. An empty glass sitting on a table is in stable equilibrium: if it were tipped over slightly—that is, with a force below a certain threshold—then it would return to its original position. This is true of a glass sitting either upright or upside-down.

Now imagine if the glass were somehow propped along the edge of a book sitting on the table, so that the bottom of the glass formed the hypotenuse of a triangle with the table as its base and the edge of the book as its other side. The glass is in equilibrium now, but unstable equilibrium, meaning that a slight disturbance—a force from which it could recover in a stable situation—would cause it to tip over.

If, on the other hand, the glass were lying on its side, then it would be in a state of neutral equilibrium. In this situation, the application of force alongside the glass will not disturb its equilibrium. The glass will not attempt to seek stable equilibrium, nor will it become more unstable; rather, all other things being equal, it will remain neutral.

CENTER OF GRAVITY. Center of gravity is the point in an object at which the weight below is equal to the weight above, the weight in front equal to the weight behind, and the weight to the left equal to the weight on the right. Every object has just one center of gravity, and if the object is suspended from that point, it will not rotate.

One interesting aspect of an object's center of gravity is that it does not necessarily have to be within the object itself. When a swimmer is poised in a diving stance, as just before the start-

A GLASS SITTING ON A TABLE IS IN A STATE OF STABLE EQUILIBRIUM. *(Photograph by John Wilkes Studio/Corbis. Reproduced by permission.)*

ing bell in an Olympic competition, the swimmer's center of gravity is to the front—some distance from his or her chest. This is appropriate, since the objective is to get into the water as quickly as possible once the race starts.

By contrast, a sprinter's stance places the center of gravity well within the body, or at least firmly surrounded by the body—specifically, at the place where the sprinter's rib cage touches the forward knee. This, too, fits with the needs of the athlete in the split-second following the starting gun. The sprinter needs to have as much traction as possible to shoot forward, rather than forward and downward, as the swimmer does.

TENSION CALCULATIONS

In the earlier discussion regarding the method of calculating tension in equilibrium, two of the three steps of this process were given: first, draw a free-body diagram, and second, resolve the forces into x and y components. The third step is to set the force components along each axis equal to zero—since, if the object is truly in equilibrium, the sum of forces will indeed equal zero. This makes it possible, finally, to solve the equations for the net tension.

IN THE STARTING BLOCKS, A SPRINTER'S CENTER OF GRAVITY IS ALIGNED ALONG THE RIB CAGE AND FORWARD KNEE, THUS MAXIMIZING THE RUNNER'S ABILITY TO SHOOT FORWARD OUT OF THE BLOCKS. *(Photograph by Ronnen Eshel/Corbis. Reproduced by permission.)*

Imagine a picture that weighs 100 lb (445 N) suspended by a wire, the left side of which may be called segment *A*, and the right side segment *B*. The wire itself is not perfectly centered on the picture-hook: *A* is at a 30° angle, and *B* on a 45° angle. It is now possible to find the tension on both.

First, one can resolve the horizontal components by the formula $F_x = T_{Bx} + T_{Ax} = 0$, meaning that the *x* component of force is equal to the product of tension for the *x* component of *B*, added to the product of tension for the *x* component of *A*, which in turn is equal to zero. Given the 30°-angle of *A*, $_{Ax} = 0.866$, which is the cosine of 30°. $_{Bx}$ is equal to cos 45°, which equals 0.707. (Recall the earlier discussion of distance, in which a square with sides 5 mi long was described: its hypotenuse was 7.07 mi, and 5/7.07 = 0.707.)

Because *A* goes off to the left from the point at which the picture is attached to the wire, this places it on the negative portion of the x axis. Therefore, the formula can now be restated as $T_B(0.707) - T_A(0.866) = 0$. Solving for T_B reveals that it is equal to $T_A(0.866/0.707) = (1.22)T_A$. This will be substituted for T_B in the formula for the total force along the y component.

However, the y-force formula is somewhat different than for x: since weight is exerted along the y axis, it must be subtracted. Thus, the formula for the y component of force is $F_y = T_{Ay} + T_{By} - w = 0$. (Note that the y components of both *A* and *B* are positive: by definition, this must be so for an object suspended from some height.)

Substituting the value for T_B obtained above, $(1.22)T_A$, makes it possible to complete the equation. Since the sine of 30° is 0.5, and the sine of 45° is 0.707—the same value as its cosine—one can state the equation thus: $T_A(0.5) + (1.22)T_A(0.707) - 100$ lb $= 0$. This can be restated as $T_A(0.5 + (1.22 \cdot 0.707)) = T_A(1.36) = 100$ lb. Hence, $T_A = (100$ lb$/1.36) = 73.53$ lb. Since $T_B = (1.22)T_A$, this yields a value of 89.71 lb for T_B.

Note that T_A and T_B actually add up to considerably more than 100 lb. This, however, is known as an algebraic sum—which is very similar to an arithmetic sum, inasmuch as algebra is simply a generalization of arithmetic. What is important here, however, is the vector sum, and the vector sum of T_A and T_B equals 100 lb.

CALCULATING CENTER OF GRAVITY. Rather than go through another lengthy calculation for center of gravity, we will explain the principles behind such calculations.

It is easy to calculate the center of gravity for a regular shape, such as a cube or sphere—assuming, of course, that the mass and therefore the weight is evenly distributed throughout the object. In such a case, the center of gravity is the geometric center. For an irregular object, however, center of gravity must be calculated.

An analogy regarding United States demographics may help to highlight the difference between geometric center and center of gravity. The geographic center of the U.S., which is analogous to geometric center, is located near the town of Castle Rock in Butte County, South Dakota. (Because Alaska and Hawaii are so far west of the other 48 states—and Alaska, with its great geographic area, is far to the north—the data is skewed in a northwestward direction. The geographic center of the 48 contiguous states is near Lebanon, in Smith County, Kansas.)

The geographic center, like the geometric center of an object, constitutes a sort of balance point in terms of physical area: there is as much U.S. land to the north of Castle Rock, South Dakota, as to the south, and as much to the east as to the west. The population center is more like the center of gravity, because it is a measure, in some sense, of "weight" rather than of volume—though in this case concentration of people is substituted for concentration of weight. Put another way, the population center is the balance point of the population, if it were assumed that every person weighed the same amount.

Naturally, the population center has been shifting westward ever since the first U.S. census in 1790, but it is still skewed toward the east: though there is far more U.S. land west of the Mississippi River, there are still more people east of it. Hence, according to the 1990 U.S. census, the geographic center is some 1,040 mi (1,664 km) in a southeastward direction from the population center: just northwest of Steelville, Missouri, and a few miles west of the Mississippi.

The United States, obviously, is an "irregular object," and calculations for either its geographic or its population center represent the mastery of numerous mathematical principles. How, then, does one find the center of gravity for a much smaller irregular object? There are a number of methods, all rather complex.

To measure center of gravity in purely physical terms, there are a variety of techniques relating to the shape of the object to be measured. There is also a mathematical formula, which involves first treating the object as a conglomeration of several more easily measured objects. Then the x components for the mass of each "sub-object" can be added, and divided by the combined mass of the object as a whole. The same can be done for the y components.

USING EQUILIBRIUM CALCULATIONS

One reason for making center of gravity calculations is to ensure that the net force on an object passes through that center. If it does not, the object will start to rotate—and for an airplane, for instance, this could be disastrous. Hence, the builders and operators of aircraft make exceedingly detailed, complicated calculations regarding center of gravity. The same is true for shipbuilders and shipping lines: if a ship's center of gravity is not vertically aligned with the focal point of the buoyant force exerted on it by the water, it may well sink.

In the case of ships and airplanes, the shapes are so irregular that center of gravity calculations require intensive analyses of the many components. Hence, a number of companies that supply measurement equipment to the aerospace and maritime industries offer center of gravity measurement instruments that enable engineers to make the necessary calculations.

On dry ground, calculations regarding equilibrium are likewise quite literally a life and death matter. In the earlier illustration, the object in equilibrium was merely a picture hanging from a wire—but what if it were a bridge or a building? The results of inaccurate estimates of net force could affect the lives of many people. Hence, structural engineers make detailed analyses of stress, once again using series of calculations that make the picture-frame illustration above look like the simplest of all arithmetic problems.

WHERE TO LEARN MORE

Beiser, Arthur. *Physics*, 5th ed. Reading, MA: Addison-Wesley, 1991.

"Determining Center of Gravity" National Aeronautics and Space Administration (Web site). <http://www.grc.nasa.gov/WWW/K-12/airplane/cg.html> (March 19, 2001).

KEY TERMS

ACCELERATION: A change in velocity.

CENTER OF GRAVITY: The point on an object at which the total weights on either side of all axes (x, y, and z) are identical. Each object has just one center of gravity, and if it is suspended from that point, it will be in a state of perfect rotational equilibrium.

COSINE: For an acute (less than 90°) angle in a right triangle, the cosine (abbreviated cos) is the ratio between the adjacent leg and the hypotenuse. Regardless of the size of the triangle, this figure is a constant for any particular angle.

DISPLACEMENT: Change in position.

EQUILIBRIUM: A state in which vector sum for all lines of force on an object is equal to zero. An object that experiences no linear acceleration is said to be in translational equilibrium, and one that experiences no rotational acceleration is referred to as being in rotational equilibrium. An object may also be in stable, unstable, or neutral equilibrium.

FORCE: The product of mass multiplied by acceleration.

FREE-BODY DIAGRAM: A sketch showing all the outside forces acting on an object in equilibrium.

HYPOTENUSE: In a right triangle, the side opposite the right angle.

RESULTANT: The sum of two or more vectors, which measures the net change in distance and direction.

RIGHT TRIANGLE: A triangle that includes a right (90°) angle. The other two angles are, by definition, acute, or less than 90°.

SCALAR: A quantity that possesses only magnitude, with no specific direction. Mass, time, and speed are all scalars. The opposite of a scalar is a vector.

SINE: For an acute (less than 90°) angle in a right triangle, the sine (abbreviated sin) is the ratio between the opposite leg and the hypotenuse. Regardless of the size of the triangle, this figure is a constant for any particular angle.

STATICS: The study of bodies at rest. Those bodies may be acted upon by a variety of forces, but as long as the vector sum for all those lines of force is equal to zero, the body itself is said to be in a state of equilibrium.

TENSION: The force exerted by a supporting object on an object in equilibrium—a force that is always equal to the amount of weight supported.

VECTOR: A quantity that possesses both magnitude and direction. Force is a vector; so too is acceleration, a component of force; and likewise weight, a variety of force. The opposite of a vector is a scalar.

VECTOR SUM: A calculation, made by different methods according to the factor being analyzed—for instance, velocity or force—that yields the net result of all the vectors applied in a particular situation.

VELOCITY: The speed of an object in a particular direction. Velocity is thus a vector quantity.

WEIGHT: A measure of the gravitational force on an object; the product of mass multiplied by the acceleration due to gravity.

"Equilibrium and Statics" (Web site). <http://www.glenbrook.k12.il.us/gbssci/phys/Class/vectors/u313c.html> (March 19, 2001).

"Exploratorium Snack: Center of Gravity." The Exploratorium (Web site). <http://www.exploratorium.edu/snacks/center_of_gravity.html> (March 19, 2001).

Faivre d'Arcier, Marima. What Is Balance? Illustrated by Volker Theinhardt. New York: Viking Kestrel, 1986.

Taylor, Barbara. Weight and Balance. Photographs by Peter Millard. New York: F. Watts, 1990.

"Where Is Your Center of Gravity?" The K-8 Aeronautics Internet Textbook (Web site). <http://wing.ucdavis.edu/Curriculums/Forces_Motion/center_howto.html> (March 19, 2001).

Wood, Robert W. Mechanics Fundamentals. Illustrated by Bill Wright. Philadelphia, PA: Chelsea House, 1997.

Zubrowski, Bernie. Mobiles: Building and Experimenting with Balancing Toys. Illustrated by Roy Doty. New York: Morrow Junior Books, 1993.

PRESSURE

CONCEPT

Pressure is the ratio of force to the surface area over which it is exerted. Though solids exert pressure, the most interesting examples of pressure involve fluids—that is, gases and liquids—and in particular water and air. Pressure plays a number of important roles in daily life, among them its function in the operation of pumps and hydraulic presses. The maintenance of ordinary air pressure is essential to human health and well-being: the body is perfectly suited to the ordinary pressure of the atmosphere, and if that pressure is altered significantly, a person may experience harmful or even fatal side-effects.

HOW IT WORKS

Force and Surface Area

When a force is applied perpendicular to a surface area, it exerts pressure on that surface equal to the ratio of F to A, where F is the force and A the surface area. Hence, the formula for pressure (p) is $p = F/A$. One interesting consequence of this ratio is the fact that pressure can increase or decrease without any change in force—in other words, if the surface becomes smaller, the pressure becomes larger, and vice versa.

If one cheerleader were holding another cheerleader on her shoulders, with the girl above standing on the shoulder blades of the girl below, the upper girl's feet would exert a certain pressure on the shoulders of the lower girl. This pressure would be equal to the upper girl's weight (F, which in this case is her mass multiplied by the downward acceleration due to gravity) divided by the surface area of her feet. Suppose, then, that

the upper girl executes a challenging acrobatic move, bringing her left foot up to rest against her right knee, so that her right foot alone exerts the full force of her weight. Now the surface area on which the force is exerted has been reduced to half its magnitude, and thus the pressure on the lower girl's shoulder is twice as great.

For the same reason—that is, that reduction of surface area increases net pressure—a well-delivered karate chop is much more effective than an open-handed slap. If one were to slap a board squarely with one's palm, the only likely result would be a severe stinging pain on the hand. But if instead one delivered a blow to the board, with the hand held perpendicular—provided, of course, one were an expert in karate—the board could be split in two. In the first instance, the area of force exertion is large and the net pressure to the board relatively small, whereas in the case of the karate chop, the surface area is much smaller—and hence, the pressure is much larger.

Sometimes, a greater surface area is preferable. Thus, snowshoes are much more effective for walking in snow than ordinary shoes or boots. Ordinary footwear is not much larger than the surface of one's foot, perfectly appropriate for walking on pavement or grass. But with deep snow, this relatively small surface area increases the pressure on the snow, and causes one's feet to sink. The snowshoe, because it has a surface area significantly larger than that of a regular shoe, reduces the ratio of force to surface area and therefore, lowers the net pressure.

The same principle applies with snow skis and water skis. Like a snowshoe, a ski makes it possible for the skier to stay on the surface of the snow, but unlike a snowshoe, a ski is long and thin, thus enabling the skier to glide more effectively down a snow-covered hill. As for skiing on water, people who are experienced at this sport can ski barefoot, but it is tricky. Most beginners require water skis, which once again reduce the net pressure exerted by the skier's weight on the surface of the water.

MEASURING PRESSURE

Pressure is measured by a number of units in the English and metric—or, as it is called in the scientific community, SI—systems. Because $p = F/A$, all units of pressure represent some ratio of force to surface area. The principle SI unit is called a pascal (Pa), or 1 N/m^2. A newton (N), the SI unit of force, is equal to the force required to accelerate 1 kilogram of mass at a rate of 1 meter per second squared. Thus, a Pascal is equal to the pressure of 1 newton over a surface area of 1 square meter.

In the English or British system, pressure is measured in terms of pounds per square inch, abbreviated as lbs./in². This is equal to $6.89 \cdot 10^3$ Pa, or 6,890 Pa. Scientists—even those in the United States, where the British system of units prevails—prefer to use SI units. However, the British unit of pressure is a familiar part of an American driver's daily life, because tire pressure in the United States is usually reckoned in terms of pounds per square inch. (The recommended tire pressure for a mid-sized car is typically 30-35 lb/in².)

Another important measure of pressure is the atmosphere (atm), which the average pressure exerted by air at sea level. In English units, this is equal to 14.7 lbs./in², and in SI units to $1.013 \cdot 10^5$ Pa—that is, 101,300 Pa. There are also two other specialized units of pressure measurement in the SI system: the bar, equal to 10^5 Pa, and the torr, equal to 133 Pa. Meteorologists, scientists who study weather patterns, use the millibar (mb), which, as its name implies, is equal to 0.001 bars. At sea level, atmospheric pressure is approximately 1,013 mb.

THE BAROMETER. The torr, once known as the "millimeter of mercury," is equal to the pressure required to raise a column of mercury (chemical symbol Hg) 1 mm. It is named for the Italian physicist Evangelista Torricelli (1608-1647), who invented the barometer, an instrument for measuring atmospheric pressure.

IN THE INSTANCE OF ONE CHEERLEADER STANDING ON ANOTHER'S SHOULDERS, THE CHEERLEADER'S FEET EXERT DOWNWARD PRESSURE ON HER PARTNER'S SHOULDERS. THE PRESSURE IS EQUAL TO THE GIRL'S WEIGHT DIVIDED BY THE SURFACE AREA OF HER FEET. *(Photograph by James L. Amos/Corbis. Reproduced by permission.)*

The barometer, constructed by Torricelli in 1643, consisted of a long glass tube filled with mercury. The tube was open at one end, and turned upside down into a dish containing more mercury: hence, the open end was submerged in mercury while the closed end at the top constituted a vacuum—that is, an area in which the pressure is much lower than 1 atm.

The pressure of the surrounding air pushed down on the surface of the mercury in the bowl, while the vacuum at the top of the tube provided an area of virtually no pressure, into which the mercury could rise. Thus, the height to which the mercury rose in the glass tube represented normal air pressure (that is, 1 atm.) Torricelli discovered that at standard atmospheric pressure, the column of mercury rose to 760 millimeters.

The value of 1 atm was thus established as equal to the pressure exerted on a column of mercury 760 mm high at a temperature of 0°C (32°F). Furthermore, Torricelli's invention eventually became a fixture both of scientific labora-

THE AIR PRESSURE ON TOP OF MOUNT EVEREST, THE WORLD'S TALLEST PEAK, IS VERY LOW, MAKING BREATH-ING DIFFICULT. MOST CLIMBERS WHO ATTEMPT TO SCALE EVEREST THUS CARRY OXYGEN TANKS WITH THEM. SHOWN HERE IS JIM WHITTAKER, THE FIRST AMERICAN TO CLIMB EVEREST. *(Photograph by Galen Rowell/Corbis. Repro-duced by permission.)*

tories and of households. Since changes in atmospheric pressure have an effect on weather patterns, many home indoor-outdoor ther-mometers today also include a barometer.

PRESSURE AND FLUIDS

In terms of physics, both gases and liquids are referred to as fluids—that is, substances that con-form to the shape of their container. Air pressure and water pressure are thus specific subjects under the larger heading of "fluid pressure." A fluid responds to pressure quite differently than a solid does. The density of a solid makes it resist-ant to small applications of pressure, but if the pressure increases, it experiences tension and, ultimately, deformation. In the case of a fluid, however, stress causes it to flow rather than to deform.

There are three significant characteristics of the pressure exerted on fluids by a container. First of all, a fluid in a container experiencing no external motion exerts a force perpendicular to the walls of the container. Likewise, the con-tainer walls exert a force on the fluid, and in

both cases, the force is always perpendicular to the walls.

In each of these three characteristics, it is assumed that the container is finite: in other words, the fluid has nowhere else to go. Hence, the second statement: the external pressure exert-ed on the fluid is transmitted uniformly. Note that the preceding statement was qualified by the term "external": the fluid itself exerts pressure whose force component is equal to its weight. Therefore, the fluid on the bottom has much greater pressure than the fluid on the top, due to the weight of the fluid above it.

Third, the pressure on any small surface of the fluid is the same, regardless of that surface's orientation. In other words, an area of fluid per-pendicular to the container walls experiences the same pressure as one parallel or at an angle to the walls. This may seem to contradict the first prin-ciple, that the force is perpendicular to the walls of the container. In fact, force is a vector quanti-ty, meaning that it has both magnitude and direction, whereas pressure is a scalar, meaning that it has magnitude but no specific direction.

REAL-LIFE APPLICATIONS

PASCAL'S PRINCIPLE AND THE HYDRAULIC PRESS

The three characteristics of fluid pressure described above have a number of implications and applications, among them, what is known as Pascal's principle. Like the SI unit of pressure, Pascal's principle is named after Blaise Pascal (1623-1662), a French mathematician and physi-cist who formulated the second of the three state-ments: that the external pressure applied on a fluid is transmitted uniformly throughout the entire body of that fluid. Pascal's principle became the basis for one of the important machines ever developed, the hydraulic press.

A simple hydraulic press of the variety used to raise a car in an auto shop typically consists of two large cylinders side by side. Each cylinder contains a piston, and the cylinders are connect-ed at the bottom by a channel containing fluid. Valves control flow between the two cylinders. When one applies force by pressing down the pis-ton in one cylinder (the input cylinder), this yields a uniform pressure that causes output in the second cylinder, pushing up a piston that raises the car.

In accordance with Pascal's principle, the pressure throughout the hydraulic press is the same, and will always be equal to the ratio between force and pressure. As long as that ratio is the same, the values of *F* and *A* may vary. In the case of an auto-shop car jack, the input cylinder has a relatively small surface area, and thus, the amount of force that must be applied is relatively small as well. The output cylinder has a relatively large surface area, and therefore, exerts a relatively large force to lift the car. This, combined with the height differential between the two cylinders (discussed in the context of mechanical advantage elsewhere in this book), makes it possible to lift a heavy automobile with a relatively small amount of effort.

THE HYDRAULIC RAM. The car jack is a simple model of the hydraulic press in operation, but in fact, Pascal's principle has many more applications. Among these is the hydraulic ram, used in machines ranging from bulldozers to the hydraulic lifts used by firefighters and utility workers to reach heights. In a hydraulic ram, however, the characteristics of the input and output cylinders are reversed from those of a car jack.

The input cylinder, called the master cylinder, has a large surface area, whereas the output cylinder (called the slave cylinder) has a small surface area. In addition—though again, this is a factor related to mechanical advantage rather than pressure, per se—the master cylinder is short, whereas the slave cylinder is tall. Owing to the larger surface area of the master cylinder compared to that of the slave cylinder, the hydraulic ram is not considered efficient in terms of mechanical advantage: in other words, the force input is much greater than the force output.

Nonetheless, the hydraulic ram is as well-suited to its purpose as a car jack. Whereas the jack is made for lifting a heavy automobile through a short vertical distance, the hydraulic ram carries a much lighter cargo (usually just one person) through a much greater vertical range—to the top of a tree or building, for instance.

EXPLOITING PRESSURE DIFFERENCES

PUMPS. A pump utilizes Pascal's principle, but instead of holding fluid in a single container, a pump allows the fluid to escape. Specifically, the pump utilizes a pressure difference, causing the fluid to move from an area of higher pressure to one of lower pressure. A very simple example of this is a siphon hose, used to draw petroleum from a car's gas tank. Sucking on one end of the hose creates an area of low pressure compared to the relatively high-pressure area of the gas tank. Eventually, the gasoline will come out of the low-pressure end of the hose. (And with luck, the person siphoning will be able to anticipate this, so that he does not get a mouthful of gasoline!)

The piston pump, more complex, but still fairly basic, consists of a vertical cylinder along which a piston rises and falls. Near the bottom of the cylinder are two valves, an inlet valve through which fluid flows into the cylinder, and an outlet valve through which fluid flows out of it. On the suction stroke, as the piston moves upward, the inlet valve opens and allows fluid to enter the cylinder. On the downstroke, the inlet valve closes while the outlet valve opens, and the pressure provided by the piston on the fluid forces it through the outlet valve.

One of the most obvious applications of the piston pump is in the engine of an automobile. In this case, of course, the fluid being pumped is gasoline, which pushes the pistons by providing a series of controlled explosions created by the spark plug's ignition of the gas. In another variety of piston pump—the kind used to inflate a basketball or a bicycle tire—air is the fluid being pumped. Then there is a pump for water, which pumps drinking water from the ground It may also be used to remove desirable water from an area where it is a hindrance, for instance, in the bottom of a boat.

BERNOULLI'S PRINCIPLE. Though Pascal provided valuable understanding with regard to the use of pressure for performing work, the thinker who first formulated general principles regarding the relationship between fluids and pressure was the Swiss mathematician and physicist Daniel Bernoulli (1700-1782). Bernoulli is considered the father of fluid mechanics, the study of the behavior of gases and liquids at rest and in motion.

While conducting experiments with liquids, Bernoulli observed that when the diameter of a pipe is reduced, the water flows faster. This suggested to him that some force must be acting upon the water, a force that he reasoned must arise from differences in pressure. Specifically, the slower-moving fluid in the wider area of pipe had a greater pressure than the portion of the fluid moving through the narrower part of the pipe. As a result, he concluded that pressure and velocity are inversely related—in other words, as one increases, the other decreases.

Hence, he formulated Bernoulli's principle, which states that for all changes in movement, the sum of static and dynamic pressure in a fluid remain the same. A fluid at rest exerts static pressure, which is commonly meant by "pressure," as in "water pressure." As the fluid begins to move, however, a portion of the static pressure—proportional to the speed of the fluid—is converted to what is known as dynamic pressure, or the pressure of movement. In a cylindrical pipe, static pressure is exerted perpendicular to the surface of the container, whereas dynamic pressure is parallel to it.

According to Bernoulli's principle, the greater the velocity of flow in a fluid, the greater the dynamic pressure and the less the static pressure: in other words, slower-moving fluid exerts greater pressure than faster-moving fluid. The discovery of this principle ultimately made possible the development of the airplane.

As fluid moves from a wider pipe to a narrower one, the volume of that fluid that moves a given distance in a given time period does not change. But since the width of the narrower pipe is smaller, the fluid must move faster (that is, with greater dynamic pressure) in order to move the same amount of fluid the same distance in the same amount of time. One way to illustrate this is to observe the behavior of a river: in a wide, unconstricted region, it flows slowly, but if its flow is narrowed by canyon walls, then it speeds up dramatically.

Bernoulli's principle ultimately became the basis for the airfoil, the design of an airplane's wing when seen from the end. An airfoil is shaped like an asymmetrical teardrop laid on its side, with the "fat" end toward the airflow. As air hits the front of the airfoil, the airstream divides, part of it passing over the wing and part passing under. The upper surface of the airfoil is curved, however, whereas the lower surface is much straighter.

As a result, the air flowing over the top has a greater distance to cover than the air flowing under the wing. Since fluids have a tendency to compensate for all objects with which they come into contact, the air at the top will flow faster to meet with air at the bottom at the rear end of the wing. Faster airflow, as demonstrated by Bernoulli, indicates lower pressure, meaning that the pressure on the bottom of the wing keeps the airplane aloft.

BUOYANCY AND PRESSURE

One hundred and twenty years before the first successful airplane flight by the Wright brothers in 1903, another pair of brothers—the Montgolfiers of France—developed another means of flight. This was the balloon, which relied on an entirely different principle to get off the ground: buoyancy, or the tendency of an object immersed in a fluid to float. As with Bernoulli's principle, however, the concept of buoyancy is related to pressure.

In the third century B.C., the Greek mathematician, physicist, and inventor Archimedes (c. 287-212 B.C.) discovered what came to be known as Archimedes's principle, which holds that the buoyant force of an object immersed in fluid is equal to the weight of the fluid displaced by the object. This is the reason why ships float: because the buoyant, or lifting, force of them is less than equal to the weight of the water they displace.

The hull of a ship is designed to displace or move a quantity of water whose weight is greater than that of the vessel itself. The weight of the displaced water—that is, its mass multiplied by the downward acceleration caused by gravity—is equal to the buoyant force that the ocean exerts on the ship. If the ship weighs less than the water it displaces, it will float; but if it weighs more, it will sink.

The factors involved in Archimedes's principle depend on density, gravity, and depth rather than pressure. However, the greater the depth within a fluid, the greater the pressure that pushes against an object immersed in the fluid. Moreover, the overall pressure at a given depth in a fluid is related in part to both density and gravity, components of buoyant force.

PRESSURE AND DEPTH. The pressure that a fluid exerts on the bottom of its container is equal to dgh, where d is density, g the acceleration due to gravity, and h the depth of the container. For any portion of the fluid, h is equal to its depth within the container, meaning that

THIS YELLOW DIVING SUIT, CALLED A "NEWT SUIT," IS SPECIALLY DESIGNED TO WITHSTAND THE ENORMOUS WATER PRESSURE THAT EXISTS AT LOWER DEPTHS OF THE OCEAN. *(Photograph by Amos Nachoum/Corbis. Reproduced by permission.)*

the deeper one goes, the greater the pressure. Furthermore, the total pressure within the fluid is equal to $dgh + p_{external}$, where $p_{external}$ is the pressure exerted on the surface of the fluid. In a piston-and-cylinder assembly, this pressure comes from the piston, but in water, the pressure comes from the atmosphere.

In this context, the ocean may be viewed as a type of "container." At its surface, the air exerts downward pressure equal to 1 atm. The density of the water itself is uniform, as is the downward acceleration due to gravity; the only variable, then, is h, or the distance below the surface. At the deepest reaches of the ocean, the pressure is incredibly great—far more than any human being could endure. This vast amount of pressure pushes upward, resisting the downward pressure of objects on its surface. At the same time, if a boat's weight is dispersed properly along its hull, the ship maximizes area and minimizes force, thus exerting a downward pressure on the surface of the water that is less than the upward pressure of the water itself. Hence, it floats.

PRESSURE AND THE HUMAN BODY

AIR PRESSURE. The Montgolfiers used the principle of buoyancy not to float on the water, but to float in the sky with a craft lighter than air. The particulars of this achievement are discussed elsewhere, in the context of buoyancy; but the topic of lighter-than-air flight suggests another concept that has been alluded to several times throughout this essay: air pressure.

Just as water pressure is greatest at the bottom of the ocean, air pressure is greatest at the surface of the Earth—which, in fact, is at the bottom of an "ocean" of air. Both air and water pressure are examples of hydrostatic pressure—the pressure that exists at any place in a body of fluid due to the weight of the fluid above. In the case of air pressure, air is pulled downward by the force of Earth's gravitation, and air along the surface has greater pressure due to the weight (a function of gravity) of the air above it. At great heights above Earth's surface, however, the gravitational force is diminished, and, thus, the air pressure is much smaller.

In ordinary experience, a person's body is subjected to an impressive amount of pressure. Given the value of atmospheric pressure discussed earlier, if one holds out one's hand—assuming that the surface is about 20 in² (0.129 m²)—the force of the air resting on it is nearly 300 lb (136 kg)! How is it, then, that one's

KEY TERMS

ATMOSPHERE: A measure of pressure, abbreviated "atm" and equal to the average pressure exerted by air at sea level. In English units, this is equal to 14.7 pounds per square inch, and in SI units to 101,300 pascals.

BAROMETER: An instrument for measuring atmospheric pressure.

BUOYANCY: The tendency of an object immersed in a fluid to float.

FLUID: Any substance, whether gas or liquid, that conforms to the shape of its container.

FLUID MECHANICS: The study of the behavior of gases and liquids at rest and in motion.

HYDROSTATIC PRESSURE: the pressure that exists at any place in a body of fluid due to the weight of the fluid above.

PASCAL: The principle SI or metric unit of pressure, abbreviated "Pa" and equal to 1 N/m^2.

PASCAL'S PRINCIPLE: A statement, formulated by French mathematician and physicist Blaise Pascal (1623–1662), which holds that the external pressure applied on a fluid is transmitted uniformly throughout the entire body of that fluid.

PRESSURE: The ratio of force to surface area, when force is applied in a direction perpendicular to that surface. The formula for pressure (p) is $p = F/A$, where F is force and A the surface area.

THE RESPONSE TO CHANGES IN AIR PRESSURE. The human body is, in fact, suited to the normal air pressure of 1 atm, and if that external pressure is altered, the body undergoes changes that may be harmful or even fatal. A minor example of this is the "popping" in the ears that occurs when one drives through the mountains or rides in an airplane. With changes in altitude come changes in pressure, and thus, the pressure in the ears changes as well.

As noted earlier, at higher altitudes, the air pressure is diminished, which makes it harder to breathe. Because air is a gas, its molecules have a tendency to be non-attractive: in other words, when the pressure is low, they tend to move away from one another, and the result is that a person at a high altitude has difficulty getting enough air into his or her lungs. Runners competing in the 1968 Olympics at Mexico City, a town in the mountains, had to train in high-altitude environments so that they would be able to breathe during competition. For baseball teams competing in Denver, Colorado (known as "the Mile-High City"), this disadvantage in breathing is compensated by the fact that lowered pressure and resistance allows a baseball to move more easily through the air.

If a person is raised in such a high-altitude environment, of course, he or she becomes used to breathing under low air pressure conditions. In the Peruvian Andes, for instance, people spend their whole lives at a height more than twice as great as that of Denver, but a person from a low-altitude area should visit such a locale only after taking precautions. At extremely great heights, of course, no human can breathe: hence airplane cabins are pressurized. Most planes are equipped with oxygen masks, which fall from the ceiling if the interior of the cabin experiences a pressure drop. Without these masks, everyone in the cabin would die.

BLOOD PRESSURE. Another aspect of pressure and the human body is blood pressure. Just as 20/20 vision is ideal, doctors recommend a target blood pressure of "120 over 80"—but what does that mean? When a person's blood pressure is measured, an inflatable cuff is wrapped around the upper arm at the same level as the heart. At the same time, a stethoscope is placed along an artery in the lower arm to monitor the sound of the blood flow. The cuff is inflated to stop the blood flow, then the pressure

hand is not crushed by all this weight? The reason is that the human body itself is under pressure, and that the interior of the body exerts a pressure equal to that of the air.

is released until the blood just begins flowing again, producing a gurgling sound in the stethoscope.

The pressure required to stop the blood flow is known as the systolic pressure, which is equal to the maximum pressure produced by the heart. After the pressure on the cuff is reduced until the blood begins flowing normally—which is reflected by the cessation of the gurgling sound in the stethoscope—the pressure of the artery is measured again. This is the diastolic pressure, or the pressure that exists within the artery between strokes of the heart. For a healthy person, systolic pressure should be 120 torr, and diastolic pressure 80 torr.

WHERE TO LEARN MORE

"Atmospheric Pressure: The Force Exerted by the Weight of Air" (Web site). <http://kids.earth.nasa.gov/archive/air_pressure/> (April 7, 2001).

Beiser, Arthur. *Physics,* 5th ed. Reading, MA: Addison-Wesley, 1991.

"Blood Pressure" (Web site). <http://www.mckinley.uiuc.edu/health-info/dis-cond/bloodpr/bloodpr.html> (April 7, 2001).

Clark, John Owen Edward. *The Atmosphere.* New York: Gloucester Press, 1992.

Cobb, Allan B. *Super Science Projects About Oceans.* New York: Rosen, 2000.

"The Physics of Underwater Diving: Pressure Lesson" (Web site). <http://www.uncwil.edu/nurc/aquarius/lessons/pressure.html> (April 7, 2001).

Provenzo, Eugene F. and Asterie Baker Provenzo. *47 Easy-to-Do Classic Experiments.* Illustrations by Peter A. Zorn, Jr. New York: Dover Publications, 1989.

"Understanding Air Pressure" *USA Today* (Web site). <http://www.usatoday.com/weather/wbarocx.html> (April 7, 2001).

Zubrowski, Bernie. *Balloons: Building and Experimenting with Inflatable Toys.* Illustrated by Roy Doty. New York: Morrow Junior Books, 1990.

ELASTICITY

CONCEPT

Unlike fluids, solids do not respond to outside force by flowing or easily compressing. The term elasticity refers to the manner in which solids respond to stress, or the application of force over a given unit area. An understanding of elasticity—a concept that carries with it a rather extensive vocabulary of key terms—helps to illuminate the properties of objects from steel bars to rubber bands to human bones.

HOW IT WORKS

CHARACTERISTICS OF A SOLID

A number of parameters distinguish solids from fluids, a term that in physics includes both gases and liquids. Solids possess a definite volume and a definite shape, whereas gases have neither; liquids have no definite shape.

At the molecular level, particles of solids tend to be precise in their arrangement and close to one another. Liquid molecules are close in proximity (though not as much so as solid molecules), and their arrangement is random, while gas molecules are both random in arrangement and far removed in proximity. Gas molecules are extremely fast-moving, and exert little or no attraction toward one another. Liquid molecules move at moderate speeds and exert a moderate attraction, but solid particles are slow-moving, and have a strong attraction to one another.

One of several factors that distinguishes solids from fluids is their relative response to pressure. Gases tend to be highly compressible, meaning that they respond well to pressure. Liquids tend to be noncompressible, yet because of their fluid characteristics, they experience exter-

nal pressure uniformly. If one applies pressure to a quantity of water in a closed container, the pressure is equal everywhere in the water. By contrast, if one places a champagne glass upright in a vise and applies pressure until it breaks, chances are that the stem or the base of the glass will be unaffected, because the pressure is not distributed equally throughout the glass.

If the surface of a solid is disturbed, it will resist, and if the force of the disturbance is sufficiently strong, it will deform—for instance, when a steel plate begins to bend under pressure. This deformation will be permanent if the force is powerful enough, as in the above example of the glass in a vise. By contrast, when the surface of a fluid is disturbed, it tends to flow.

TYPES OF STRESS

Deformation occurs as a result of stress, whether that stress be in the form of tension, compression, or shear. Tension occurs when equal and opposite forces are exerted along the ends of an object. These operate on the same line of action, but away from each other, thus stretching the object. A perfect example of an object under tension is a rope in the middle of a tug-of-war competition. The adjectival form of "tension" is "tensile": hence the term "tensile stress," which will be discussed later.

Earlier, stress was defined as the application of force over a given unit area, and in fact, the formula for stress can be written as F/A, where F is force and A area. This is also the formula for pressure, though in order for an object to be under pressure, the force must be applied in a direction perpendicular to—and in the same direction as—its surface. The one form of stress that clearly matches these parameters is compression, produced by the action of equal and opposite forces, whose effect is to reduce the length of a material. Thus compression (for example, crushing an aluminum can in one's hand) is both a form of stress and a form of pressure.

Note that compression was defined as reducing length, yet the example given involved a reduction in what most people would call the "width" or diameter of the aluminum can. In fact, width and height are the same as length, for the purposes of most discussions in physics. Length is, along with time, mass, and electric current, one of the fundamental units of measure used to express virtually all other physical quantities. Width and height are simply length expressed in terms of other planes, and within the subject of elasticity, it is not important to distinguish between these varieties of length. (By contrast, when discussing gravitational attraction—which is always vertical—it is obviously necessary to distinguish between "vertical length," or height, and horizontal length.)

The third variety of stress is shear, which occurs when a solid is subjected to equal and opposite forces that do not act along the same line, and which are parallel to the surface area of the object. If a thick hardbound book is lying flat, and a person places a finger on the spine and pushes the front cover away from the spine so that the covers and pages no longer constitute parallel planes, this is an example of shear. Stress resulting from shear is called shearing stress.

HOOKE'S LAW AND ELASTIC LIMIT

To sum up the three varieties of stress, tension stretches an object, compression shrinks it, and shear twists it. In each case, the object is deformed to some degree. This deformation is expressed in terms of strain, or the ratio between change in dimension and the original dimensions of the object. The formula for strain is $\delta L/L_o$, where δL is the change in length (δ, the Greek letter delta, means "change" in scientific notation) and L_o the original length.

Hooke's law, formulated by English physicist Robert Hooke (1635-1703), relates strain to stress. Hooke's law can be stated in simple terms as "the strain is proportional to the stress," and can also be expressed in a formula, $F = ks$, where F is the applied force, s, the resulting change in

dimension, and k, a constant whose value is related to the nature and size of the object under stress. The harder the material, the higher the value of k; furthermore, the value of k is directly proportional to the object's cross-sectional area or thickness.

The elastic limit of a given solid is the maximum stress to which it can be subjected without experiencing permanent deformation. Elastic limit will be discussed in the context of several examples below; for now, it is important merely to know that Hooke's law is applicable only as long as the material in question has not reached its elastic limit. The same is true for any modulus of elasticity, or the ratio between a particular type of applied stress and the strain that results. (The term "modulus," whose plural is "moduli," is Latin for "small measure.")

MODULI OF ELASTICITY

In cases of tension or compression, the modulus of elasticity is Young's modulus. Named after English physicist Thomas Young (1773-1829), Young's modulus is simply the ratio between F/A and $\delta L/L_o$—in other words, stress divided by strain. There are also moduli describing the behavior of objects exposed to shearing stress (shear modulus), and of objects exposed to compressive stress from all sides (bulk modulus).

Shear modulus is the relationship of shearing stress to shearing strain. This can be expressed as the ratio between F/A and ϕ. The latter symbol, the Greek letter phi, stands for the angle of shear—that is, the angle of deformation along the sides of an object exposed to shearing stress. The greater the amount of surface area A, the less that surface will be displaced by the force F. On the other hand, the greater the amount of force in proportion to A, the greater the value of ϕ, which measures the strain of an object exposed to shearing stress. (The value of ϕ, however, will usually be well below 90°, and certainly cannot exceed that magnitude.)

With tensile and compressive stress, A is a surface perpendicular to the direction of applied force, but with shearing stress, A is parallel to F. Consider again the illustration used above, of a thick hardbound book lying flat. As noted, when one pushes the front cover from the side so that the covers and pages no longer constitute parallel planes, this is an example of shear. If one pulled the spine and the long end of the pages away

THE MACHINE PICTURED HERE ROLLS OVER STEEL IN ORDER TO BEND IT INTO PIPES. BECAUSE OF ITS ELASTIC NATURE, STEEL CAN BE BENT WITHOUT BREAKING. *(Photograph by Vince Streano/Corbis. Reproduced by permission.)*

from one another, that would be tensile stress, whereas if one pushed in on the sides of the pages and spine, that would be compressive stress. Shearing stress, by contrast, would stress only the front cover, which is analogous to A for any object under shearing stress.

The third type of elastic modulus is bulk modulus, which occurs when an object is subjected to compression from all sides—that is, volume stress. Bulk modulus is the relationship of volume stress to volume strain, expressed as the ratio between F/A and $\delta V/V_o$, where δV is the change in volume and V_o is the original volume.

REAL-LIFE APPLICATIONS

ELASTIC AND PLASTIC DEFORMATION

As noted earlier, the elastic limit is the maximum stress to which a given solid can be subjected without experiencing permanent deformation, referred to as plastic deformation. Plastic deformation describes a permanent change in shape or size as a result of stress; by contrast, elastic deformation is only a temporary change in dimension.

A classic example of elastic deformation, and indeed, of highly elastic behavior, is a rubber band: it can be deformed to a length many times its original size, but upon release, it returns to its original shape. Examples of plastic deformation, on the other hand, include the bending of a steel rod under tension or the breaking of a glass under compression. Note that in the case of the steel rod, the object is deformed without rupturing—that is, without breaking or reducing to pieces. The breaking of the glass, however, is obviously an instance of rupturing.

METALS AND ELASTICITY

Metals, in fact, exhibit a number of interesting characteristics with regard to elasticity. With the notable exception of cast iron, metals tend to possess a high degree of ductility, or the ability to be deformed beyond their elastic limits without experiencing rupture. Up to a certain point, the ratio of tension to elongation for metals is high: in other words, a high amount of tension produces only a small amount of elongation. Beyond the elastic limit, however, the ratio is much lower: that is, a relatively small amount of tension produces a high degree of elongation.

Because of their ductility, metals are highly malleable, and, therefore, capable of experienc-

RUBBER BANDS, LIKE THE ONES SHOWN HERE FORMED INTO A BALL, ARE A CLASSIC EXAMPLE OF ELASTIC DEFORMATION. *(Photograph by Matthew Klein/Corbis. Reproduced by permission.)*

ing mechanical deformation through metallurgical processes, such as forging, rolling, and extrusion. Cold extrusion involves the application of high pressure—that is, a high bulk modulus—to a metal without heating it, and is used on materials such as tin, zinc, and copper to change their shape. Hot extrusion, on the other hand, involves heating a metal to a point of extremely high malleability, and then reshaping it. Metals may also be melted for the purposes of casting, or pouring the molten material into a mold.

ULTIMATE STRENGTH. The tension that a material can withstand is called its ultimate strength, and due to their ductile properties, most metals possess a high value of ultimate strength. It is possible, however, for a metal to break down due to repeated cycles of stress that are well below the level necessary to rupture it. This occurs, for instance, in metal machines such as automobile engines that experience a high frequency of stress cycles during operation.

The high ultimate strength of metals, both in tension and compression, makes them useful in a number of structural capacities. Steel has an ultimate compressive strength 25 times as great as concrete, and an ultimate tensile strength 250 times as great. For this reason, when concrete is poured for building bridges or other large struc-

tures, steel rods are inserted in the concrete. Called "rebar" (for "reinforced bars"), the steel rods have ridges along them in order to bond more firmly with the concrete as it dries. As a result, reinforced concrete has a much greater ability than plain concrete to withstand tension and compression.

STEEL BARS AND RUBBER BANDS UNDER STRESS

CRYSTALLINE MATERIALS. Metals are crystalline materials, meaning that they are composed of solids called crystals. Particles of crystals are highly ordered, with a definite geometric arrangement repeated in all directions, rather like a honeycomb. (It should be noted, however, that the crystals are not necessarily as uniform in size as the "cells" of the honeycomb.) The atoms of a crystal are arranged in orderly rows, bound to one another by strongly attractive forces that act like microscopic springs.

Just as a spring tends to return to its original length, the highly attractive atoms in a steel bar, when it is stretched, tend to restore it to its original dimensions. Likewise, it takes a great deal of force to pull apart the atoms. When the metal is subjected to plastic deformation, the atoms move

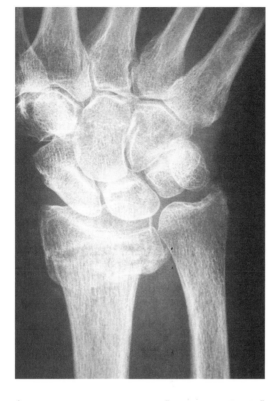

A HUMAN BONE HAS A GREATER "ULTIMATE STRENGTH" THAN THAT OF CONCRETE. (*Ecoscene/Corbis. Reproduced by permission.*)

to new positions and form new bonds. The atoms are incapable of forming bonds; however, when the metal has been subjected to stress exceeding its ultimate strength, at that point, the metal breaks.

The crystalline structure of metal influences its behavior under high temperatures. Heat causes atoms to vibrate, and in the case of metals, this means that the "springs" are stretching and compressing. As temperature increases, so do the vibrations, thus increasing the average distance between atoms. For this reason, under extremely high temperature, the elastic modulus of the metal decreases, and the metal becomes less resistant to stress.

POLYMERS AND ELASTOMERS. Rubber is so elastic in behavior that in everyday life, the term "elastic" is most often used for objects containing rubber: the waistband on a pair of underwear, for instance. The long, thin molecules of rubber, which are arranged side-by-side, are called "polymers," and the super-elastic polymers in rubber are called "elastomers." The chemical bonds between the atoms in a polymer

are flexible, and tend to rotate, producing kinks along the length of the molecule.

When a piece of rubber is subjected to tension, as, for instance, if one pulls a rubber band by the ends, the kinks and loops in the elastomers straighten. Once the stress is released, however, the elastomers immediately return to their original shape. The more "kinky" the polymers, the higher the elastic modulus, and hence, the more capable the item is of stretching and rebounding.

It is interesting to note that steel and rubber, materials that are obviously quite different, are both useful in part for the same reason: their high elastic modulus when subjected to tension, and their strength under stress. But a rubber band exhibits behaviors under high temperatures that are quite different from that of a metal: when heated, rubber contracts. It does so quite suddenly, in fact, suggesting that the added energy of the heat allows the bonds in the elastomers to begin rotating again, thus restoring the kinked shape of the molecules.

BONES

The tensile strength in bone fibers comes from the protein collagen, while the compressive strength is largely due to the presence of inorganic (non-living) salt crystals. It may be hard to believe, but bone actually has an ultimate strength—both in tension and compression—greater than that of concrete!

The ultimate strength of most materials is rendered in factors of 10^8 N/m$_2$—that is, 100,000,000 newtons (the metric unit of force) per square meter. For concrete under tensile stress, the ultimate strength is 0.02, whereas for bone, it is 1.3. Under compressive stress, the values are 0.2 and 1.7, respectively. In fact, the ultimate tensile strength of bone is close to that of cast iron (1.7), though the ultimate compressive strength of cast iron (5.5) is much higher than for bone.

Even with these figures, it may be hard to understand how bone can be stronger than concrete, but that is largely because the volume of concrete used in most situations is much greater than the volume of any bone in the body of a human being. By way of explanation, consider a piece of concrete no bigger than a typical bone: under relatively small amounts of stress, it would crumble.

KEY TERMS

ANGLE OF SHEAR: The angle of deformation on the sides of an object exposed to shearing stress. Its symbol is φ (the Greek letter phi), and its value will usually be well below 90°.

BULK MODULUS: The modulus of elasticity for a material subjected to compression on all surfaces—that is, volume stress. Bulk modulus is the relationship of volume stress to volume strain, expressed as the ratio between F/A and dV/V_o, where dV is the change in volume and V_o is the original volume.

COMPRESSION: A form of stress produced by the action of equal and opposite forces, whose effect is to reduce the length of a material. Compression is a form of pressure. When compressive stress is applied to all surfaces of a material, this is known as volume stress.

DUCTILITY: A property whereby a material is capable of being deformed far beyond its elastic limit without experiencing rupture—that is, without breaking. Most metals other than cast iron are highly ductile.

ELASTIC DEFORMATION: A temporary change in shape or size experienced by a solid subjected to stress. Elastic deformation is thus less severe than plastic deformation.

ELASTIC LIMIT: The maximum stress to which a given solid can be subjected without experiencing plastic deformation—that is, without being permanently deformed.

ELASTICITY: The response of solids to stress.

HOOKE'S LAW: A principle of elasticity formulated by English physicist Robert Hooke (1635-1703), who discovered that strain is proportional to stress. Hooke's law can be written as a formula, $F = ks$, where F is the applied force, s the resulting change in dimension, and k a constant whose value is related to the nature and size of the object being subjected to stress. Hooke's law applies only when the elastic limit has not been exceeded.

LENGTH: In discussions of elasticity, "length" refers to an object's dimensions on any given plane, thus, it can be used not only to refer to what is called length in everyday language, but also to width or height.

MODULUS OF ELASTICITY: The ratio between a type of applied stress (that is, tension, compression, and shear) and the strain that results in the object to which stress has been applied. Elastic moduli—including Young's modulus, shearing modulus, and bulk modulus—are applicable only as long as the object's elastic limit has not been reached.

PLASTIC DEFORMATION: A permanent change in shape or size experienced by a solid subjected to stress. Plastic deformation is thus more severe than elastic deformation.

PRESSURE: The ratio of force to surface area, when force is applied in a direction perpendicular to, and in the same direction as, that surface.

KEY TERMS CONTINUED

SHEAR: A form of stress resulting from equal and opposite forces that do not act along the same line. If a thick hardbound book is lying flat, and one pushes the front cover from the side so that the covers and pages no longer constitute parallel planes, this is an example of shear.

SHEAR MODULUS: The modulus of elasticity for an object exposed to shearing stress. It is expressed as the ratio between F/A and ϕ, where ϕ (the Greek letter phi) stands for the angle of shear.

STRAIN: The ratio between the change in dimension experienced by an object that has been subjected to stress, and the original dimensions of the object. The formula for strain is dL/L_o, where dL is the change in length and L_o the original length. Hooke's law, as well as the various moduli of elasticity, relates strain to stress.

STRESS: In general terms, stress is any attempt to deform a solid. Types of stress include tension, compression, and shear. More specifically, stress is the ratio of force to unit area, F/A, where F is force and A

area. Thus, it is similar to pressure, and indeed, compression is a form of pressure.

TENSION: A form of stress produced by a force which acts to stretch a material. The adjectival form of "tension" is "tensile": hence the terms "tensile stress" and "tensile strain."

ULTIMATE STRENGTH: The tension that a material can withstand without rupturing. Due to their high levels of ductility, most metals have a high value of ultimate strength.

VOLUME STRESS: The stress that occurs in a material when it is subjected to compression from all sides. The modus of elasticity for volume stress is the bulk modulus.

YOUNG'S MODULUS: A modulus of elasticity describing the relationship between stress to strain for objects under either tension or compression. Named after English physicist Thomas Young (1773-1829), Young's modulus is simply the ratio between F/A and $\delta L/L_o$—in other words, stress divided by strain.

WHERE TO LEARN MORE

Beiser, Arthur. *Physics,* 5th ed. Reading, MA: Addison-Wesley, 1991.

"Dictionary of Metallurgy" Steelmill.com: The Polish Steel Industry Directory (Web site). <http://www.steelmill.com/DICTIONARY/Diction-ary.htm> (April 9, 2001).

"Engineering Processes." eFunda.com (Web site). <http://www.efunda.com/processes/processes_home/process.cfm> (April 9, 2001).

Gibson, Gary. *Making Shapes.* Illustrated by Tony Kenyon. Brookfield, CT: Copper Beech Books, 1996.

"Glossary of Materials Testing Terms" (Web site). <http://www.instron.com/apps/glossary> (April 9, 2001).

Goodwin, Peter H. *Engineering Projects for Young Scientists.* New York: Franklin Watts, 1987.

Johnston, Tom. *The Forces with You!* Illustrated by Sarah Pooley. Milwaukee, WI: Gareth Stevens Publishing, 1988.

WORK AND ENERGY

MECHANICAL ADVANTAGE AND
SIMPLE MACHINES

ENERGY

MECHANICAL ADVANTAGE AND SIMPLE MACHINES

CONCEPT

When the term machine is mentioned, most people think of complex items such as an automobile, but, in fact, a machine is any device that transmits or modifies force or torque for a specific purpose. Typically, a machine increases either the force of the person operating it—an aspect quantified in terms of mechanical advantage—or it changes the distance or direction across which that force can be operated. Even a humble screw is a machine; so too is a pulley, and so is one of the greatest machines ever invented: the wheel. Virtually all mechanical devices are variations on three basic machines: the lever, the inclined plane, and the hydraulic press. From these three, especially the first two, arose literally hundreds of machines that helped define history, and which still permeate daily life.

HOW IT WORKS

MACHINES AND CLASSICAL MECHANICS

There are four known types of force in the universe: gravitational, electromagnetic, weak nuclear, and strong nuclear. This was the order in which the forces were identified, and the number of machines that use each force descends in the same order. The essay that follows will make little or no reference to nuclear-powered machines. Somewhat more attention will be paid to electrical machines; however, to trace in detail the development of forces in that context would require a new and somewhat cumbersome vocabulary.

Instead, the machines presented for consideration here depend purely on gravitational force and the types of force explainable purely in a gravitational framework. This is the realm of classical physics, a term used to describe the studies of physicists from the time of Galileo Galilei (1564-1642) to the end of the nineteenth century. During this era, physicists were primarily concerned with large-scale interactions that were easily comprehended by the senses, as opposed to the atomic behaviors that have become the subject of modern physics.

Late in the classical era, the Scottish physicist James Clerk Maxwell (1831-1879)—building on the work of many distinguished predecessors—identified electromagnetic force. For most of the period, however, the focus was on gravitational force and mechanics, or the study of matter, motion, and forces. Likewise, the majority of machines invented and built during most of the classical period worked according to the mechanical principles of plain gravitational force.

This was even true to some extent with the steam engine, first developed late in the seventeenth century and brought to fruition by Scotland's James Watt (1736-1819.) Yet the steam engine, though it involved ordinary mechanical processes in part, represented a new type of machine, which used thermal energy. This is also true of the internal-combustion engine; yet both steam- and gas-powered engines to some extent borrowed the structure of the hydraulic press, one of the three basic types of machine. Then came the development of electronic power, thanks to Thomas Edison (1847-1931) and others, and machines became increasingly divorced from basic mechanical laws.

THE LEVER, LIKE THIS HYDROELECTRIC ENGINE LEVER, IS A SIMPLE MACHINE THAT PERFECTLY ILLUSTRATES THE CONCEPT OF MECHANICAL ADVANTAGE. *(Photograph by E.O. Hoppe/Corbis. Reproduced by permission.)*

The heyday of classical mechanics—when classical studies in mechanics represented the absolute cutting edge of experimentation—was in the period from the beginning of the seventeenth century to the beginning of the nineteenth. One figure held a dominant position in the world of physics during those two centuries, and indeed was the central figure in the history of physics between Galileo and Albert Einstein (1879-1955). This was Sir Isaac Newton (1642-1727), who discerned the most basic laws of physical reality—laws that govern everyday life, including the operation of simple machines.

Newton and his principles are essential to the study that follows, but one other figure deserves "equal billing": the Greek mathematician, physicist, and inventor Archimedes (c. 287-212 B.C.). Nearly 2,000 years before Newton, Archimedes explained and improved a number of basic machines, most notably the lever. Describing the powers of the lever, he is said to have promised, "Give me a lever long enough and a place to stand, and I will move the world." This he demonstrated, according to one story, by moving a fully loaded ship single-handedly with the use of a lever, while remaining seated some distance away.

A common trait runs through all forms of machinery: mechanical advantage, or the ratio of force output to force input. In the case of the lever, a simple machine that will be discussed in detail below, mechanical advantage is high. In some machines, however, mechanical advantage is actually less than 1, meaning that the resulting force is less than the applied force.

This does not necessarily mean that the machine itself has a flaw; on the contrary, it can mean that the machine has a different purpose than that of a lever. One example of this is the screw: a screw with a high mechanical advantage—that is, one that rewarded the user's input of effort by yielding an equal or greater output—would be useless. In this case, mechanical advantage could only be achieved if the screw backed out from the hole in which it had been placed, and that is clearly not the purpose of a screw.

Here a machine offers an improvement in terms of direction rather than force; likewise with scissors or a fishing rod, both of which will be discussed below, an improvement with regard to distance or range of motion is bought at the expense of force. In these and many more cases, mechanical advantage alone does not measure the benefit. Thus, it is important to keep in mind what was previously stated: a machine either increases force output, or changes the force's distance or direction of operation.

Most machines, however, work best when mechanical advantage is maximized. Yet mechanical advantage—whether in theoretical terms or real-life instances—can only go so high, because there are factors that limit it. For one thing, the operator must give some kind of input to yield an output; furthermore, in most situations friction greatly diminishes output. Hence, in the operation of a car, for instance, one-quarter of the vehicle's energy is expended simply on overcoming the resistance of frictional forces.

For centuries, inventors have dreamed of creating a mechanism with an almost infinite mechanical advantage. This is the much-sought-after "perpetual motion machine," that would only require a certain amount of initial input; after that, the machine would simply run on its own forever. As output compounded over the years, its ratio to input would become so high that the figure for mechanical advantage would approach infinity.

A number of factors, most notably the existence of friction, prevent the perpetual motion machine from becoming anything other than a pipe dream. In outer space, however, the near-absence of friction makes a perpetual motion machine viable: hence, a space probe launched from Earth can travel indefinitely unless or until it enters the gravitational field of some other body in deep space.

The concept of a perpetual motion machine, at least on Earth, is only an idealization; yet idealization does have its place in physics. Physicists discuss most concepts in terms of an idealized state. For instance, when illustrating the acceleration due to gravity experienced by a body in free fall, it is customary to treat such an event as though it were taking place under conditions divorced from reality. To consider the effects of friction, air resistance, and other factors on the body's fall would create an impossibly complicated problem—yet real-world situations are just that complicated.

In light of this tendency to discuss physical processes in idealized terms, it should be noted that there are two types of mechanical advantage: theoretical and actual. Efficiency, as applied to machines in its most specific scientific sense, is the ratio of actual to theoretical mechanical advantage. This in some ways resembles the formula for mechanical advantage itself: once again, what is being measured is the relationship between "output" (the real behavior of the machine) and "input" (the planned behavior of the machine).

As with other mechanical processes, the actual mechanical advantage of a machine is a much more complicated topic than the theoretical mechanical advantage. The gulf between the two, indeed, is enormous. It would be almost impossible to address the actual behavior of machines within an environment framework that includes complexities such as friction.

Each real-world framework—that is, each physical event in the real world—is just a bit different from every other one, due to the many varieties of factors involved. By contrast, the idealized machines of physics problems behave exactly the same way in one imaginary situation after another, assuming outside conditions are the same. Therefore, the only form of mechanical advantage that a physicist can easily discuss is theoretical. For that reason, the term "efficiency"

will henceforth be used as a loose synonym for mechanical advantage—even though the technical definition is rather different.

TYPES OF MACHINES

The term "simple machine" is often used to describe the labor-saving devices known to the ancient world, most of which consisted of only one or two essential parts. Historical sources vary regarding the number of simple machines, but among the items usually listed are levers, pulleys, winches, wheels and axles, inclined planes, wedges, and screws. The list, though long, can actually be reduced to just two items: levers and inclined planes. All the items listed after the lever and before the inclined plane—including the wheel and axle—are merely variations on the lever. The same goes for the wedge and the screw, with regard to the inclined plane.

In fact, all machines are variants on three basic devices: the lever, the inclined plane, and the hydraulic press. Each transmits or modifies force or torque, producing an improvement in force, distance, or direction. The first two, which will receive more attention here, share several aspects not true of the third. First of all, the lever and the inclined plane originated at the beginning of civilization, whereas the hydraulic press is a much more recent invention.

The lever appeared as early as 5000 B.C. in the form of a simple balance scale, and within a few thousand years, workers in the Near East and India were using a crane-like lever called the shaduf to lift containers of water. The shaduf, introduced in Mesopotamia in about 3000 B.C., consisted of a long wooden pole that pivoted on two upright posts. At one end of the lever was a counterweight, and at the other a bucket. The operator pushed down on the pole to fill the bucket with water, and then used the counterweight to assist in lifting the bucket.

The inclined plane made its appearance in the earliest days of civilization, when the Egyptians combined it with rollers in the building of their monumental structures, the pyramids. Modern archaeologists generally believe that Egyptian work gangs raised the huge stone blocks of the pyramids through the use of sloping earthen ramps. These were most probably built up alongside the pyramid itself, and then removed when the structure was completed.

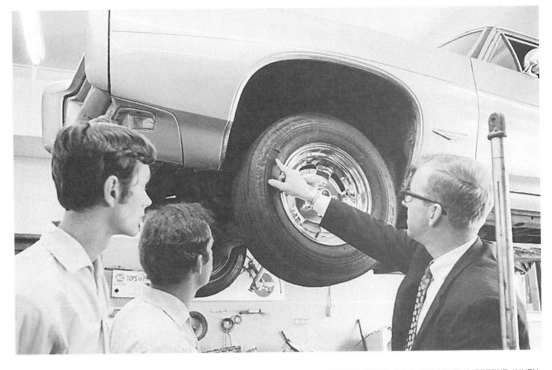

THE HYDRAULIC PRESS, LIKE THE LIFT HOLDING UP THIS CAR IN A REPAIR SHOP, IS A RELATIVELY RECENT INVENTION THAT IS USED IN MANY MODERN DEVICES, FROM CAR JACKS TO TOILETS. *(Photograph by Charles E. Rotkin/Corbis. Reproduced by permission.)*

In contrast to the distant origins of the other two machines, the hydraulic press seems like a mere youngster. Also, the first two clearly arose as practical solutions first: by the time Archimedes achieved a conceptual understanding of these machines, they had been in use a long, long time. The hydraulic press, on the other hand, first emerged from the theoretical studies of the brilliant French mathematician and physicist Blaise Pascal (1623-1662.) Nearly 150 years passed before the English inventor Joseph Bramah (1748-1814) developed a workable hydraulic press in 1796, by which time Watt had already introduced his improved steam engine. The Industrial Revolution was almost underway.

REAL-LIFE APPLICATIONS

THE LEVER

In its most basic form, the lever consists of a rigid bar supported at one point, known as the fulcrum. One of the simplest examples of a lever is a crowbar, which one might use to move a heavy object, such as a rock. In this instance, the fulcrum could be the ground, though a more rigid

"artificial" fulcrum (such as a brick) would probably be more effective.

As the operator of the crowbar pushes down on its long shaft, this constitutes an input of force, variously termed applied force, effort force, or merely effort. Newton's third law of motion shows that there is no such thing as an unpaired force in the universe: every input of force in one area will yield an output somewhere else. In this case, the output is manifested by dislodging the stone—that is, the output force, resistance force, or load.

Use of the lever gives the operator much greater lifting force than that available to a person who tried to lift with only the strength of his or her own body. Like all machines, the lever links input to output, harnessing effort to yield beneficial results—in this case, by translating the input effort into the output effort of a dislodged stone. Note, however, the statement at the beginning of this paragraph: proper use of a lever actually gives a person much greater force than he or she would possess unaided. How can this be?

There is a close relationship between the behavior of the lever and the concept of torque, as, for example, the use of a wrench to remove a lug nut. A wrench, in fact, is a sort of lever (Class

I—a distinction that will be explored below.) In any object experiencing torque, the distance from the pivot point (the lug nut, in this case), to the area where force is being applied is called the moment arm. On the wrench, this is the distance from the lug nut to the place where the operator is pushing on the wrench handle. Torque is the product of force multiplied by moment arm, and the greater the torque, the greater the tendency of the object to be put into rotation. As with machines in general, the greater the input, the greater the output.

The fact that torque is the product of force and moment arm means that if one cannot increase force, it is still possible to gain greater torque by increasing the moment arm. This is the reason why, when one tries and fails to disengage a stubborn lug nut, it is a good idea to get a longer wrench. Likewise with a lever, greater leverage can be gained without applying more force: all one needs is a longer lever arm.

As one might suspect, the lever arm is the distance from the force input to the fulcrum, or from the fulcrum to the force output. If a carpenter is using a nail-puller on the head of a hammer to extract a nail from a board, the lever arm of force input would be from the carpenter's hand gripping the hammer handle to the place where the hammerhead rests against the board. The lever arm of force output would be from the hammerhead to the end of the nail-puller.

With a lever, the input force (that of the carpenter's hand pulling back on the hammer handle) multiplied by the input lever arm is always equal to the output force (that of the nail-puller pulling up the nail) multiplied by the output lever arm. The relationship between input and output force and lever arm then makes it possible to determine a formula for the lever's mechanical advantage.

Since $F_{in}L_{in} = F_{out}L_{out}$ (where F = force and L = lever arm), it is possible to set up an equation for a lever's mechanical advantage. Once again, mechanical advantage is always F_{out}/F_{in}, but with a lever, it is also L_{in}/L_{out}. Hence, the mechanical advantage of a lever is always the same as the inverse ratio of the lever arm. If the input arm is 5 units long and the output arm is 1 unit long, the mechanical advantage will be 5, but if the positions are reversed, it will be 0.2.

CLASSES OF LEVERS. Levers are divided into three classes, depending on the rela-

tive positions of the input lever arm, the fulcrum, and the output arm or load. In a Class I lever, such as the crowbar and the wrench, the fulcrum is between the input arm and the output arm. By contrast, a Class II lever, for example, a wheelbarrow, places the output force (the load carried in the barrow itself) between the input force (the action of the operator lifting the handles) and the fulcrum, which in this case is the wheel.

Finally, there is the Class III lever, which is the reverse of a Class II. Here, the input force is between the output force and the fulcrum. The human arm itself is an example of a Class III lever: if one grasps a weight in one's hand, one's bent elbow is the fulcrum, the arm raising the weight is the input force, and the weight held in the hand—now rising—is the output force. The Class III lever has a mechanical advantage of less than 1, but what it loses in force output in gains in range of motion.

The world abounds with levers. Among the Class I varieties in common use are a nail puller on a hammerhead, described earlier, as well as postal scales and pliers. A handcart, though it might seem at first like a wheelbarrow, is actually a Class I lever, because the wheel or fulcrum is between the input effort—the force of a person's hands gripping the handles—and the output, which is the lifting of the load in the handcart itself. Scissors constitute an interesting type of Class I lever, because the force of the output (the cutting blades) is reduced in order to create a greater lever arm for the input, in this case the handles gripped when cutting.

A handheld bottle opener provides an excellent illustration of a Class II lever. Here the fulcrum is on the far end of the opener, away from the operator: the top end of the opener ring, which rests atop the bottle cap. The cap itself is the load, and one provides input force by pulling up on the opener handle, thus prying the cap from the bottle with the lower end of the opener ring. Nail clippers represent a type of combination lever: the handle that one operates is a Class II, while the cutting blades are a Class III.

Whereas Class II levers maximize force at the expense of range of motion, Class III levers operate in exactly the opposite fashion. When using a fishing rod to catch a fish, the fisherman's left hand (assuming he is right-handed) constitutes the fulcrum as it holds the rod just below the reel assembly. The right hand supplies the effort, jerking upward, while the fish is the load. The purpose here is not to raise a heavy object (one reason why a fishing rod may break if one catches too large a fish) but rather to use the increased lever arm for one's advantage in catching an object at some distance. Similarly, a hammer, which constitutes a Class III lever, with the operator's wrist as fulcrum, magnifies the motion of the operator's hand with a hammerhead that cuts a much wider arc.

Many machines that arose in the Industrial Age are a combination of many levers—that is, a compound lever. In a manual typewriter or piano, for instance, each key is a complex assembly of levers designed for a given task. An automobile, too, uses multiple levers—most notably, a special variety known as the wheel and axle.

THE WHEEL AND AXLE. A wheel is a variation on a Class I lever, but it represents such a stunning technological advancement that it deserves to be considered on its own. When driving a car, the driver places input force on the rim of the steering wheel, whose fulcrum is the center of the wheel. The output force is translated along the steering column to the driveshaft.

The combination of wheel and axle overcomes one factor that tends to limit the effectiveness of most levers, regardless of class: limited range of motion. An axle is really a type of wheel, though it has a smaller radius, which means that the output lever arm is correspondingly smaller. Given what was already said about the equation for mechanical advantage in a lever, this presents a very fortunate circumstance.

When a wheel turns, it has a relatively large lever arm (the rim), that turns a relatively small lever arm, the axle. Because the product of input force and input lever arm must equal output force multiplied by output lever arm, this means that the output force will be higher than the input force. Therefore, the larger the wheel in proportion to the axle, the greater the mechanical advantage.

It is for this reason that large vehicles without power steering often have very large steering wheels, which have a larger range of motion and, thus, a greater torque on the axle—that is, the steering column. Some common examples of the wheel-and-axle principle in operation today include a doorknob and a screwdriver; however, long before the development of the wheel and axle, there were wheels alone.

There is nothing obvious about the wheel, and in fact, it is not nearly as old an invention as most people think. Until the last few centuries, most peoples in sub-Saharan Africa, remote parts of central and northern Asia, the Americas, and the Pacific Islands remained unaware of it. This did not necessarily make them "primitive": even the Egyptians who built the Great Pyramid of Cheops in about 2550 B.C. had no concept of the wheel.

What the Egyptians did have, however, were rollers—-most often logs, onto which a heavy object was hoisted using a lever. From rollers developed the idea of a sledge, a sled-like device for sliding large loads atop a set of rollers. A sledge appears in a Sumerian illustration from about 3500 B.C., the oldest known representation of a wheel-like object.

The transformation from the roller-and-sledge assembly to wheeled vehicles is not as easy as it might seem, and historians still disagree as to the connection. Whatever the case, it appears that the first true wheels originated in Sumer (now part of Iraq) in about 3500 B.C. These were tripartite wheels, made by attaching three pieces of wood and then cutting out a circle. This made a much more durable wheel than a sawed-off log, and also overcame the fact that few trees are perfectly round.

During the early period of wheeled transportation, from 3500 to 2000 B.C., donkeys and oxen rather than horses provided the power, in part because wheeled vehicles were not yet made for the speeds that horses could achieve. Hence, it was a watershed event when wheelmakers began fashioning axles as machines separate from the wheel. Formerly, axles and wheels were made up of a single unit; separating them made carts much more stable, especially when making turns—and, as noted earlier, greatly increased the mechanical advantage of the wheels themselves.

Transportation entered a new phase in about 2000 B.C., when improvements in technology made possible the development of spoked wheels. By heat-treating wood, it became possible to bend the material slightly, and to attach spokes between the rim and hub of the wheel. When the wood cooled, the tension created a much stronger wheel—capable of carrying heavier loads faster and over greater distances.

In China during the first century B.C., a new type of wheeled vehicle—identified earlier as a

Class II lever—was born in the form of the wheelbarrow, or "wooden ox." The wheelbarrow, whose invention the Chinese attributed to a semi-legendary figure named Ko Yu, was of such value to the imperial army for moving arms and military equipment that China's rulers kept its design secret for centuries.

In Europe, around the same time, chariots were dying out, and it was a long time before the technology of wheeled transport improved. The first real innovation came during the 1500s, with the development of the horsedrawn coach. By 1640, a German family was running a regular stagecoach service, and, in 1667, a new, light, two-wheeled carriage called a cabriolet made its first appearance. Later centuries, of course, saw the development of increasingly more sophisticated varieties of wheeled vehicles powered in turn by human effort (the bicycle), steam (the locomotive), and finally, the internal combustion engine (the automobile.)

But the wheel was never just a machine for transport: long before the first wheeled carts came into existence, potters had been using wheels that rotated in place to fashion perfectly round objects, and in later centuries, wheels gained many new applications. By 500 B.C., farmers in Greece and other parts of the Mediterranean world were using rotary mills powered by donkeys. These could grind grain much faster than a person working with a hand-powered grindstone could hope to do, and in time, the Greeks found a means of powering their mills with a force more useful than donkeys: water.

The first waterwheels, turned by human or animal power, included a series of buckets along the rim that made it possible to raise water from the river below and disperse it to other points. By about 70 B.C., however, Roman engineers recognized that they could use the power of water itself to turn wheels and grind grain. Thus, the waterwheel became one of the first two rotor mechanisms in which an inanimate source (as opposed to the effort of humans or animals) created power to spin a shaft.

In this way, the waterwheel was a prototype for the engine developed many centuries later. Indeed, in the first century A.D., Hero of Alexandria—who discovered the concept of steam power some 1,700 years before anyone took up the idea and put it to use—proposed what has been considered a prototype for the turbine engine. However, for a variety of complex reasons, the ancient world was simply not ready for the technological leap portended by such an invention; and so, in terms of significant progress in the development of machines, Europe was asleep for more than a millennium.

The other significant form of wheel powered by an "inanimate" source was the windmill, first mentioned in 85 B.C. by Antipater of Thessalonica, who commented on a windmill he saw in northern Greece. In this early version of the windmill, the paddle wheel moved on a horizontal plane. However, the windmill did not take hold in Europe during ancient times, and, in fact, its true origins lie further east, and it did not become widespread until much later.

In the seventh century A.D., windmills began to appear in the region of modern Iran and Afghanistan, and the concept spread to the Arab world. Europeans in the Near East during the Crusades (1095-1291) observed the windmill, and brought the idea back to Europe with them. By the twelfth century Europeans had developed the more familiar vertical mill.

Finally, there was a special variety of wheel that made its appearance as early as 500 B.C.: the toothed gearwheel. By 300 B.C., it was in use throughout Egypt, and by about 270 B.C., Ctesibius of Alexandria (fl. c. 270-250 B.C.) had applied the gear in devising a constant-flow water clock called a clepsydra.

Some 2,100 years after Ctesibius, toothed gears became a critical component of industrialization. The most common type is a spur gear, in which the teeth of the wheel are parallel to the axis of rotation. Helical gears, by contrast, have curved teeth in a spiral pattern at an angle to their rotational axes. This means that several teeth of one gearwheel are always in contact with several teeth of the adjacent wheel, thus providing greater torque.

In bevel gears, the teeth are straight, as with a spur gear, but they slope at a 45°-angle relative to their axes so that two gearwheels can fit together at up to 90°-angles to one another without a change in speed. Finally, planetary gears are made such that one or more smaller gearwheels can fit within a larger gearwheel, which has teeth cut on the inside rather than the outside.

Similar in concept to the gearwheel is the V belt drive, which consists of two wheels side by side, joined with a belt. Each of the wheels has grooves cut in it for holding the belt, making this a modification of the pulley, and the grooves provide much greater gripping power for holding the belt in place. One common example of a V belt drive, combined with gearwheels, is a bicycle chain assembly.

PULLEYS. A pulley is essentially a grooved wheel on an axle attached to a frame, which in turn is attached to some form of rigid support such as a ceiling. A rope runs along the grooves of the pulley, and one end is attached to a load while the other is controlled by the operator.

In several instances, it has been noted that a machine may provide increased range of motion or position rather than power. So, this simplest kind of pulley, known as a single or fixed pulley, only offers the advantage of direction rather than improved force. When using Venetian blinds, there is no increase in force; the advantage of the machine is simply that it allows one to move objects upward and downward. Thus, the theoretical mechanical advantage of a fixed pulley is 1 (or almost certainly less under actual conditions, where friction is a factor).

Here it is appropriate to return to Archimedes, whose advancements in the understanding of levers translated to improvements in pulleys. In the case of the lever, it was Archimedes who first recognized that the longer the effort arm, the less effort one had to apply in raising the load. Likewise, with pulleys and related devices—cranes and winches—he explained and improved the way these machines worked.

The first crane device dates to about 1000 B.C., but evidence from pictures suggests that pulleys may have been in use as early as seven thousand years before. Several centuries before Archimedes's time, the Greeks were using compound pulleys that contained several wheels and thus provided the operator with much greater mechanical advantage than a fixed pulley. Archimedes, who was also the first to recognize the relationship between pulleys and levers, created the first fully realized block-and-tackle system using compound pulleys and cranes. In the late modern era, compound pulley systems were used in applications such as elevators and escalators.

A compound pulley consists of two or more wheels, with at least one attached to the support while the other wheel or wheels lift the load. A rope runs from the support pulley down to the load-bearing wheel, wraps around that pulley and comes back up to a fixed attachment on the upper pulley. Whereas the upper pulley is fixed, the load-bearing pulley is free to move, and raises the load as the rope is pulled below.

The simplest kind of compound pulley, with just two wheels, has a mechanical advantage of 2. On a theoretical level, at least, it is possible to calculate the mechanical advantage of a compound pulley with more wheels: the number is equal to the segments of rope between the lower pulleys and the upper, or support pulley. In reality, however, friction, which is high as ropes rub against the pulley wheels, takes its toll. Thus mechanical advantage is never as great as it might be.

A block-and-tackle, like a compound pulley, uses just one rope with a number of pulley wheels. In a block-and-tackle, however, the wheels are arranged along two axles, each of which includes multiple pulley wheels that are free to rotate along the axle. The upper row is attached to the support, and the lower row to the load. The rope connects them all, running from the first pulley in the upper set to the first in the lower set, then to the second in the upper set, and so on. In theory, at least, the mechanical advantage of a block-and-tackle is equal to the number of wheels used, which must be an even number—but again, friction diminishes the theoretical mechanical advantage.

THE INCLINED PLANE

To the contemporary mind, it is difficult enough to think of a lever as a "machine"—but levers at least have more than one part, unlike an inclined plane. The latter, by contrast, is exactly what it seems to be: a ramp. Yet it was just such a ramp structure, as noted earlier, that probably enabled the Egyptians to build the pyramids—a feat of engineering so stunning that even today, some people refuse to believe that the ancient Egyptians could have achieved it on their own.

Surely, as anti-scientific proponents of various fantastic theories often insist, the building of the pyramids could only have been done with machines provided by super-intelligent, extraterrestrial beings. Even in ancient times, the Greek historian Herodotus (c. 484-c. 424 B.C.) speculated that the Egyptians must have used huge cranes that had long since disappeared.

These bizarre guesses concerning the technology for raising the pyramid's giant blocks serve to highlight the brilliance of a gloriously simple machine that, in essence, doubles force. If one needs to move a certain weight to a certain height, there are two options. One can either raise the weight straight upward, expending an enormous amount of effort, even with a pulley system; or one can raise the weight gradually along an inclined plane. The inclined plane is a much wiser choice, because it requires half the effort.

Why half? Imagine an inclined plane sloping evenly upward to the right. The plane exists in a sort of frame that is equal in both length and height to the dimensions of the plane itself. As we can easily visualize, the plane takes up exactly half of the frame, and this is true whether the slope is more than, less than, or equal to 45°. For any plane in which the slope is more than 45°, however, the mechanical advantage will be less than 1, and it is indeed hard to imagine why anyone would use such a plane unless forced to do so by limitations on their horizontal space—for example, when lifting a heavy object from a narrow canyon.

The mechanical advantage of an inclined plane is equal to the ratio between the distance over which input force is applied and the distance of output; or, more simply, the ratio of length to height. If a man is pushing a crate up a ramp 4 ft high and 8 ft long (1.22 m by 2.44 m), 8 ft is the input distance and 4 ft the output distance; hence, the mechanical advantage is 2. If the ramp length were doubled to 16 ft (4.88 m), the mechanical advantage would likewise double to 4, and so on.

The concept of work, in terms of physics, has specific properties that are a subject unto themselves; however, it is important here only to recognize that work is the product of force (that is, effort) multiplied by distance. This means that if one increases the distance, a much smaller quantity of force is needed to achieve the same amount of work.

On an everyday level, it is easy to see this in action. Walking or running up a gentle hill, obviously, is easier than going up a steep hill. Therefore, if one's primary purpose is to conserve effort, it is best to choose the gentler hill. On the other hand, one may wish to minimize distance—or, if moving for the purpose of exercise,

to maximize force input to burn calories. In either case, the steeper hill would be the better option.

WEDGES. The type of inclined plane discussed thus far is a ramp, but there are a number of much smaller varieties of inclined plane at work in the everyday world. A knife is an excellent example of one of the most common types, a wedge. Again, the mechanical advantage of a wedge is the ratio of length to height, which, in the knife, would be the depth of the blade compared to its cross-sectional width. Due to the ways in which wedges are used, however, friction plays a much greater role, therefore greatly reducing the theoretical mechanical advantage.

Other types of wedges may be used with a lever, as a form of fulcrum for raising objects. Or, a wedge may be placed under objects to stabilize them, as for instance, when a person puts a folded matchbook under the leg of a restaurant table to stop it from wobbling. Wedges also stop other objects from moving: a triangular piece of wood under a door will keep it from closing, and a more substantial wedge under the front wheels of a car will stop it from rolling forward.

Variations of the wedge are everywhere. Consider all the types of cutting or chipping devices that exist: scissors, chisels, ice picks, axes, splitting wedges (used with a mallet to split a log down the center), saws, plows, electric razors, etc. Then there are devices that use a complex assembly of wedges working together. The part of a key used to open a lock is really just a row of wedges for moving the pins inside the lock to the proper position for opening the door. Similarly, each tooth in a zipper is a tiny wedge that fits tightly with the adjacent teeth.

SCREWS. As the wheel is, without a doubt, the greatest conceptual variation on the lever, so the screw may be identified as a particularly cunning adaptation of an inclined plane. The uses of screws today are many and obvious, but as with wheels and axles, these machines have more applications than are commonly recognized. Not only are there screws for holding things together, but there are screws such as those on vises, clamps, or monkey wrenches for applying force to objects.

A screw is an inclined plane in the shape of a helix, wrapped around an axis or cylinder. In order to determine its mechanical advantage, one must first find the pitch, which is the distance

A SCREW, LIKE THIS CORKSCREW USED TO OPEN A BOTTLE OF WINE, IS AN INCLINED PLANE IN THE SHAPE OF A HELIX, WRAPPED AROUND AN AXIS OR CYLINDER. *(Ecoscene/Corbis. Reproduced by permission.)*

between adjacent threads. The other variable is lever arm, which with a screwdriver is the radius, or on a wrench, the length from the crescent or clamp to the area of applied force. Obviously, the lever arm is much greater for a wrench, which explains why a wrench is sometimes preferable to a screwdriver, when removing a highly resistant material screw or bolt.

When one rotates a screw of a given pitch, the applied force describes a circle whose area may be calculated as $2\pi L$, where L is the lever arm. This figure, when divided by the pitch, is the same as the ratio between the distance of force input to force output. Either number is equal to the mechanical advantage for a screw. As suggested earlier, that mechanical advantage is usually low, because force input (screwing in the screw) takes place in a much greater range of motion than force output (the screw working its way into the surface). But this is exactly what the screw is designed to do, and what it lacks in mechanical advantage, it more than makes up in its holding power.

As with the lever and pulley, Archimedes did not invent the screw, but he did greatly improve human understanding of it. Specifically, he developed a mathematical formula for a simple spiral, and translated this into the highly practi-

cal Archimedes screw, a device for lifting water. The invention consists of a metal pipe in a corkscrew shape, which draws water upward as it revolves. It proved particularly useful for lifting water that had seeped into the lower parts of a ship, and in many countries today, it remains in use as a simple pump for drawing water out of the ground.

Some historians maintain that Archimedes did not invent the screw-type pump, but rather saw an example of it in Egypt. In any case, he clearly developed a practical version of the device, and it soon gained application throughout the ancient world. Archaeologists discovered a screw-driven olive press in the ruins of Pompeii, destroyed by the eruption of Mt. Vesuvius in A.D. 79, and Hero of Alexandria later mentioned the use of a screw-type machine in his *Mechanica.*

Yet, Archimedes is the figure most widely associated with the development of this wondrous device. Hence, in 1837, when the Swedish-American engineer John Ericsson (1803-1899) demonstrated the use of a screw-driven ship's propeller, he did so on a craft he named the Archimedes.

From screws planted in wood to screws that drive ships at sea, the device is everywhere in modern life. Faucets, corkscrews, drills, and meat grinders are obvious examples. Though many types of jacks used for lifting an automobile or a house are levers, others are screw assemblies on which one rotates the handle along a horizontal axis. In fact, the jack is a particularly interesting device. Versions of the jack represent all three types of simple machine: lever, inclined plane, and hydraulic press.

THE HYDRAULIC PRESS

As noted earlier, the hydraulic press came into existence much, much later than the lever or inclined plane, and its birth can be seen within the context of a larger movement toward the use of water power, including steam. A little more than a quarter-century after Pascal created the theoretical framework for hydraulic power, his countryman Denis Papin (1647-1712) introduced the steam digester, a prototype for the pressure cooker. In 1687, Papin published a work describing a machine in which steam operated a piston—an early model for the steam engine.

Papin's concept, which was on the absolute cutting edge of technological development at that time, utilized not only steam power but also the very hydraulic concept that Pascal had identified a few decades earlier. Indeed, the assembly of pistons and cylinders that forms the central component of the internal-combustion engine reflects this hydraulic rule, discovered by Pascal in 1653. It was then that he formulated what is known as Pascal's principle: that the external pressure applied on a fluid is transmitted uniformly throughout the entire body of that fluid.

Inside a piston and cylinder assembly, one of the most basic varieties of hydraulic press, the pressure is equal to the ratio of force to the horizontal area of pressure. A simple hydraulic press of the variety that might be used to raise a car in an auto shop typically consists of two large cylinders side by side, connected at the bottom by a channel in which valves control flow. When one applies force over a given area of input—that is, by pressing down on one cylinder—this yields a uniform pressure that causes output in the second cylinder.

Once again, mechanical advantage is equal to the ratio of force output to force input, and for a hydraulic press, this can also be measured as the ratio of area output to area input. Just as there is an inverse relationship between lever arm and force in a lever, and between length and height in an inclined plane, so there is such a relationship between horizontal area and force in a hydraulic pump. Consequently, in order to increase force, one should minimize area.

However, there is another factor to consider: height. The mechanical advantage of a hydraulic pump is equal to the vertical distance to which the input force is applied, divided by that of the output force. Hence, the greater the height of the input cylinder compared to the output cylinder, the greater the mechanical advantage. And since these three factors—height, area, and force—work together, it is possible to increase the lifting force and area by minimizing the height.

Consider once again the auto-shop car jack. Typically, the input cylinder will be relatively tall and thin, and the output cylinder short and squat. Because the height of the input cylinder is large, the area of input will be relatively small, as will the force of input. But according to Pascal's principle, whatever the force applied on the input, the pressure will be the same on the output. At the output end, where the car is raised, one needs a large amount of force and a relatively large lifting area. Therefore, height is minimized to increase force and area. If the output area is 10 times the size of the input area, an input force of 1 unit will produce an output force of 10 units—but in order to raise the weight by 1 unit of height, the input piston must move downward by 10 units.

This type of car jack provides a basic model of the hydraulic press in operation, but, in fact, hydraulic technology has many more applications. A hydraulic pump, whether for pumping air into a tire or water from a basement, uses very much the same principle as the hydraulic jack. So too does the hydraulic ram, used in machines ranging from bulldozers to the hydraulic lifts used by firefighters and utility workers to reach great heights.

In a hydraulic ram, however, the characteristics of the input and output cylinders are reversed from those of a car jack. The input cylinder, called the master cylinder, is short and squat, whereas the output cylinder—the slave

COMPOUND LEVER: A machine that combines multiple levers to accomplish its task. An example is a piano or manual typewriter.

CLASS I LEVER: A lever in which the fulcrum is between the input force and output force. Examples include a crowbar, a nail puller, and scissors.

CLASS II LEVER: A lever in which the output force is between the input force and the fulcrum. Class II levers, of which wheelbarrows and bottle openers are examples, maximize output force at the expense of range of motion.

CLASS III LEVER: A lever in which the input force is between the output force and the fulcrum. Class III levers, of which a fishing rod is an example, maximize range of motion at the expense of output force.

EFFICIENCY: The ratio of actual mechanical advantage to theoretical mechanical advantage.

FRICTION: The force that resists motion when the surface of one object comes into contact with the surface of another.

FULCRUM: The support point of a lever.

INERTIA: The tendency of an object in motion to remain in motion, and of an object at rest to remain at rest.

INPUT: The effort supplied by the operator of a machine. In a Class I lever such as

a crowbar, input would be the energy one expends by pushing down on the bar. Input force is often called applied force, effort force, or simply effort.

LEVER: One of the three basic varieties of machine, a lever consists of a rigid bar supported at one point, known as the fulcrum.

LEVER ARM: On a lever, the distance from the input force or the output force to the fulcrum.

MACHINE: A device that transmits or modifies force or torque for a specific purpose.

MECHANICAL ADVANTAGE: The ratio of force output to force input for a machine.

MOMENT ARM: For an object experiencing torque, moment arm is the distance from the pivot or balance point to the vector on which force is being applied. Moment arm is always perpendicular to the direction of force.

OUTPUT: The results achieved from the operation of a machine. In a Class I lever such as a crowbar, output is the moving of a stone or other heavy load dislodged by the crowbar. Output force is often called the load or resistance force.

TORQUE: In general terms, torque is turning force; in scientific terms, it is the product of moment arm multiplied by force.

cylinder—is tall and thin. The reason for this change is that in objects using a hydraulic ram, height is more important than force output: they

are often raising people rather than cars. When the slave cylinder exerts pressure on the stabilizer ram above it (the bucket containing the firefighter, for example), it rises through a much larger range of vertical motion than that of the fluid flowing from the master cylinder.

As noted earlier, the pistons of a car engine are hydraulic pumps—specifically, reciprocating hydraulic pumps, so named because they all work together. In scientific terms, "fluid" can mean either a liquid or a gas such as air; hence, there is an entire subset of hydraulic machines that are pneumatic, or air-powered. Among these are power brakes in a car, pneumatic drills, and even hovercrafts. As with the other two varieties of simple machine, the hydraulic press is in evidence throughout virtually every nook and cranny of daily life. The pump in a toilet tank is a type of hydraulic press, as is the inner chamber of a pen. So too are aerosol cans, fire extinguishers, scuba tanks, water meters, and pressure gauges.

WHERE TO LEARN MORE

Archimedes (Web site). <http://www.mcs.drexel.edu/~crorres/Archimedes/contents.html> (March 9, 2001.)

Bains, Rae. *Simple Machines.* Mahwah, N.J.: Troll Associates, 1985.

Beiser, Arthur. *Physics,* 5th ed. Reading, MA: Addison-Wesley, 1991.

Canizares, Susan. *Simple Machines.* New York: Scholastic, 1999.

Haslam, Andrew and David Glover. *Machines.* Photography by Jon Barnes. Chicago: World Book, 1998.

"Inventors Toolbox: The Elements of Machines" (Web site). <http://www.mos.org/sln/Leonardo/InventorsToolbox.html> (March 9, 2001).

Macaulay, David. *The New Way Things Work.* Boston: Houghton Mifflin, 1998.

"Machines" (Web site). <http://www.galaxy.net:80/~k12/machines/index.s.html> (March 9, 2001).

"Motion, Energy, and Simple Machines" (Web site). <http://www.necc.mass.edu/MRVIS/MR3_13/start.html> (March 9, 2001).

O'Brien, Robert, and the Editors of Life. *Machines.* New York: Time-Life Books, 1964.

"Simple Machines" (Web site). <http://sln.fi.edu/qa97/spotlight3/spotlight3.html> (March 9, 2001).

Singer, Charles Joseph, et al., editors. *A History of Technology* (8 vols.) Oxford, England: Clarendon Press, 1954-84.

ENERGY

CONCEPT

As with many concepts in physics, energy—along with the related ideas of work and power—has a meaning much more specific, and in some ways quite different, from its everyday connotation. According to the language of physics, a person who strains without success to pull a rock out of the ground has done no work, whereas a child playing on a playground produces a great deal of work. Energy, which may be defined as the ability of an object to do work, is neither created nor destroyed; it simply changes form, a concept that can be illustrated by the behavior of a bouncing ball.

HOW IT WORKS

In fact, it might actually be more precise to say that energy is the ability of "a thing" or "something" to do work. Not only tangible objects (whether they be organic, mechanical, or electromagnetic) but also non-objects may possess energy. At the subatomic level, a particle with no mass may have energy. The same can be said of a magnetic force field.

One cannot touch a force field; hence, it is not an object—but obviously, it exists. All one has to do to prove its existence is to place a natural magnet, such as an iron nail, within the magnetic field. Assuming the force field is strong enough, the nail will move through space toward it—and thus the force field will have performed work on the nail.

Work: What It Is and Is Not

Work may be defined in general terms as the exertion of force over a given distance. In order for work to be accomplished, there must be a displacement in space—or, in colloquial terms, something has to be moved from point A to point B. As noted earlier, this definition creates results that go against the common-sense definition of "work."

A person straining, and failing, to pull a rock from the ground has performed no work (in terms of physics) because nothing has been moved. On the other hand, a child on a playground performs considerable work: as she runs from the slide to the swing, for instance, she has moved her own weight (a variety of force) across a distance. She is even working when her movement is back-and-forth, as on the swing. This type of movement results in no net displacement, but as long as displacement has occurred at all, work has occurred.

Similarly, when a man completes a full push-up, his body is in the same position—parallel to the floor, arms extended to support him—as he was before he began it; yet he has accomplished work. If, on the other hand, he at the end of his energy, his chest is on the floor, straining but failing, to complete just one more push-up, then he is not working. The fact that he feels as though he has worked may matter in a personal sense, but it does not in terms of physics.

CALCULATING WORK. Work can be defined more specifically as the product of force and distance, where those two vectors are exerted in the same direction. Suppose one were to drag a block of a certain weight across a given distance of floor. The amount of force one exerts parallel to the floor itself, multiplied by the distance, is equal to the amount of work exerted. On the other hand, if one pulls up on the block in a

position perpendicular to the floor, that force does not contribute toward the work of dragging the block across the floor, because it is not par allel to distance as defined in this particular situation.

Similarly, if one exerts force on the block at an angle to the floor, only a portion of that force counts toward the net product of work—a portion that must be quantified in terms of trigonometry. The line of force parallel to the floor may be thought of as the base of a triangle, with a line perpendicular to the floor as its second side. Hence there is a 90°-angle, making it a right triangle with a hypotenuse. The hypotenuse is the line of force, which again is at an angle to the floor.

The component of force that counts toward the total work on the block is equal to the total force multiplied by the cosine of the angle. A cosine is the ratio between the leg adjacent to an acute (less than 90°) angle and the hypotenuse. The leg adjacent to the acute angle is, of course, the base of the triangle, which is parallel to the floor itself. Sizes of triangles may vary, but the ratio expressed by a cosine (abbreviated cos) does not. Hence, if one is pulling on the block by a rope that makes a 30°-angle to the floor, then

force must be multiplied by cos 30°, which is equal to 0.866.

Note that the cosine is less than 1; hence when multiplied by the total force exerted, it will yield a figure 13.4% smaller than the total force. In fact, the larger the angle, the smaller the cosine; thus for 90°, the value of cos = 0. On the other hand, for an angle of 0°, cos = 1. Thus, if total force is exerted parallel to the floor—that is, at a 0°-angle to it—then the component of force that counts toward total work is equal to the total force. From the standpoint of physics, this would be a highly work-intensive operation.

GRAVITY AND OTHER PECU-LIARITIES OF WORK. The above discussion relates entirely to work along a horizontal plane. On the vertical plane, by contrast, work is much simpler to calculate due to the presence of a constant downward force, which is, of course, gravity. The force of gravity accelerates objects at a rate of 32 ft (9.8 m)/sec^2. The mass (m) of an object multiplied by the rate of gravitational acceleration (g) yields its weight, and the formula for work done against gravity is equal to weight multiplied by height (h) above some lower reference point: mgh.

Distance and force are both vectors—that is, quantities possessing both magnitude and direction. Yet work, though it is the product of these two vectors, is a scalar, meaning that only the magnitude of work (and not the direction over which it is exerted) is important. Hence mgh can refer either to the upward work one exerts against gravity (that is, by lifting an object to a certain height), or to the downward work that gravity performs on the object when it is dropped. The direction of h does not matter, and its value is purely relative, referring to the vertical distance between one point and another.

The fact that gravity can "do work"—and the irrelevance of direction—further illustrates the truth that work, in the sense in which it is applied by physicists, is quite different from "work" as it understood in the day-to-day world. There is a highly personal quality to the everyday meaning of the term, which is completely lacking from its physics definition.

If someone carried a heavy box up five flights of stairs, that person would quite naturally feel justified in saying "I've worked." Certainly he or she would feel that the work expended was far greater than that of someone who had simply allowed the the elevator to carry the box up those five floors. Yet in terms of work done against gravity, the work done on the box by the elevator is exactly the same as that performed by the person carrying it upstairs. The identity of the "worker"—not to mention the sweat expended or not expended—is irrelevant from the standpoint of physics.

MEASUREMENT OF WORK AND POWER

In the metric system, a newton (N) is the amount of force required to accelerate 1 kg of mass by 1 meter per second squared (m/s^2). Work is measured by the joule (J), equal to 1 newton-meter (N • m). The British unit of force is the pound, and work is measured in foot-pounds, or the work done by a force of 1 lb over a distance of one foot.

Power, the rate at which work is accomplished over time, is the same as work divided by time. It can also be calculated in terms of force multiplied by speed, much like the force-multiplied-by-distance formula for work. However, as with work, the force and speed must be in the same direction. Hence, the formula for power in these terms is F • cos θ • v, where F=force, v=speed, and cos θ is equal to the cosine of the angle θ (the Greek letter theta) between F and the direction of v.

The metric-system measure of power is the watt, named after James Watt (1736-1819), the Scottish inventor who developed the first fully viable steam engine and thus helped inaugurate the Industrial Revolution. A watt is equal to 1 joule per second, but this is such a small unit that it is more typical to speak in terms of kilowatts, or units of 1,000 watts.

Ironically, Watt himself—like most people in the British Isles and America—lived in a world that used the British system, in which the unit of power is the foot-pound per second. The latter, too, is very small, so for measuring the power of his steam engine, Watt suggested a unit based on something quite familiar to the people of his time: the power of a horse. One horsepower (hp) is equal to 550 foot-pounds per second.

SORTING OUT METRIC AND BRITISH UNITS. The British system, of course, is horridly cumbersome compared to the metric system, and thus it long ago fell out of favor with the international scientific community. The British system is the product of loosely developed conventions that emerged over time: for instance, a foot was based on the length of the reigning king's foot, and in time, this became standardized. By contrast, the metric system was

IN THIS 1957 PHOTOGRAPH, ITALIAN OPERA SINGER LUIGI INFANTINO TRIES TO BREAK A WINE GLASS BY SINGING A HIGH "C" NOTE. CONTRARY TO POPULAR BELIEF, THE NOTE DOES NOT HAVE TO BE A PARTICULARLY HIGH ONE TO BREAK THE GLASS: RATHER, THE NOTE SHOULD BE ON THE SAME WAVELENGTH AS THE GLASS'S OWN VIBRATIONS. WHEN THIS OCCURS, SOUND ENERGY IS TRANSFERRED DIRECTLY TO THE GLASS, WHICH IS SHATTERED BY THIS SUDDEN NET INTAKE OF ENERGY. *(Hulton-Deutsch Collection/Corbis. Reproduced by permission.)*

created quite deliberately over a matter of just a few years following the French Revolution, which broke out in 1789. The metric system was adopted ten years later.

During the revolutionary era, French intellectuals believed that every aspect of existence could and should be treated in highly rational, scientific terms. Out of these ideas arose much folly—especially after the supposedly "rational" leaders of the revolution began chopping off people's heads—but one of the more positive outcomes was the metric system. This system, based entirely on the number 10 and its exponents, made it easy to relate one figure to another: for instance, there are 100 centimeters in a meter and 1,000 meters in a kilometer. This is vastly more convenient than converting 12 inches to a foot, and 5,280 feet to a mile.

For this reason, scientists—even those from the Anglo-American world—use the metric system for measuring not only horizontal space, but volume, temperature, pressure, work, power, and so on. Within the scientific community, in fact, the metric system is known as SI, an abbreviation of the French *Système International d'Unités*—that is, "International System of Units."

Americans have shown little interest in adopting the SI system, yet where power is concerned, there is one exception. For measuring the power of a mechanical device, such as an automobile or even a garbage disposal, Americans use the British horsepower. However, for measuring electrical power, the SI kilowatt is used. When an electric utility performs a meter reading on a family's power usage, it measures that usage in terms of electrical "work" performed for the family, and thus bills them by the kilowatt-hour.

THREE TYPES OF ENERGY

KINETIC AND POTENTIAL ENERGY FORMULAE. Earlier, energy was defined as the ability of an object to accomplish work—a definition that by this point has acquired a great deal more meaning. There are three types of energy: kinetic energy, or the energy that something possesses by virtue of its motion; potential energy, the energy it possesses by virtue of its position; and rest energy, the energy it possesses by virtue of its mass.

The formula for kinetic energy is KE = ½ mv^2. In other words, for an object of mass m, kinetic energy is equal to half the mass multiplied by the square of its speed v. The actual derivation of this formula is a rather detailed process, involving reference to the second of the three laws of motion formulated by Sir Isaac Newton (1642-1727.) The second law states that $F = ma$, in other words, that force is equal to mass multiplied by acceleration. In order to understand kinetic energy, it is necessary, then, to understand the formula for uniform acceleration. The latter is $v_f^2 = v_0^2 + 2as$, where v_f^2 is the final speed of the object, v_0^2 its initial speed, a acceleration and s distance. By substituting values within these equations, one arrives at the formula of ½ mv^2 for kinetic energy.

The above is simply another form of the general formula for work—since energy is, after all, the ability to perform work. In order to produce an amount of kinetic energy equal to ½ mv^2 within an object, one must perform an amount of work on it equal to Fs. Hence, kinetic energy also equals Fs, and thus the preceding paragraph simply provides a means for translating that into more specific terms.

The potential energy (PE) formula is much simpler, but it also relates to a work formula given earlier: that of work done against gravity. Potential energy, in this instance, is simply a function of gravity and the distance h above some reference point. Hence, its formula is the same as that for work done against gravity, mgh or wh, where w stands for weight. (Note that this refers to potential energy in a gravitational field; potential energy may also exist in an electromagnetic field, in which case the formula would be different from the one presented here.)

REST ENERGY AND ITS INTRIGUING FORMULA. Finally, there is rest energy, which, though it may not sound very exciting, is in fact the most intriguing—and the most complex—of the three. Ironically, the formula for rest energy is far, far more complex in derivation than that for potential or even kinetic energy, yet it is much more well-known within the popular culture.

Indeed, $E = mc^2$ is perhaps the most famous physics formula in the world—even more so than the much simpler $F = ma$. The formula for rest energy, as many people know, comes from the man whose Theory of Relativity invalidated certain specifics of the Newtonian framework: Albert Einstein (1879-1955). As for what the formula actually means, that will be discussed later.

REAL-LIFE APPLICATIONS

FALLING AND BOUNCING BALLS

One of the best—and most frequently used—illustrations of potential and kinetic energy involves standing at the top of a building, holding a baseball over the side. Naturally, this is not an experiment to perform in real life. Due to its relatively small mass, a falling baseball does not have a great amount of kinetic energy, yet in the real world, a variety of other conditions (among them inertia, the tendency of an object to maintain its state of motion) conspire to make a hit on the head with a baseball potentially quite serious.

If dropped from a great enough height, it could be fatal.

When one holds the baseball over the side of the building, potential energy is at a peak, but once the ball is released, potential energy begins to decrease in favor of kinetic energy. The relationship between these, in fact, is inverse: as the value of one decreases, that of the other increases in exact proportion. The ball will only fall to the point where its potential energy becomes 0, the same amount of kinetic energy it possessed before it was dropped. At the same point, kinetic energy will have reached maximum value, and will be equal to the potential energy the ball possessed at the beginning. Thus the sum of kinetic energy and potential energy remains constant, reflecting the conservation of energy, a subject discussed below.

It is relatively easy to understand how the ball acquires kinetic energy in its fall, but potential energy is somewhat more challenging to comprehend. The ball does not really "possess" the potential energy: potential energy resides within an entire system comprised by the ball, the space through which it falls, and the Earth. There is thus no "magic" in the reciprocal relationship between potential and kinetic energy: both are part of a single system, which can be envisioned by means of an analogy.

Imagine that one has a 20-dollar bill, then buys a pack of gum. Now one has, say, $19.20. The positive value of dollars has decreased by $0.80, but now one has increased "non-dollars" or "anti-dollars" by the same amount. After buying lunch, one might be down to $12.00, meaning that "anti-dollars" are now up to $8.00. The same will continue until the entire $20.00 has been spent. Obviously, there is nothing magical about this: the 20-dollar bill was a closed system, just like the one that included the ball and the ground. And just as potential energy decreased while kinetic energy increased, so "non-dollars" increased while dollars decreased.

BOUNCING BACK. The example of the baseball illustrates one of the most fundamental laws in the universe, the conservation of energy: within a system isolated from all other outside factors, the total amount of energy remains the same, though transformations of energy from one form to another take place. An interesting example of this comes from the case

of another ball and another form of vertical motion.

This time instead of a baseball, the ball should be one that bounces: any ball will do, from a basketball to a tennis ball to a superball. And rather than falling from a great height, this one is dropped through a range of motion ordinary for a human being bouncing a ball. It hits the floor and bounces back—during which time it experiences a complex energy transfer.

As was the case with the baseball dropped from the building, the ball (or more specifically, the system involving the ball and the floor) possesses maximum potential energy prior to being released. Then, in the split-second before its impact on the floor, kinetic energy will be at a maximum while potential energy reaches zero.

So far, this is no different than the baseball scenario discussed earlier. But note what happens when the ball actually hits the floor: it stops for an infinitesimal fraction of a moment. What has happened is that the impact on the floor (which in this example is assumed to be perfectly rigid) has dented the surface of the ball, and this saps the ball's kinetic energy just at the moment when the energy had reached its maximum value. In accordance with the energy conservation law, that energy did not simply disappear: rather, it was transferred to the floor.

Meanwhile, in the wake of its huge energy loss, the ball is motionless. An instant later, however, it reabsorbs kinetic energy from the floor, undents, and rebounds. As it flies upward, its kinetic energy begins to diminish, but potential energy increases with height. Assuming that the person who released it catches it at exactly the same height at which he or she let it go, then potential energy is at the level it was before the ball was dropped.

WHEN A BALL LOSES ITS BOUNCE. The above, of course, takes little account of energy "loss"—that is, the transfer of energy from one body to another. In fact, a part of the ball's kinetic energy will be lost to the floor because friction with the floor will lead to an energy transfer in the form of thermal, or heat, energy. The sound that the ball makes when it bounces also requires a slight energy loss; but friction—a force that resists motion when the surface of one object comes into contact with the surface of another—is the principal culprit where energy transfer is concerned.

Of particular importance is the way the ball responds in that instant when it hits bottom and stops. Hard rubber balls are better suited for this purpose than soft ones, because the harder the rubber, the greater the tendency of the molecules to experience only elastic deformation. What this means is that the spacing between molecules changes, yet their overall position does not.

If, however, the molecules change positions, this causes them to slide against one another, which produces friction and reduces the energy that goes into the bounce. Once the internal friction reaches a certain threshold, the ball is "dead"—that is, unable to bounce. The deader the ball is, the more its kinetic energy turns into heat upon impact with the floor, and the less energy remains for bouncing upward.

VARIETIES OF ENERGY IN ACTION

The preceding illustration makes several references to the conversion of kinetic energy to thermal energy, but it should be stressed that there are only three fundamental varieties of energy: potential, kinetic, and rest. Though heat is often discussed as a form unto itself, this is done only because the topic of heat or thermal energy is complex: in fact, thermal energy is simply a result of the kinetic energy between molecules.

To draw a parallel, most languages permit the use of only three basic subject-predicate constructions: first person ("I"), second person ("you"), and third person ("he/she/it.") Yet within these are endless varieties such as singular and plural nouns or various temporal orientations of verbs: present ("I go"); present perfect ("I have gone"); simple past ("I went"); past perfect ("I had gone.") There are even "moods," such as the subjunctive or hypothetical, which permit the construction of complex thoughts such as "I would have gone." Yet for all this variety in terms of sentence pattern—actually, a degree of variety much greater than for that of energy types—all subject-predicate constructions can still be identified as first, second, or third person.

One might thus describe thermal energy as a manifestation of energy, rather than as a discrete form. Other such manifestations include electromagnetic (sometimes divided into electrical and magnetic), sound, chemical, and nuclear. The principles governing most of these are similar: for instance, the positive or negative attraction between two electromagnetically charged particles is analogous to the force of gravity.

MECHANICAL ENERGY. One term not listed among manifestations of energy is mechanical energy, which is something different altogether: the sum of potential and kinetic energy. A dropped or bouncing ball was used as a convenient illustration of interactions within a larger system of mechanical energy, but the example could just as easily have been a roller coaster, which, with its ups and downs, quite neatly illustrates the sliding scale of kinetic and potential energy.

Likewise, the relationship of Earth to the Sun is one of potential and kinetic energy transfers: as with the baseball and Earth itself, the planet is pulled by gravitational force toward the larger body. When it is relatively far from the Sun, it possesses a higher degree of potential energy, whereas when closer, its kinetic energy is highest. Potential and kinetic energy can also be illustrated within the realm of electromagnetic, as opposed to gravitational, force: when a nail is some distance from a magnet, its potential energy is high, but as it moves toward the magnet, kinetic energy increases.

ENERGY CONVERSION IN A DAM. A dam provides a beautiful illustration of energy conversion: not only from potential to kinetic, but from energy in which gravity provides the force component to energy based in electromagnetic force. A dam big enough to be used for generating hydroelectric power forms a vast steel-and-concrete curtain that holds back millions of tons of water from a river or other body. The water nearest the top—the "head" of the dam—thus has enormous potential energy.

Hydroelectric power is created by allowing controlled streams of this water to flow downward, gathering kinetic energy that is then transferred to powering turbines. Dams in popular vacation spots often release a certain amount of water for recreational purposes during the day. This makes it possible for rafters, kayakers, and others downstream to enjoy a relatively fast-flowing river. (Or, to put it another way, a stream with high kinetic energy.) As the day goes on, however, the sluice-gates are closed once again to build up the "head." Thus when night comes, and energy demand is relatively high as people retreat to their homes, vacation cabins, and hotels, the dam is ready to provide the power they need.

OTHER MANIFESTATIONS OF ENERGY. Thermal and electromagnetic energy are much more readily recognizable manifestations of energy, yet sound and chemical energy are two forms that play a significant part as well. Sound, which is essentially nothing more than the series of pressure fluctuations within a medium such as air, possesses enormous energy: consider the example of a singer hitting a certain note and shattering a glass.

Contrary to popular belief, the note does not have to be particularly high: rather, the note should be on the same wavelength as the glass's own vibrations. When this occurs, sound energy is transferred directly to the glass, which is shattered by this sudden net intake of energy. Sound waves can be much more destructive than that: not only can the sound of very loud music cause permanent damage to the ear drums, but also, sound waves of certain frequencies and decibel levels can actually drill through steel. Indeed, sound is not just a by-product of an explosion; it is part of the destructive force.

As for chemical energy, it is associated with the pull that binds together atoms within larger molecular structures. The formation of water molecules, for instance, depends on the chemical bond between hydrogen and oxygen atoms. The combustion of materials is another example of chemical energy in action.

With both chemical and sound energy, however, it is easy to show how these simply reflect the larger structure of potential and kinetic energy discussed earlier. Hence sound, for instance, is potential energy when it emerges from a source, and becomes kinetic energy as it moves toward a receiver (for example, a human ear). Furthermore, the molecules in a combustible material contain enormous chemical potential energy, which becomes kinetic energy when released in a fire.

REST ENERGY AND ITS NUCLEAR MANIFESTATION

Nuclear energy is similar to chemical energy, though in this instance, it is based on the binding of particles within an atom and its nucleus. But it is also different from all other kinds of energy, because its force component is neither gravitational nor electromagnetic, but based on one of two other known varieties of force: strong nuclear and weak nuclear. Furthermore, nuclear energy—to a much greater extent than thermal or chemical energy—involves not only kinetic and potential energy, but also the mysterious, extraordinarily powerful, form known as rest energy.

Throughout this discussion, there has been little mention of rest energy; yet it is ever-present. Kinetic and potential energy rise and fall with respect to one another; but rest energy changes little. In the baseball illustration, for instance, the ball had the same rest energy at the top of the building as it did in flight—the same rest energy, in fact, that it had when sitting on the ground. And its rest energy is enormous.

NUCLEAR WARFARE. This brings back the subject of the rest energy formula: $E = mc^2$, famous because it made possible the creation of the atomic bomb. The latter, which fortunately has been detonated in warfare only twice in history, brought a swift end to World War II when the United States unleashed it against Japan in August 1945. From the beginning, it was clear that the atom bomb possessed staggering power, and that it would forever change the way nations conducted their affairs in war and peace.

Yet the atom bomb involved only nuclear fission, or the splitting of an atom, whereas the hydrogen bomb that appeared just a few years after the end of World War II used an even more powerful process, the nuclear fusion of atoms. Hence, the hydrogen bomb upped the ante to a much greater extent, and soon the two nuclear superpowers—the United States and the Soviet Union—possessed the power to destroy most of the life on Earth.

The next four decades were marked by a superpower struggle to control "the bomb" as it came to be known—meaning any and all nuclear weapons. Initially, the United States controlled all atomic secrets through its heavily guarded Manhattan Project, which created the bombs used against Japan. Soon, however, spies such as Julius and Ethel Rosenberg provided the Soviets with U.S. nuclear secrets, ensuring that the dictatorship of Josef Stalin would possess nuclear capabilities as well. (The Rosenbergs were executed for treason, and their alleged innocence became a celebrated cause among artists and intellectuals; however, Soviet documents released since the collapse of the Soviet empire make it clear that they were guilty as charged.)

KEY TERMS

CONSERVATION OF ENERGY: A law of physics which holds that within a system isolated from all other outside factors, the total amount of energy re-mains the same, though transformations of energy from one form to another take place.

COSINE: For an acute (less than 90°) in a right triangle, the cosine (abbreviated cos) is the ratio between the adjacent leg and the hypotenuse. Regardless of the size of the triangle, this figure is a constant for any particular angle.

ENERGY: The ability of an object (or in some cases a non-object, such as a magnetic force field) to accomplish work.

FRICTION: The force that resists motion when the surface of one object comes into contact with the surface of another.

HORSEPOWER: The British unit of power, equal to 550 foot-pounds per second.

HYPOTENUSE: In a right triangle, the side opposite the right angle.

JOULE: The SI measure of work. One joule (1 J) is equal to the work required to accelerate 1 kilogram of mass by 1 meter per second squared (1 m/s^2) over a distance of 1 meter. Due to the small size of the joule, however, it is often replaced by the kilowatt-hour, equal to 3.6 million (3.6 · 10^6) J.

KINETIC ENERGY: The energy that an object possesses by virtue of its motion.

MATTER: Physical substance that occupies space, has mass, is composed of atoms (or in the case of subatomic particles, is part of an atom), and is convertible into energy.

MECHANICAL ENERGY: The sum of potential energy and kinetic energy within a system.

POTENTIAL ENERGY: The energy that an object possesses by virtue of its position.

POWER: The rate at which work is accomplished over time, a figure rendered mathematically as work divided by time.

Both nations began building up missile arsenals. It was not, however, just a matter of the United States and the Soviet Union. By the 1970s, there were at least three other nations in the "nuclear club": Britain, France, and China. There were also other countries on the verge of developing nuclear bombs, among them India and Israel. Furthermore, there was a great threat that a terrorist leader such as Libya's Muammar al-Qaddafi would acquire nuclear weapons and do the unthinkable: actually use them.

Though other nations acquired nuclear weapons, however, the scale of the two super-power arsenals dwarfed all others. And at the heart of the U.S.-Soviet nuclear competition was

a sort of high-stakes chess game—to use a metaphor mentioned frequently during the 1970s. Soviet leaders and their American counterparts both recognized that it would be the end of the world if either unleashed their nuclear weapons; yet each was determined to be able to meet the other's ever-escalating nuclear threat.

United States President Ronald Reagan earned harsh criticism at home for his nuclear buildup and his hard line in negotiations with Soviet President Mikhail Gorbachev; but as a result of this one-upmanship, he put the Soviets into a position where they could no longer compete. As they put more and more money into nuclear weapons, they found themselves less and

The SI unit of power is the watt, while the British unit is the foot-pound per second. The latter, because it is small, is usually reckoned in terms of horsepower.

REST ENERGY: The energy an object possesses by virtue of its mass.

RIGHT TRIANGLE: A triangle that includes a right (90°) angle. The other two angles are, by definition, acute or less than 90°.

SCALAR: A quantity that possesses only magnitude, with no specific direction.

SI: An abbreviation of the French *Système International d'Unités*, which means "International System of Units." This is the term within the scientific community for the entire metric system, as applied to a wide variety of quantities ranging from length, weight and volume to work and power, as well as electromagnetic units.

SYSTEM: In discussions of energy, the term "system" refers to a closed set of inter-actions free from interference by outside factors. An example is the baseball dropped from a height to illustrate potential energy and kinetic energy the ball, the space through which it falls, and the ground below together form a system.

VECTOR: A quantity that possesses both magnitude and direction.

WATT: The metric unit of power, equal to 1 joule per second. Because this is such a small unit, scientists and engineers typically speak in terms of kilowatts, or units of 1,000 watts.

WORK: The exertion of force over a given distance. Work is the product of force and distance, where force and distance are exerted in the same direction. Hence the actual formula for work is $F \cdot \cos \theta \cdot s$, where F = force, s = distance, and $\cos \theta$ is equal to the cosine of the angle θ (the Greek letter theta) between F and s. In the metric or SI system, work is measured by the joule (J), and in the British system by the foot-pound.

---

less able to uphold their already weak economic system. This was precisely Reagan's purpose in using American economic might to outspend the Soviets—or, in the case of the proposed multi-trillion-dollar Strategic Defense Initiative (SDI or "Star Wars")—threatening to outspend them. The Soviets expended much of their economic energy in competing with U.S. military strength, and this (along with a number of other complex factors), spelled the beginning of the end of the Communist empire.

$E = mc^2$. The purpose of the preceding historical brief is to illustrate the epoch-making significance of a single scientific formula: $E = mc^2$. It ended World War II and ensured that no war like it would ever happen again—but brought on the specter of global annihilation. It created a superpower struggle—yet it also ultimately helped bring about the end of Soviet totalitarianism, thus opening the way for a greater level of peace and economic and cultural exchange than the world has ever known. Yet nuclear arsenals still remain, and the nuclear threat is far from over.

So just what is this literally earth-shattering formula? E stands for rest energy, m for mass, and c for the speed of light, which is 186,000 mi (297,600 km) per second. Squared, this yields an almost unbelievably staggering number.

Hence, even an object of insignificant mass possesses an incredible amount of rest energy. The baseball, for instance, weighs only about 0.333 lb, which—on Earth, at least—converts to 0.15 kg. (The latter is a unit of mass, as opposed to weight.) Yet when factored into the rest energy equation, it yields about 3.75 billion kilowatt-hours—enough to provide an American home with enough electrical power to last it more than 156,000 years!

How can a mere baseball possess such energy? It is not the baseball in and of itself, but its mass; thus every object with mass of any kind possesses rest energy. Often, mass energy can be released in very small quantities through purely thermal or chemical processes: hence, when a fire burns, an almost infinitesimal portion of the matter that went into making the fire is converted into energy. If a stick of dynamite that weighed 2.2 lb (1 kg) exploded, the portion of it that "disappeared" would be equal to 6 parts out of 100 billion; yet that portion would cause a blast of considerable proportions.

As noted much earlier, the derivation of Einstein's formula—and, more to the point, how he came to recognize the fundamental principles involved—is far beyond the scope of this essay. What is important is the fact, hypothesized by Einstein and confirmed in subsequent experiments, that matter is convertible to energy, a fact that becomes apparent when matter is accelerated to speeds close to that of light.

Physicists do not possess a means for propelling a baseball to a speed near that of light—or of controlling its behavior and capturing its energy. Instead, atomic energy—whether of the wartime or peacetime varieties (that is, in power plants)—involves the acceleration of mere atomic particles. Nor is any atom as good as another. Typically physicists use uranium and other extremely rare minerals, and often, they further process these minerals in highly specialized ways. It is the rarity and expense of those minerals, incidentally—not the difficulty of actually putting atomic principles to work—that has kept smaller nations from developing their own nuclear arsenals.

WHERE TO LEARN MORE

Beiser, Arthur. *Physics,* 5th ed. Reading, MA: Addison-Wesley, 1991.

Berger, Melvin. *Sound, Heat and Light: Energy at Work.* Illustrated by Anna DiVito. New York: Scholastic, 1992.

Gardner, Robert. *Energy Projects for Young Scientists.* New York: F. Watts, 1987.

"Kinetic and Potential Energy" Thinkquest (Web site). <http://library.thinkquest.org/2745/data/ke.htm> (March 12, 2001).

Snedden, Robert. *Energy.* Des Plaines, IL: Heinemann, Library, 1999.

Suplee, Curt. *Everyday Science Explained.* Washington, D.C.: National Geographic Society, 1996.

"Work and Energy" (Web site). <http://www.glenbrook.k12.il.us/gbssci/phys/Class/energy/energtoc.html> (March 12, 2001).

World of Coasters (Web site). <http://www.worldofcoasters.com> (March 12, 2001).

Zubrowski, Bernie. *Raceways: Having Fun with Balls and Tracks.* Illustrated by Roy Doty. New York: Morrow, 1985.

THERMODYNAMICS

GAS LAWS

CONCEPT

Gases respond more dramatically to temperature and pressure than do the other three basic types of matter (liquids, solids and plasma). For gases, temperature and pressure are closely related to volume, and this allows us to predict their behavior under certain conditions. These predictions can explain mundane occurrences, such as the fact that an open can of soda will soon lose its fizz, but they also apply to more dramatic, life-and-death situations.

HOW IT WORKS

Ordinary air pressure at sea level is equal to 14.7 pounds per square inch, a quantity referred to as an atmosphere (atm). Because a pound is a unit of force and a kilogram a unit of mass, the metric equivalent is more complex in derivation. A newton (N), or 0.2248 pounds, is the metric unit of force, and a pascal (Pa)—1 newton per square meter—the unit of pressure. Hence, an atmosphere, expressed in metric terms, is 1.013×10^5 Pa.

GASES VS. SOLIDS AND LIQ-UIDS: A STRIKINGLY DIFFERENT RESPONSE

Regardless of the units you use, however, gases respond to changes in pressure and temperature in a remarkably different way than do solids or liquids. Using a small water sample, say, 0.2642 gal (1 l), an increase in pressure from 1-2 atm will decrease the volume of the water by less than 0.01%. A temperature increase from 32° to 212°F (0 to 100°C) will increase its volume by only 2% The response of a solid to these changes is even

less dramatic; however, the reaction of air (a combination of oxygen, nitrogen, and other gases) to changes in pressure and temperature is radically different.

For air, an equivalent temperature increase would result in a volume increase of 37%, and an equivalent pressure increase will decrease the volume by a whopping 50%. Air and other gases also have a boiling point below room temperature, whereas the boiling point for water is higher than room temperature and that of solids is much higher. The reason for this striking difference in response can be explained by comparing all three forms of matter in terms of their overall structure, and in terms of their molecular behavior. (Plasma, a gas-like state found, for instance, in stars and comets' tails, does not exist on Earth, and therefore it will not be included in the comparisons that follow.)

MOLECULAR STRUCTURE DETER-MINES REACTION

Solids possess a definite volume and a definite shape, and are relatively noncompressible: for instance, if you apply extreme pressure to a steel plate, it will bend, but not much. Liquids have a definite volume, but no definite shape, and tend to be noncompressible. Gases, on the other hand, possess no definite volume or shape, and are compressible.

At the molecular level, particles of solids tend to be definite in their arrangement and close in proximity—indeed, part of what makes a solid "solid," in the everyday meaning of that term, is the fact that its constituent parts are basically immovable. Liquid molecules, too, are close in proximity, though random in arrangement. Gas molecules, too, are random in arrangement, but tend to be more widely spaced than liquid molecules. Solid particles are slow moving, and have a strong attraction to one another, whereas gas particles are fast-moving, and have little or no attraction. (Liquids are moderate in both regards.)

Given these interesting characteristics of gases, it follows that a unique set of parameters—collectively known as the "gas laws"—are needed to describe and predict their behavior. Most of the gas laws were derived during the eighteenth and nineteenth centuries by scientists whose work is commemorated by the association of their names with the laws they discovered. These men include the English chemists Robert Boyle (1627-1691), John Dalton (1766-1844), and William Henry (1774-1836); the French physicists and chemists J. A. C. Charles (1746-1823) and Joseph Gay-Lussac (1778-1850), and the Italian physicist Amedeo Avogadro (1776-1856).

BOYLE'S, CHARLES'S, AND GAY-LUSSAC'S LAWS

Boyle's law holds that in isothermal conditions (that is, a situation in which temperature is kept constant), an inverse relationship exists between the volume and pressure of a gas. (An inverse relationship is a situation involving two variables, in which one of the two increases in direct proportion to the decrease in the other.) In this case, the greater the pressure, the less the volume and vice versa. Therefore the product of the volume multiplied by the pressure remains constant in all circumstances.

Charles's law also yields a constant, but in this case the temperature and volume are allowed to vary under isobarometric conditions—that is, a situation in which the pressure remains the same. As gas heats up, its volume increases, and when it cools down, its volume reduces accordingly. Hence, Charles established that the ratio of temperature to volume is constant.

By now a pattern should be emerging: both of the aforementioned laws treat one parameter (temperature in Boyle's, pressure in Charles's) as unvarying, while two other factors are treated as variables. Both in turn yield relationships between the two variables: in Boyle's law, pressure and volume are inversely related, whereas in Charles's law, temperature and volume are directly related.

In Gay-Lussac's law, a third parameter, volume, is treated as a constant, and the result is a constant ratio between the variables of pressure and temperature. According to Gay-Lussac's law, the pressure of a gas is directly related to its absolute temperature.

Absolute temperature refers to the Kelvin scale, established by William Thomson, Lord Kelvin (1824-1907). Drawing on Charles's discovery that gas at 0°C (32°F) regularly contracted by about 1/273 of its volume for every Celsius degree drop in temperature, Thomson derived the value of absolute zero (-273.15°C or -459.67°F). Using the Kelvin scale of absolute temperature, Gay-Lussac found that at lower temperatures, the pressure of a gas is lower, while at higher temperatures its pressure is higher. Thus, the ratio of pressure to temperature is a constant.

AVOGADRO'S LAW

Gay-Lussac also discovered that the ratio in which gases combine to form compounds can be expressed in whole numbers: for instance, water is composed of one part oxygen and two parts hydrogen. In the language of modern science, this would be expressed as a relationship between molecules and atoms: one molecule of water contains one oxygen atom and two hydrogen atoms.

In the early nineteenth century, however, scientists had yet to recognize a meaningful distinction between atoms and molecules. Avogadro was the first to achieve an understanding of the difference. Intrigued by the whole-number relationship discovered by Gay-Lussac, Avogadro reasoned that one liter of any gas must contain the same number of particles as a liter of another gas. He further maintained that gas consists of particles—which he called molecules—that in turn consist of one or more smaller particles.

In order to discuss the behavior of molecules, it was necessary to establish a large quantity as a basic unit, since molecules themselves are very small. For this purpose, Avogadro established the mole, a unit equal to 6.022137×10^{23} (more than 600 billion trillion) molecules. The term "mole" can be used in the same way we use the word "dozen." Just as "a dozen" can refer to twelve cakes or twelve chickens, so "mole" always describes the same number of molecules.

Just as one liter of water, or one liter of mercury, has a certain mass, a mole of any given substance has its own particular mass, expressed in grams. The mass of one mole of iron, for instance, will always be greater than that of one mole of oxygen. The ratio between them is exactly the same as the ratio of the mass of one iron atom to one oxygen atom. Thus the mole makes if possible to compare the mass of one element or one compound to that of another.

Avogadro's law describes the connection between gas volume and number of moles. According to Avogadro's law, if the volume of gas is increased under isothermal and isobarometric conditions, the number of moles also increases. The ratio between volume and number of moles is therefore a constant.

THE IDEAL GAS LAW

Once again, it is easy to see how Avogadro's law can be related to the laws discussed earlier, since they each involve two or more of the four parameters: temperature, pressure, volume, and quantity of molecules (that is, number of moles). In fact, all the laws so far described are brought together in what is known as the ideal gas law, sometimes called the combined gas law.

The ideal gas law can be stated as a formula, $pV = nRT$, where p stands for pressure, V for volume, n for number of moles, and T for temperature. R is known as the universal gas constant, a figure equal to 0.0821 atm · liter/mole · K. (Like most terms in physics, this one is best expressed in metric rather than English units.)

Given the equation $pV = nRT$ and the fact that R is a constant, it is possible to find the value of any one variable—pressure, volume, number of moles, or temperature—as long as one knows the value of the other three. The ideal gas law also makes it possible to discern certain relations: thus if a gas is in a relatively cool state, the product of its pressure and volume is proportionately low; and if heated, its pressure and volume product increases correspondingly. Thus

$$\frac{p_1 V_1}{T_1} = \frac{p_2 V_2}{T_2},$$

where $p_1 V_1$ is the product of its initial pressure and its initial volume, T_1 its initial temperature,

A FIRE EXTINGUISHER CONTAINS A HIGH-PRESSURE MIXTURE OF WATER AND CARBON DIOXIDE THAT RUSHES OUT OF THE SIPHON TUBE, WHICH IS OPENED WHEN THE RELEASE VALVE IS DEPRESSED. *(Photograph by Craig Lovell/ Corbis. Reproduced by permission.)*

$p_2 V_2$ the product of its final volume and final pressure, and T_2 its final temperature.

FIVE POSTULATES REGARDING THE BEHAVIOR OF GASES

Five postulates can be applied to gases. These more or less restate the terms of the earlier discussion, in which gases were compared to solids and liquids; however, now those comparisons can be seen in light of the gas laws.

First, the size of gas molecules is minuscule in comparison to the distance between them, making gas highly compressible. In other words, there is a relatively high proportion of empty space between gas molecules.

Second, there is virtually no force attracting gas molecules to one another.

Third, though gas molecules move randomly, frequently colliding with one another, their net effect is to create uniform pressure.

A HOT-AIR BALLOON FLOATS BECAUSE THE AIR INSIDE IT IS NOT AS DENSE THAN THE AIR OUTSIDE. THE WAY IN WHICH THE DENSITY OF THE AIR IN THE BALLOON IS REDUCED REFLECTS THE GAS LAWS. *(Duomo/Corbis. Reproduced by permission.)*

Fourth, the elastic nature of the collisions results in no net loss of kinetic energy, the energy that an object possesses by virtue of its motion. If a stone is dropped from a height, it rapidly builds kinetic energy, but upon hitting a nonelastic surface such as pavement, most of that kinetic energy is transferred to the pavement. In the case of two gas molecules colliding, however, they simply bounce off one another, only to collide with other molecules and so on, with no kinetic energy lost.

Fifth, the kinetic energy of all gas molecules is directly proportional to the absolute temperature of the gas.

LAWS OF PARTIAL PRESSURE

Two gas laws describe partial pressure. Dalton's law of partial pressure states that the total pressure of a gas is equal to the sum of its partial pressures—that is, the pressure exerted by each component of the gas mixture. As noted earlier, air is composed mostly of nitrogen and oxygen. Along with these are small components carbon dioxide and gases collectively known as the rare or noble gases: argon, helium, krypton, neon, radon, and xenon. Hence, the total pressure of a given quantity of air is equal to the sum of the pressures exerted by each of these gases.

Henry's law states that the amount of gas dissolved in a liquid is directly proportional to the partial pressure of the gas above the surface of the solution. This applies only to gases such as oxygen and hydrogen that do not react chemically to liquids. On the other hand, hydrochloric acid will ionize when introduced to water: one or more of its electrons will be removed, and its atoms will convert to ions, which are either positive or negative in charge.

REAL-LIFE APPLICATIONS

PRESSURE CHANGES

OPENING A SODA CAN. Inside a can or bottle of carbonated soda is carbon dioxide gas (CO_2), most of which is dissolved in the drink itself. But some of it is in the space (sometimes referred to as "head space") that makes up the difference between the volume of the soft drink and the volume of the container.

At the bottling plant, the soda manufacturer adds high-pressure carbon dioxide to the head space in order to ensure that more CO_2 will be absorbed into the soda itself. This is in accordance with Henry's law: the amount of gas (in this case CO_2) dissolved in the liquid (soda) is directly proportional to the partial pressure of

the gas above the surface of the solution—that is, the CO_2 in the head space. The higher the pressure of the CO_2 in the head space, the greater the amount of CO_2 in the drink itself; and the greater the CO_2 in the drink, the greater the "fizz" of the soda.

Once the container is opened, the pressure in the head space drops dramatically. Once again, Henry's law indicates that this drop in pressure will be reflected by a corresponding drop in the amount of CO_2 dissolved in the soda. Over a period of time, the soda will release that gas, and will eventually go "flat."

FIRE EXTINGUISHERS. A fire extinguisher consists of a long cylinder with an operating lever at the top. Inside the cylinder is a tube of carbon dioxide surrounded by a quantity of water, which creates pressure around the CO_2 tube. A siphon tube runs vertically along the length of the extinguisher, with one opening near the bottom of the water. The other end opens in a chamber containing a spring mechanism attached to a release valve in the CO_2 tube.

The water and the CO_2 do not fill the entire cylinder: as with the soda can, there is "head space," an area filled with air. When the operating lever is depressed, it activates the spring mechanism, which pierces the release valve at the top of the CO_2 tube. When the valve opens, the CO_2 spills out in the "head space," exerting pressure on the water. This high-pressure mixture of water and carbon dioxide goes rushing out of the siphon tube, which was opened when the release valve was depressed. All of this happens, of course, in a fraction of a second—plenty of time to put out the fire.

AEROSOL CANS. Aerosol cans are similar in structure to fire extinguishers, though with one important difference. As with the fire extinguisher, an aerosol can includes a nozzle that depresses a spring mechanism, which in turn allows fluid to escape through a tube. But instead of a gas cartridge surrounded by water, most of the can's interior is made up of the product (for instance, deodorant), mixed with a liquid propellant.

The "head space" of the aerosol can is filled with highly pressurized propellant in gas form, and in accordance with Henry's law, a corresponding proportion of this propellant is dissolved in the product itself. When the nozzle is depressed, the pressure of the propellant forces the product out through the nozzle.

A propellant, as its name implies, propels the product itself through the spray nozzle when the latter is depressed. In the past, chlorofluorocarbons (CFCs)—manufactured compounds containing carbon, chlorine, and fluorine atoms—were the most widely used form of propellant. Concerns over the harmful effects of CFCs on the environment, however, has led to the development of alternative propellants, most notably hydrochlorofluorocarbons (HCFCs), CFC-like compounds that also contain hydrogen atoms.

WHEN THE TEMPERATURE CHANGES

A number of interesting things, some of them unfortunate and some potentially lethal, occur when gases experience a change in temperature. In these instances, it is possible to see the gas laws—particularly Boyle's and Charles's—at work.

There are a number of examples of the disastrous effects that result from an increase in the temperature of a product containing combustible gases, as with natural gas and petroleum-based products. In addition, the pressure on the gases in aerosol cans makes the cans highly explosive—so much so that discarded cans at a city dump may explode on a hot summer day. Yet there are other instances when heating a gas can produce positive effects.

A hot-air balloon, for instance, floats because the air inside it is not as dense than the air outside. By itself, this fact does not depend on any of the gas laws, but rather reflects the concept of buoyancy. However, the way in which the density of the air in the balloon is reduced does indeed reflect the gas laws.

According to Charles's law, heating a gas will increase its volume. Also, as noted in the first and second propositions regarding the behavior of gases, gas molecules are highly nonattractive to one another, and therefore, there is a great deal of space between them. The increase in volume makes that space even greater, leading to a significant difference in density between the air in the balloon and the air outside. As a result, the balloon floats, or becomes buoyant.

Although heating a gas can be beneficial, cooling a gas is not always a wise idea. If someone were to put a bag of potato chips into a freezer, thinking this would preserve their flavor, he would be in for a disappointment. Much of what maintains the flavor of the chips is the pressurization of the bag, which ensures a consistent internal environment in which preservative chemicals, added during the manufacture of the chips, can keep them fresh. Placing the bag in the freezer causes a reduction in pressure, as per Gay-Lussac's law, and the bag ends up a limp version of its earlier self.

Propane tanks and tires offer an example of the pitfalls that may occur by either allowing a gas to heat up or cool down by too much. Because most propane tanks are made according to strict regulations, they are generally safe, but it is not entirely inconceivable that an extremely hot summer day could cause a defective tank to burst. Certainly the laws of physics are there: an increase in temperature leads to an increase in pressure, in accordance with Gay-Lussac's law, and could lead to an explosion.

Because of the connection between heat and pressure, propane trucks on the highways during the summer are subjected to weight tests to ensure that they are not carrying too much of the gas. On the other hand, a drastic reduction in temperature could result in a loss in gas pressure. If a propane tank from Florida were transported by truck during the winter to northern Canada, the pressure would be dramatically reduced by the time it reached its destination.

GAS REACTIONS THAT MOVE AND STOP A CAR

In operating a car, we experience two examples of gas laws in operation. One of these, common to everyone, is that which makes the car run: the combustion of gases in the engine. The other is, fortunately, a less frequent phenomenon—but it can and does save lives. This is the operation of an air bag, which, though it is partly related to laws of motion, depends also on the behaviors explained in Charles's law.

With regard to the engine, when the driver pushes down on the accelerator, this activates a throttle valve that sprays droplets of gasoline mixed with air into the engine. (Older vehicles used a carburetor to mix the gasoline and air, but most modern cars use fuel-injection, which sprays the air-gas combination without requiring an intermediate step.) The mixture goes into the

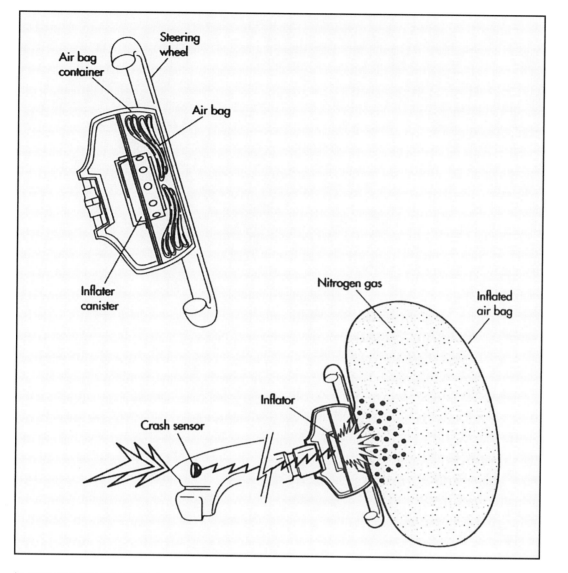

IN CASE OF A CAR COLLISION, A SENSOR TRIGGERS THE AIR BAG TO INFLATE RAPIDLY WITH NITROGEN GAS. BEFORE YOUR BODY REACHES THE BAG, HOWEVER, IT HAS ALREADY BEGUN DEFLATING. *(Illustration by Hans & Cassidy. The Gale Group.)*

cylinder, where the piston moves up, compressing the gas and air.

While the mixture is still compressed (high pressure, high density), an electric spark plug produces a flash that ignites it. The heat from this controlled explosion increases the volume of air, which forces the piston down into the cylinder. This opens an outlet valve, causing the piston to rise and release exhaust gases.

As the piston moves back down again, an inlet valve opens, bringing another burst of gasoline-air mixture into the chamber. The piston, whose downward stroke closed the inlet valve, now shoots back up, compressing the gas and air to repeat the cycle. The reactions of the gasoline

and air are what move the piston, which turns a crankshaft that causes the wheels to rotate.

So much for moving—what about stopping? Most modern cars are equipped with an airbag, which reacts to sudden impact by inflating. This protects the driver and front-seat passenger, who, even if they are wearing seatbelts, may otherwise be thrown against the steering wheel or dashboard..

But an airbag is much more complicated than it seems. In order for it to save lives, it must deploy within 40 milliseconds (0.04 seconds). Not only that, but it has to begin deflating before the body hits it. An airbag does not inflate if a car simply goes over a bump; it only operates in sit-

KEY TERMS

ABSOLUTE TEMPERATURE: Temperature in relation to absolute zero (-273.15°C or -459.67°F). Its unit is the Kelvin (K), named after William Thomson, Lord Kelvin (1824-1907), who created the scale. The Kelvin and Celsius scales are directly related; hence, Celsius temperatures can be converted to Kelvins (for which neither the word or symbol for "degree" are used) by adding 273.15.

AVOGADRO'S LAW: A statement, derived by the Italian physicist Amedeo Avogadro (1776-1856), which holds that as the volume of gas increases under isothermal and isobarometric conditions, the number of molecules (expressed in terms of mole number), increases as well. Thus the ratio of volume to mole number is a constant.

BOYLE'S LAW: A statement, derived by English chemist Robert Boyle (1627-1691), which holds that for gases in isothermal conditions, an inverse relationship exists between the volume and pressure of a gas. This means that the greater the pressure, the less the volume and vice versa, and therefore the product of pressure multiplied by volume yields a constant figure.

CHARLES'S LAW: A statement, derived by French physicist and chemist J. A. C. Charles (1746-1823), which holds that for gases in isobarometric conditions, the ratio between the volume and temperature of a gas is constant. This means that the greater the temperature, the greater the volume and vice versa.

DALTON'S LAW OF PARTIAL PRESSURE: A statement, derived by the English chemist John Dalton (1766-1844), which holds that the total pressure of a gas is equal to the sum of its partial pres-

uations when the vehicle experiences extreme deceleration. When this occurs, there is a rapid transfer of kinetic energy to rest energy, as with the earlier illustration of a stone hitting concrete. And indeed, if you were to smash against a fully inflated airbag, it would feel like hitting concrete—with all the expected results.

The airbag's sensor contains a steel ball attached to a permanent magnet or a stiff spring. The spring holds it in place through minor mishaps in which an airbag would not be warranted—for instance, if a car were simply to be "tapped" by another in a parking lot. But in a case of sudden deceleration, the magnet or spring releases the ball, sending it down a smooth bore. It flips a switch, turning on an electrical circuit. This in turn ignites a pellet of sodium azide, which fills the bag with nitrogen gas.

The events described in the above illustration take place within 40 milliseconds—less time than it takes for your body to come flying forward; and then the airbag has to begin deflating before the body reaches it. At this point, the highly pressurized nitrogen gas molecules begin escaping through vents. Thus as your body hits the bag, the deflation of the latter is moving it in the same direction that your body is going—only much, much more slowly. Two seconds after impact, which is an eternity in terms of the processes involved, the pressure inside the bag has returned to 1 atm.

WHERE TO LEARN MORE

Beiser, Arthur. *Physics,* 5th ed. Reading, MA: Addison-Wesley, 1991.

"*Chemistry Units: Gas Laws.*" (Web site). <http://bio.bio.rpi.edu/MS99/ausemaW/chem/gases.hmtl> (February 21, 2001).

sures—that is, the pressure exerted by each component of the gas mixture.

GAY-LUSSAC'S LAW: A statement, derived by the French physicist and chemist Joseph Gay-Lussac (1778-1850), which holds that the pressure of a gas is directly related to its absolute temperature. Hence the ratio of pressure to absolute temperature is a constant.

HENRY'S LAW: A statement, derived by the English chemist William Henry (1774-836), which holds that the amount of gas dissolved in a liquid is directly proportional to the partial pressure of the gas above the solution. This holds true only for gases, such as hydrogen and oxygen, that are capable of dissolving in water without undergoing ionization.

IDEAL GAS LAW: A proposition, also known as the combined gas law, that draws on all the gas laws. The ideal gas law can be expressed as the formula $pV = nRT$, where p stands for pressure, V for volume, n for number of moles, and T for temperature. R is known as the universal gas constant, a figure equal to 0.0821 atm · liter/mole · K.

INVERSE RELATIONSHIP: A situation involving two variables, in which one of the two increases in direct proportion to the decrease in the other.

IONIZATION: A reaction in which an atom or group of atoms loses one or more electrons. The atoms are then converted to ions, which are either wholly positive or negative in charge.

ISOTHERMAL: Referring to a situation in which temperature is kept constant.

ISOBAROMETRIC: Referring to a situation in which pressure is kept constant.

MOLE: A unit equal to 6.022137×10^{23} molecules.

Laws of Gases. New York: Arno Press, 1981.

Macaulay, David. *The New Way Things Work.* Boston: Houghton Mifflin, 1998.

Mebane, Robert C. and Thomas R. Rybolt. *Air and Other Gases.* Illustrations by Anni Matsick. New York: Twenty-First Century Books, 1995.

"Tutorials—6." <http://www.chemistrycoach.com/tutorials-6.html> (February 21, 2001).

MOLECULAR DYNAMICS

CONCEPT

Physicists study matter and motion, or matter in motion. These forms of matter may be large, or they may be far too small to be seen by the most high-powered microscopes available. Such is the realm of molecular dynamics, the study and simulation of molecular motion. As its name suggests, molecular dynamics brings in aspects of dynamics, the study of why objects move as they do, as well as thermodynamics, the study of the relationships between heat, work, and energy. Existing at the borders between physics and chemistry, molecular dynamics provides understanding regarding the properties of matter—including phenomena such as the liquefaction of gases, in which one phase of matter is transformed into another.

HOW IT WORKS

Molecules

The physical world is made up of matter, physical substance that has mass; occupies space; is composed of atoms; and is, ultimately, convertible to energy. On Earth, three principal phases of matter exist, namely solid, liquid, and gas. The differences between these three are, on the surface at least, easily perceivable. Clearly, water is a liquid, just as ice is a solid and steam a gas. Yet, the ways in which various substances convert between phases are often complex, as are the interrelations between these phases. Ultimately, understanding of the phases depends on an awareness of what takes place at the molecular level.

An atom is the smallest particle of a chemical element. It is not, however, the smallest thing in the universe; atoms are composed of subatomic particles, including protons, neutrons, and electrons. These subatomic particles are discussed in the context of the structure of matter elsewhere in this volume, where they are examined largely with regard to their electromagnetic properties. In the present context, the concern is primarily with the properties of atomic and molecular particles, in terms of mechanics, the study of bodies in motion, and thermodynamics.

An atom must, by definition, represent one and only one chemical element, of which 109 have been identified and named. It should be noted that the number of elements changes with continuing research, and that many of the elements, particularly those discovered relatively recently—as, for instance, meitnerium (No. 109), isolated in the 1990s—are hardly part of everyday experience. So, perhaps 100 would be a better approximation; in any case, consider the multitude of possible ways in which the elements can be combined.

Musicians have only seven tones at their disposal, and artists only seven colors—yet they manage to create a seemingly infinite variety of mutations in sound and sight, respectively. There are only 10 digits in the numerical system that has prevailed throughout the West since the late Middle Ages, yet it is possible to use that system to create such a range of numbers that all the books in all the libraries in the world could not contain them. This gives some idea of the range of combinations available using the hundred-odd chemical elements nature has provided—in other words, the number of possible molecular combinations that exist in the universe.

THIS HUGE LIQUEFIED NATURAL GAS CONTAINER WILL BE INSTALLED ON A SHIP. THE VOLUME OF THE LIQUEFIED GAS IS FAR LESS THAN IT WOULD BE IF THE GAS WERE IN A VAPORIZED STATE, THUS ENABLING EASE AND ECONOMY IN TRANSPORT. *(Photograph by James L. Amos/Corbis. Reproduced by permission.)*

THE STRUCTURE OF MOLE-CULES. A molecule is a group of atoms joined in a single structure. Often, these atoms come from different elements, in which case the molecule represents a particular chemical compound, such as water, carbon dioxide, sodium chloride (salt), and so on. On the other hand, a molecule may consist only of one type of atom: oxygen molecules, for instance, are formed by the joining of two oxygen atoms.

As much as scientists understand about molecules and their structure, there is much that they do not know. Molecules of water are fairly easy to understand, because they have a simple, regular structure that does not change. A water molecule is composed of one oxygen atom joined by two hydrogen atoms, and since the oxygen atom is much larger than the two hydrogens, its shape can be compared to a basketball with two soft-balls attached. The scale of the molecule, of course, is so small as to boggle the mind: to borrow an illustration from American physicist Richard Feynman (1918-1988), if a basketball were blown up to the size of Earth, the molecules inside of it would not even be as large as an ordinary-sized basketball.

As for the water molecule, scientists know a number of things about it: the distance between the two hydrogen atoms (measured in units called an angstrom), and even the angle at which they join the oxygen atom. In the case of salt, however, the molecular structure is not nearly as uniform as that of water: atoms join together, but not always in regular ways. And then there are compounds far more complex than water or salt, involving numerous elements that fit together in precise and complicated ways. But, once that discussion is opened, one has stepped from the realm of physics into that of chemistry, and that is not the intention here. Rather, the purpose of the foregoing and very cursory discussion of molecular structure is to point out that molecules are at the heart of all physical existence—and that the things we cannot see are every bit as complicated as those we can.

THE MOLE. Given the tiny—to use an understatement—size of molecules, how do scientists analyze their behavior? Today, physicists have at their disposal electron microscopes and other advanced forms of equipment that make it possible to observe activity at the atomic and molecular levels. The technology that makes this possible is beyond the scope of the present dis-

HOW SMALL ARE MOLECULES? IF THIS BASKETBALL WERE BLOWN UP TO THE SIZE OF EARTH, THE MOLECULES INSIDE IT WOULD NOT BE AS BIG AS A REAL BASKETBALL. *(Photograph by Dimitri Iundt/Corbis. Reproduced by permission.)*

cussion. On the other hand, consider a much simpler question: how do physicists weigh molecules?

Obviously "a bunch" of iron (an element known by the chemical symbol Fe) weighs more than "a bunch" of oxygen, but what exactly is "a bunch"? Italian physicist Amedeo Avogadro (1776-1856), the first scientist to clarify the distinction between atoms and molecules, created a unit that made it possible to compare the masses of various molecules. This is the mole, also known as "Avogadro's number," a unit equal to 6.022137×10^{23} (more than 600 billion trillion) molecules.

The term "mole" can be used in the same way that the word "dozen" is used. Just as "a dozen" can refer to twelve cakes or twelve chickens, so "mole" always describes the same number of molecules. A mole of any given substance has its own particular mass, expressed in grams. The mass of one mole of iron, for instance, will always be greater than that of one mole of oxygen. The ratio between them is exactly the same as the ratio of the mass of one iron atom to one oxygen atom. Thus, the mole makes it possible to compare the mass of one element or compound to that of another.

MOLECULAR ATTRACTION AND MOTION

Molecular dynamics can be understood primarily in terms of the principles of motion, identified by Sir Isaac Newton (1642-1727), principles that receive detailed discussion at several places in this volume. However, the attraction between particles at the atomic and molecular level cannot be explained by reference to gravitational force, also identified by Newton. For more than a century, gravity was the only type of force known to physicists, yet the pull of gravitation alone was too weak to account for the strong pull between atoms and molecules.

During the eighteenth century and early nineteenth centuries, however, physicists and other scientists became increasingly aware of another form of interaction at work in the world—one that could not be explained in gravitational terms. This was the force of electricity and magnetism, which Scottish physicist James Clerk Maxwell (1831-1879) suggested were different manifestations of a "new" kind of force, electromagnetism. All subatomic particles possess either a positive, negative, or neutral electrical charge. An atom usually has a neutral charge, meaning that it is composed of an equal number of protons (positive) and electrons (negative). In certain situations, however, it may lose one or more electrons and, thus, acquire a net charge, making it an ion.

Positive and negative charges attract one another, much as the north and south poles of two different magnets attract. (In fact, magnetism is simply an aspect of electromagnetic force.) Not only do the positive and negative elements of an atom attract one another, but positive elements in atoms attract negative elements in other atoms, and vice versa. These interactions are much more complex than the preceding discussion suggests, of course; the important point is that a force other than gravitation draws matter together at the atomic and molecular levels. On the other hand, the interactions that are critical to the study of molecular dynamics are primarily mechanical, comprehensible from the standpoint of Newtonian dynamics.

MOLECULAR BEHAVIOR AND PHASES OF MATTER. All molecules are in motion, and the rate of that motion is affected by the attraction between them. This attraction or repulsion can be though of like a spring connecting two molecules, an analogy that works best for solids, but in a limited way for liquids. Most molecular motion in liquids and gases is caused by collisions with other molecules; even in solids, momentum is transferred from one molecule to the next along the "springs," but ultimately the motion is caused by collisions. Hence molecular collisions provide the mechanism by which heat is transferred between two bodies in contact.

The rate at which molecules move in relation to one another determines phase of matter—that is, whether a particular item can be described as solid, liquid, or gas. The movement of molecules means that they possess kinetic energy, or the energy of movement, which is manifested as thermal energy and measured by temperature. Temperature is really nothing more than molecules in motion, relative to one another: the faster they move, the greater the kinetic energy, and the greater the temperature.

When the molecules in a material move slowly in relation to one another, they tend to be close in proximity, and hence the force of attraction between them is strong. Such a material is called a solid. In molecules of liquid, by contrast, the rate of relative motion is higher, so the molecules tend to be a little more spread out, and

therefore the force between them is weaker. A material substance whose molecules move at high speeds, and therefore exert little attraction toward one another, is known as a gas. All forms of matter possess a certain (very large) amount of energy due to their mass; thermal energy, however, is—like phase of matter—a function of the attractions between particles. Hence, solids generally have less energy than liquids, and liquids less energy than gases.

REAL-LIFE APPLICATIONS

KINETIC THEORIES OF MATTER

English chemist John Dalton (1766-1844) was the first to recognize that nature is composed of tiny particles. In putting forward his idea, Dalton adopted a concept from the Greek philosopher Democritus (c. 470-380 B.C.), who proposed that matter is made up of tiny units he called atomos, or "indivisible."

Dalton recognized that the structure of atoms in a particular element or compound is uniform, and maintained that compounds are made up of compound atoms: in other words, that water, for instance, is composed of "water atoms." Soon after Dalton, however, Avogadro clarified the distinction between atoms and molecules. Neither Dalton nor Avogadro offered much in the way of a theory regarding atomic or molecular behavior; but another scientist had already introduced the idea that matter at the smallest levels is in a constant state of motion.

This was Daniel Bernoulli (1700-1782), a Swiss mathematician and physicist whose studies of fluids—a term which encompasses both gases and liquids—provided a foundation for the field of fluid mechanics. (Today, Bernoulli's principle, which relates the velocity and pressure of fluids, is applied in the field of aerodynamics, and explains what keeps an airplane aloft.) Bernoulli published his fluid mechanics studies in *Hydrodynamica* (1700-1782), a work in which he provided the basis for what came to be known as the kinetic theory of gases.

BROWNIAN MOTION. Because he came before Dalton and Avogadro, and, thus, did not have the benefit of their atomic and molecular theories, Bernoulli was not able to develop his kinetic theory beyond the seeds of an idea. The subsequent elaboration of kinetic theory, which is applied not only to gases but (with somewhat less effectiveness) to liquids and solids, in fact, resulted from an accidental discovery.

In 1827, Scottish botanist Robert Brown (1773-1858) was studying pollen grains under a microscope, when he noticed that the grains underwent a curious zigzagging motion in the water. The pollen assumed the shape of a colloid, a pattern that occurs when particles of one substance are dispersed—but not dissolved—in another substance. Another example of a colloidal pattern is a puff of smoke.

At first, Brown assumed that the motion had a biological explanation—that is, that it resulted from life processes within the pollen—but later, he discovered that even pollen from long-dead plants behaved in the same way. He never understood what he was witnessing. Nor did a number of other scientists, who began noticing other examples of what came to be known as Brownian motion: the constant but irregular zigzagging of colloidal particles, which can be seen clearly through a microscope.

MAXWELL, BOLTZMANN, AND THE MATURING OF KINETIC THEORY. A generation after Brown's time, kinetic theory came to maturity through the work of Maxwell and Austrian physicist Ludwig E. Boltzmann (1844-1906). Working independently, the two men developed a theory, later dubbed the Maxwell-Boltzmann theory of gases, which described the distribution of molecules in a gas. In 1859, Maxwell described the distribution of molecular velocities, work that became the foundation of statistical mechanics—the study of large systems—by examining the behavior of their smallest parts.

A year later, in 1860, Maxwell published a paper in which he presented the kinetic theory of gases: the idea that a gas consists of numerous molecules, relatively far apart in space, which interact by colliding. These collisions, he proposed, are responsible for the production of thermal energy, because when the velocity of the molecules increases—as it does after collision—the temperature increases as well. Eight years later, in 1868, Boltzmann independently applied statistics to the kinetic theory, explaining the behavior of gas molecules by means of what would come to be known as statistical mechanics.

Kinetic theory offered a convincing explanation of the processes involved in Brownian motion. According to the kinetic view, what Brown observed had nothing to do with the pollen particles; rather, the movement of those particles was simply the result of activity on the part of the water molecules. Pollen grains are many thousands of times larger than water molecules, but since there are so many molecules in even one drop of water, and their motion is so constant but apparently random, the water molecules are bound to move a pollen grain once every few thousand collisions.

In 1905, Albert Einstein (1879-1955) analyzed the behavior of particles subjected to Brownian motion. His work, and the confirmation of his results by French physicist Jean Baptiste Perrin (1870-1942), finally put an end to any remaining doubts concerning the molecular structure of matter. The kinetic explanation of molecular behavior, however, remains a theory.

KINETIC THEORY AND GASES

Maxwell's and Boltzmann's work helped explain characteristics of matter at the molecular level, but did so most successfully with regard to gases. Kinetic theory fits with a number of behaviors exhibited by gases: their tendency to fill any container by expanding to fit its interior, for instance, and their ability to be easily compressed.

This, in turn, concurs with the gas laws (discussed in a separate essay titled "Gas Laws")—for instance, Boyle's law, which maintains that pressure decreases as volume increases, and vice versa. Indeed, the ideal gas law, which shows an inverse relationship between pressure and volume, and a proportional relationship between temperature and the product of pressure and volume, is an expression of kinetic theory.

THE GAS LAWS ILLUSTRATED. The operations of the gas laws are easy to visualize by means of kinetic theory, which portrays gas molecules as though they were millions upon billions of tiny balls colliding at random. Inside a cube-shaped container of gas, molecules are colliding with every possible surface, but the net effect of these collisions is the same as though the molecules were divided into thirds, each third colliding with opposite walls inside the cube.

If the cube were doubled in size, the molecules bouncing back and forth between two sets of walls would have twice as far to travel between each collision. Their speed would not change, but the time between collisions would double, thus, cutting in half the amount of pressure they would exert on the walls. This is an illustration of Boyle's law: increasing the volume by a factor of two leads to a decrease in pressure to half of its original value.

On the other hand, if the size of the container were decreased, the molecules would have less distance to travel from collision to collision. This means they would be colliding with the walls more often, and, thus, would have a higher degree of energy—and, hence, a higher temperature. This illustrates another gas law, Charles's law, which relates volume to temperature: as one of the two increases or decreases, so does the other. Thus, it can be said, in light of kinetic theory, that the average kinetic energy produced by the motions of all the molecules in a gas is proportional to the absolute temperature of the gas.

GASES AND ABSOLUTE TEMPERATURE. The term "absolute temperature" refers to the Kelvin scale, established by William Thomson, Lord Kelvin (1824-1907). Drawing on Charles's discovery that gas at 0°C (32°F) regularly contracts by about 1/273 of its volume for every Celsius degree drop in temperature, Thomson derived the value of absolute zero (-273.15°C or -459.67°F). The Kelvin and Celsius scales are directly related; hence, Celsius temperatures can be converted to Kelvins by adding 273.15.

The Kelvin scale measures temperature in relation to absolute zero, or 0K. (Units in the Kelvin system, known as Kelvins, do not include the word or symbol for degree.) But what is absolute zero, other than a very cold temperature? Kinetic theory provides a useful definition: the temperature at which all molecular movement in a gas ceases. But this definition requires some qualification.

First of all, the laws of thermodynamics show the impossibility of actually reaching absolute zero. Second, the vibration of atoms never completely ceases: rather, the vibration of the average atom is zero. Finally, one element—helium—does not freeze, even at temperatures near absolute zero. Only the application of pressure will push helium past the freezing point.

CHANGES OF PHASE

Kinetic theory is more successful when applied to gases than to liquids and solids, because liquid and solid molecules do not interact nearly as frequently as gas particles do. Nonetheless, the proposition that the internal energy of any substance—gas, liquid, or solid—is at least partly related to the kinetic energies of its molecules helps explain much about the behavior of matter.

The thermal expansion of a solid, for instance, can be clearly explained in terms of kinetic theory. As discussed in the essay on elasticity, many solids are composed of crystals, regular shapes composed of molecules joined to one another, as though on springs. A spring that is pulled back, just before it is released, is an example of potential energy: the energy that an object possesses by virtue of its position. For a crystalline solid at room temperature, potential energy and spacing between molecules are relatively low. But as temperature increases and the solid expands, the space between molecules increases—as does the potential energy in the solid.

An example of a liquid displaying kinetic behavior is water in the process of vaporization. The vaporization of water, of course, occurs in boiling, but water need not be anywhere near the boiling point to evaporate. In either case, the process is the same. Speeds of molecules in any substance are distributed along a curve, meaning that a certain number of molecules have speeds well below, or well above, the average. Those whose speeds are well above the average have enough energy to escape the surface, and once they depart, the average energy of the remaining liquid is less than before. As a result, evaporation leads to cooling. (In boiling, of course, the continued application of thermal energy to the entire water sample will cause more molecules to achieve greater energy, even as highly energized molecules leave the surface of the boiling water as steam.)

THE PHASE DIAGRAM

The vaporization of water is an example of a change of phase—the transition from one phase of matter to another. The properties of any substance, and the points at which it changes phase, are plotted on what is known as a phase diagram. The latter typically shows temperature along the x-axis, and pressure along the y-axis. It is also possible to construct a phase diagram that plots volume against temperature, or volume against pressure, and there are even three-dimensional phase diagrams that measure the relationship between all three—volume, pressure, and temperature. Here we will consider the simpler two-dimensional diagram we have described.

For simple substances such as water and carbon dioxide, the solid form of the substance appears at a relatively low temperature, and at pressures anywhere from zero upward. The line between solids and liquids, indicating the temperature at which a solid becomes a liquid at any pressure above a certain level, is called the fusion curve. Though it appears to be a line, it is indeed curved, reflecting the fact that at high pressures, a solid well below the normal freezing point for that substance may be melted to create a liquid.

Liquids occupy the area of the phase diagram corresponding to relatively high temperatures and high pressures. Gases or vapors, on the other hand, can exist at very low temperatures, but only if the pressure is also low. Above the melting point for the substance, gases exist at higher pressures and higher temperatures. Thus, the line between liquids and gases often looks almost like a 45° angle. But it is not a straight line, as its name, the vaporization curve, indicates. The curve of vaporization reflects the fact that at relatively high temperatures and high pressures, a substance is more likely to be a gas than a liquid.

CRITICAL POINT AND SUBLIMATION. There are several other interesting phenomena mapped on a phase diagram. One is the critical point, which can be found at a place of very high temperature and pressure along the vaporization curve. At the critical point, high temperatures prevent a liquid from remaining a liquid, no matter how high the pressure. At the same time, the pressure causes gas beyond that point to become more and more dense, but due to the high temperatures, it does not condense into a liquid. Beyond the critical point, the substance cannot exist in anything other than the gaseous state. The temperature component of the critical point for water is 705.2°F (374°C)—at 218 atm, or 218 times ordinary atmospheric pressure. For helium, however, critical temperature is just a few degrees above absolute zero. This is why helium is rarely seen in forms other than a gas.

There is also a certain temperature and pressure, called the triple point, at which some substances—water and carbon dioxide are examples—will be a liquid, solid, and gas all at once. Another interesting phenomenon is the sublimation curve, or the line between solid and gas. At certain very low temperatures and pressures, a substance may experience sublimation, meaning that a gas turns into a solid, or a solid into a gas, without passing through a liquid stage. A well-known example of sublimation occurs when "dry ice," which is made of carbon dioxide, vaporizes at temperatures above (-109.3°F [-78.5°C]). Carbon dioxide is exceptional, however, in that it experiences sublimation at relatively high pressures, such as those experienced in everyday life: for most substances, the sublimation point occurs at such a low pressure point that it is seldom witnessed outside of a laboratory.

LIQUEFACTION OF GASES

One interesting and useful application of phase change is the liquefaction of gases, or the change of gas into liquid by the reduction in its molecular energy levels. There are two important properties at work in liquefaction: critical temperature and critical pressure. Critical temperature is that temperature above which no amount of pressure will cause a gas to liquefy. Critical pressure is the amount of pressure required to liquefy the gas at critical temperature.

Gases are liquefied by one of three methods: (1) application of pressure at temperatures below critical; (2) causing the gas to do work against external force, thus, removing its energy and changing it to the liquid state; or (3) causing the gas to do work against some internal force. The second option can be explained in terms of the operation of a heat engine, as explored in the Thermodynamics essay.

In a steam engine, an example of a heat engine, water is boiled, producing energy in the form of steam. The steam is introduced to a cylinder, in which it pushes on a piston to drive some type of machinery. In pushing against the piston, the steam loses energy, and as a result, changes from a gas back to a liquid.

As for the use of internal forces to cool a gas, this can be done by forcing the vapor through a small nozzle or porous plug. Depending on the temperature and properties of the gas, such an

ABSOLUTE ZERO: The temperature, defined as 0K on the Kelvin scale, at which the motion of molecules in a solid virtually ceases. Absolute zero is equal to -459.67°F (-273.15°C).

ATOM: The smallest particle of a chemical element. An atom can exist either alone or in combination with other atoms in a molecule.

BROWNIAN MOTION: The constant but irregular zigzagging of colloidal particles, which can be seen clearly through a microscope. The phenomenon is named after Scottish botanist Robert Brown (1773-1858), who first witnessed it but was not able to explain it. The behavior exhibited in Brownian motion provides evidence for the kinetic theory of matter.

CHANGE OF PHASE: The transition from one phase of matter to another.

CHEMICAL COMPOUND: A substance made up of atoms of more than one chemical element. These atoms are usually joined in molecules.

CHEMICAL ELEMENT: A substance made up of only one kind of atom.

COLLOID: A pattern that occurs when particles of one substance are dispersed—but not dissolved—in another substance. A puff of smoke in the air is an example of a colloid, whose behavior is typically characterized by Brownian motion.

CRITICAL POINT: A coordinate, plotted on a phase diagram, above which a substance cannot exist in anything other than the gaseous state. Located at a position of very high temperature and pressure, the critical point marks the termination of the vaporization curve.

DYNAMICS: The study of why objects move as they do. Dynamics is an element of mechanics.

FLUID: Any substance, whether gas or liquid, which tends to flow, and which conforms to the shape of its container. Unlike solids, fluids are typically uniform in molecular structure: for instance, one molecule of water is the same as another water molecule.

FUSION CURVE: The boundary between solid and liquid for any given substance, as plotted on a phase diagram.

GAS: A phase of matter in which molecules exert little or no attraction toward one another, and, therefore, move at high speeds.

HEAT: Internal thermal energy that flows from one body of matter to another.

KELVIN SCALE: Established by William Thomson, Lord Kelvin (1824-1907), the Kelvin scale measures temperature in relation to absolute zero, or 0K. (Units in the Kelvin system, known as Kelvins, do not include the word or symbol for degree.) The Kelvin and Celsius scales are directly related; hence, Celsius temperatures can be converted to Kelvins by adding 273.15.

KINETIC ENERGY: The energy that an object possesses by virtue of its motion.

KINETIC THEORY OF GASES: The idea that a gas consists of numerous molecules, relatively far apart in space, which interact by colliding. These collisions are responsible for the production of thermal energy, because when the velocity of the molecules increases—as it does after collision—the temperature increases as well.

KINETIC THEORY OF MATTER: The application of the kinetic theory of gases to all forms of matter. Since particles of liquids and solids move much more slowly than do gas particles, kinetic theory is not as successful in this regard; however, the proposition that the internal energy of any substance is at least partly related to the kinetic energies of its molecules helps explain much about the behavior of matter.

LIQUID: A phase of matter in which molecules exert moderate attractions toward one another, and, therefore, move at moderate speeds.

MATTER: Physical substance that has mass; occupies space; is composed of atoms; and is ultimately convertible to energy. There are several phases of matter, including solids, liquids, and gases.

MECHANICS: The study of bodies in motion.

MOLE: A unit equal to 6.022137×10^{23} (more than 600 billion trillion) molecules. Since their size makes it impossible to weigh molecules in relatively small quantities; hence, the mole, devised by Italian physicist Amedeo Avogadro (1776-1856), facilitates comparisons of mass between substances.

MOLECULAR DYNAMICS: The study and simulation of molecular motion.

MOLECULE: A group of atoms, usually of more than one chemical element, joined in a structure.

PHASE DIAGRAM: A chart, plotted for any particular substance, identifying the particular phase of matter for that substance at a given temperature and pressure level. A phase diagram usually shows temperature along the x-axis, and pressure along the y-axis.

PHASES OF MATTER: The various forms of material substance (matter),

operation may be enough to remove energy sufficient for liquefaction to take place. Sometimes, the process must be repeated before the gas fully condenses into a liquid.

HISTORICAL BACKGROUND. Like the steam engine itself, the idea of gas liquefaction is a product of the early Industrial Age. One of the pioneering figures in the field was the brilliant English physicist Michael Faraday (1791-1867), who liquefied a number of high-critical temperature gases, such as carbon dioxide.

Half a century after Faraday, French physicist Louis Paul Cailletet (1832-1913) and Swiss chemist Raoul Pierre Pictet (1846-1929) developed the nozzle and porous-plug methods of liquefaction. This, in turn, made it possible to liq-

uefy gases with much lower critical temperatures, among them oxygen, nitrogen, and carbon monoxide.

By the end of the nineteenth century, physicists were able to liquefy the gases with the lowest critical temperatures. James Dewar of Scotland (1842-1923) liquefied hydrogen, whose critical temperature is -399.5°F (-239.7°C). Some time later, Dutch physicist Heike Kamerlingh Onnes (1853-1926) successfully liquefied the gas with the lowest critical temperature of them all: helium, which, as mentioned earlier, becomes a gas at almost unbelievably low temperatures. Its critical temperature is -449.9°F (-267.7°C), or just 5.3K.

APPLICATIONS OF GAS LIQUEFACTION. Liquefied natural gas (LNG) which are defined primarily in terms of the behavior exhibited by their atomic or molecular structures. On Earth, three principal phases of matter exist, namely solid, liquid, and gas.

POTENTIAL ENERGY: The energy an object possesses by virtue of its position.

SOLID: A phase of matter in which molecules exert strong attractions toward one another, and, therefore, move slowly.

STATISTICAL MECHANICS: A realm of the physical sciences devoted to the study of large systems by examining the behavior of their smallest parts.

SUBLIMATION CURVE: The boundary between solid and gas for any given substance, as plotted on a phase diagram.

SYSTEM: In physics, the term "system" usually refers to any set of physical interactions isolated from the rest of the universe. Anything outside of the system, including all factors and forces irrelevant to a discussion of that system, is known as the environment.

TEMPERATURE: A measure of the average kinetic energy—or molecular translational energy in a system. Differences in temperature determine the direction of internal energy flow between two systems when heat is being transferred.

THERMAL ENERGY: Heat energy, a form of kinetic energy produced by the movement of atomic or molecular particles. The greater the movement of these particles, the greater the thermal energy.

THERMODYNAMICS: The study of the relationships between heat, work, and energy.

VAPORIZATION CURVE: The boundary between liquid and gas for any given substance as plotted on a phase diagram.

and liquefied petroleum gas (LPG), the latter a mixture of by-products obtained from petroleum and natural gas, are among the examples of liquefied gas in daily use. In both cases, the volume of the liquefied gas is far less than it would be if the gas were in a vaporized state, thus enabling ease and economy in transport.

Liquefied gases are used as heating fuel for motor homes, boats, and homes or cabins in remote areas. Other applications of liquefied gases include liquefied oxygen and hydrogen in rocket engines, and liquefied oxygen and petroleum used in welding. The properties of liquefied gases also figure heavily in the science of producing and studying low-temperature environments. In addition, liquefied helium is used in studying the behavior of matter at temperatures close to absolute zero.

A "NEW" FORM OF MATTER?

Physicists at a Colorado laboratory in 1995 revealed a highly interesting aspect of atomic motion at temperatures approaching absolute zero. Some 70 years before, Einstein had predicted that, at extremely low temperatures, atoms would fuse to form one large "superatom." This hypothesized structure was dubbed the Bose-Einstein Condensate after Einstein and Satyendranath Bose (1894-1974), an Indian physicist whose statistical methods contributed to the development of quantum theory.

Because of its unique atomic structure, the Bose-Einstein Condensate has been dubbed a "new" form of matter. It represents a quantum mechanical effect, relating to a cutting-edge area of physics devoted to studying the properties of subatomic particles and the interaction of matter with radiation. Thus it is not directly related to molecular dynamics; nonetheless, the Bose-Einstein Condensate is mentioned here as an example of the exciting work being performed at a level beyond that addressed by molecular dynamics. Its existence may lead to a greater understanding of quantum mechanics, and on an everyday level, the "superatom" may aid in the design of smaller, more powerful computer chips.

WHERE TO LEARN MORE

Cooper, Christopher. *Matter.* New York: DK Publishing, 1999.

"Kinetic Theory of Gases: A Brief Review" University of Virginia Department of Physics (Web site). <http://www.phys.virginia.edu/classes/252/kinetic_theory.html> (April 15, 2001).

"The Kinetic Theory Page" (Web site). <http://comp.uark.edu/~jgeabana/mol_dyn/> (April 15, 2001).

Medoff, Sol and John Powers. *The Student Chemist Explores Atoms and Molecules.* Illustrated by Nancy Lou Gahan. New York: R. Rosen Press, 1977.

"Molecular Dynamics" (Web site). <http://www.biochem.vt.edu/courses/modeling/molecular_dynamics.html> (April 15, 2001).

"Molecular Simulation Molecular Dynamics Page" (Web site). <http://www.phy.bris.ac.uk/research/theory/simulation/md.html> (April 15, 2001).

Santrey, Laurence. *Heat.* Illustrated by Lloyd Birmingham. Mahwah, NJ: Troll Associates, 1985.

Strasser, Ben. *Molecules in Motion.* Illustrated by Vern Jorgenson. Pasadena, CA: Franklin Publications, 1967.

Van, Jon. *"U.S. Scientists Create a 'Superatom.'"* *Chicago Tribune,* July 14, 1995, p. 3.

STRUCTURE OF MATTER

CONCEPT

The physical realm is made up of matter. On Earth, matter appears in three clearly defined forms—solid, liquid, and gas—whose visible and perceptible structure is a function of behavior that takes place at the molecular level. Though these are often referred to as "states" of matter, it is also useful to think of them as phases of matter. This terminology serves as a reminder that any one substance can exist in any of the three phases. Water, for instance, can be ice, liquid, or steam; given the proper temperature and pressure, it may be solid, liquid, and gas all at once! But the three definite earthbound states of matter are not the sum total of the material world: in outer space a fourth phase, plasma, exists—and there may be still other varieties in the physical universe.

HOW IT WORKS

MATTER AND ENERGY

Matter can be defined as physical substance that has mass; occupies space; is composed of atoms; and is ultimately convertible to energy. A significant conversion of matter to energy, however, occurs only at speeds approaching that of the speed of light, a fact encompassed in the famous statement formulated by Albert Einstein (1879-1955), $E = mc^2$.

Einstein's formula means that every item possesses a quantity of energy equal to its mass multiplied by the squared speed of light. Given the fact that light travels at 186,000 mi (297,600 km) per second, the quantities of energy available from even a tiny object traveling at that speed are massive indeed. This is the basis for both nuclear power and nuclear weaponry, each of which uses some of the smallest particles in the known universe to produce results that are both amazing and terrifying.

The forms of matter that most people experience in their everyday lives, of course, are traveling at speeds well below that of the speed of light. Even so, transfers between matter and energy take place, though on a much, much smaller scale. For instance, when a fire burns, only a tiny fraction of its mass is converted to energy. The rest is converted into forms of mass different from that of the wood used to make the fire. Much of it remains in place as ash, of course, but an enormous volume is released into the atmosphere as a gas so filled with energy that it generates not only heat but light. The actual mass converted into energy, however, is infinitesimal.

CONSERVATION AND CONVERSION. The property of energy is, at all times and at all places in the physical universe, conserved. In physics, "to conserve" something means "to result in no net loss of" that particular component—in this case, energy. Energy is never destroyed: it simply changes form. Hence, the conservation of energy, a law of physics stating that within a system isolated from all other outside factors, the total amount of energy remains the same, though transformations of energy from one form to another take place.

Whereas energy is perfectly conserved, matter is only approximately conserved, as shown with the example of the fire. Most of the matter from the wood did indeed turn into more mat-

AN ICEBERG FLOATS BECAUSE THE DENSITY OF ICE IS LOWER THAN WATER, WHILE ITS VOLUME IS GREATER, MAKING THE ICEBERG BUOYANT. *(Photograph by Ric Engenbright/Corbis. Reproduced by permission.)*

ter—that is, vapor and ash. Yet, as also noted, a tiny quantity of matter—too small to be perceived by the senses—turned into energy.

The conservation of mass holds that total mass is constant, and is unaffected by factors such as position, velocity, or temperature, in any system that does not exchange any matter with its environment. This, however, is a qualified statement: at speeds well below c (the speed of light), it is essentially true, but for matter approaching c and thus, turning into energy, it is not.

Consider an item of matter moving at the speed of 100 mi (160 km)/sec. This is equal to 360,000 MPH (576,000 km/h) and in terms of the speeds to which humans are accustomed, it seems incredibly fast. After all, the fastest any human beings have ever traveled was about 25,000 MPH (40,000 km/h), in the case of the astronauts aboard *Apollo 11* in May 1969, and the speed under discussion is more than 14 times greater. Yet 100 mi/sec is a snail's pace compared to c: in fact, the proportional difference between an actual snail's pace and the speed of a human walking is not as great. Yet even at this leisurely gait, equal to 0.00054c, a portion of mass equal to 0.0001% (one-millionth of the total mass) converts to energy.

MATTER AT THE ATOMIC LEVEL

In his brilliant work *Six Easy Pieces*, American physicist Richard Feynman (1918-1988) asked his readers, "If, in some cataclysm, all of scientific knowledge were to be destroyed, and only one sentence passed on to the next generations of creatures, what statement would contain the most information in the fewest words? I believe it is the atomic hypothesis (or the atomic fact, or whatever you wish to call it) that all things are made of atoms—little articles that move around in perpetual motion, attracting each other when they are a little distance apart, but repelling upon being squeezed into one another. In that sentence, you will see, there is an enormous amount of information about the world, if just a little imagination and thinking are applied."

Feynman went on to offer a powerful series of illustrations concerning the size of atoms relative to more familiar objects: if an apple were magnified to the size of Earth, for instance, the atoms in it would each be about the size of a regular apple. Clearly atoms and other atomic particles are far too small to be glimpsed by even the most highly powered optical microscope. Yet, it is the behavior of particles at the atomic level that defines the shape of the entire physical world. Viewed from this perspective, it becomes easy to understand how and why matter is convertible to energy. Likewise, the interaction between atoms and other particles explains why some types of matter are solid, others liquid, and still others, gas.

ATOMS AND MOLECULES. An atom is the smallest particle of a chemical element. It is not, however, the smallest particle in the universe; atoms are composed of subatomic particles, including protons, neutrons, and electrons. But at the subatomic level, it is meaningless to refer to, for instance, "an oxygen electron": electrons are just electrons. An atom, then, is the fundamental unit of matter. Most of the substances people encounter in the world, however, are not pure elements, such as oxygen or iron; they are chemical compounds, in which atoms of more than one element join together to form molecules.

One of the most well-known molecular forms in the world is water, or H_2O, composed of two hydrogen atoms and one oxygen atom. The arrangement is extremely precise and never varies: scientists know, for instance, that the two hydrogen atoms join the oxygen atom (which is much larger than the hydrogen atoms) at an angle of 105° 3'. Other molecules are much more complex than those of water—some of them much, much more complex, which is reflected in the sometimes unwieldy names required to identify their chemical components.

ATOMIC AND MOLECULAR THEORY. The idea of atoms is not new. More than 24 centuries ago, the Greek philosopher Democritus (c. 470-380 B.C.) proposed that matter is composed of tiny particles he called *atomos*, or "indivisible." Democritus was not, however, describing matter in a concrete, scientific way: his "atoms" were idealized, philosophical constructs, not purely physical units.

Yet, he came amazingly close to identifying the fundamental structure of physical reality—much closer than any number of erroneous theories (such as the "four elements" of earth, air, fire, and water) that prevailed until modern times. English chemist John Dalton (1766-1844) was the first to identify what Feynman later called the "atomic hypothesis": that nature is composed of tiny particles. In putting forward his idea, Dalton adopted Democritus's word "atom" to describe these basic units.

Dalton recognized that the structure of atoms in a particular element or compound is uniform. He maintained that compounds are made up of compound atoms: in other words, that water, for instance, is a compound of "water atoms." Water, however, is not an element, and thus, it was necessary to think of its atomic composition in a different way—in terms of molecules rather than atoms. Dalton's contemporary Amedeo Avogadro (1776-1856), an Italian physicist, was the first scientist to clarify the distinction between atoms and molecules.

THE MOLE. Obviously, it is impractical to weigh a single molecule, or even several thousand; what was needed, then, was a number large enough to make possible practical comparisons of mass. Hence, the mole, a quantity equal to "Avogadro's number." The latter, named after Avogadro though not derived by him, is equal to 6.022137×10^{23} (more than 600 billion trillion) molecules.

The term "mole" can be used in the same way that the word "dozen" is used. Just as "a dozen" can refer to twelve cakes or twelve chickens, so "mole" always describes the same number of molecules. A mole of any given substance has its own particular mass, expressed in grams. The mass of one mole of iron, for instance, will always be greater than that of one mole of oxygen. The ratio between them is exactly the same as the ratio of the mass of one iron atom to one oxygen atom. Thus, the mole makes it possible to compare the mass of one element or compound to that of another.

BROWNIAN MOTION AND KINETIC THEORY. Contemporary to both Dalton and Avogadro was Scottish naturalist Robert Brown (1773-1858), who in 1827 stumbled upon a curious phenomenon. While studying pollen grains under a microscope, Brown noticed that the grains underwent a curious zigzagging motion in the water. At first, he assumed that the motion had a biological explanation—that is, it resulted from life processes within the pollen—but later he discovered that even pollen from long-dead plants behaved in the same way.

Brown never understood what he was witnessing. Nor did a number of other scientists, who began noticing other examples of what came to be known as Brownian motion: the constant but irregular zigzagging of particles in a puff of smoke, for instance. Later, however, Scottish physicist James Clerk Maxwell (1831-1879) and others were able to explain it by what came to be known as the kinetic theory of matter.

The kinetic theory, which is discussed in depth elsewhere in this book, is based on the idea that molecules are constantly in motion: hence, the water molecules were moving the pollen grains Brown observed. Pollen grains are many thousands of times as large as water molecules, but there are so many molecules in just one drop of water, and their motion is so constant but apparently random, that they are bound to move a pollen grain once every few thousand collisions.

GROWTH IN UNDERSTANDING THE ATOM. Einstein, who was born the year Maxwell died, published a series of papers in which he analyzed the behavior of particles subjected to Brownian motion. His work, and the confirmation of his results by French physicist Jean Baptiste Perrin (1870-1942), finally put an end to any remaining doubts concerning the molecular structure of matter.

It may seem amazing that the molecular and atomic ideas were still open to question in the early twentieth century; however, the vast majority of what is known today concerning the atom emerged after World War I. At the end of the nineteenth century, scientists believed the atom to be indivisible, but growing evidence concerning electrical charges in atoms brought with it the awareness that there must be something smaller creating those charges.

Eventually, physicists identified protons and electrons, but the neutron, with no electrical charge, was harder to discover: it was not identified until 1932. After that point, scientists were convinced that just three types of subatomic particles existed. However, subsequent activity among physicists—particularly those in the field of quantum mechanics—led to the discovery of other elementary particles, such as the photon. However, in this discussion, the only subatomic particles whose behavior is reviewed are the proton, electron, and neutron.

MOTION AND ATTRACTION IN ATOMS AND MOLECULES

At the molecular level, every item of matter in the world is in motion. This may be easy enough to imagine with regard to air or water, since both tend to flow. But what about a piece of paper, or a glass, or a rock? In fact, all molecules are in constant motion, and depending on the particular phase of matter, this motion may vary from a mere vibration to a high rate of speed.

Molecular motion generates kinetic energy, or the energy of movement, which is manifested as heat or thermal energy. Indeed, heat is really nothing more than molecules in motion relative to one another: the faster they move, the greater the kinetic energy, and the greater the heat.

The movement of atoms and molecules is always in a straight line and at a constant velocity, unless acted upon by some outside force. In fact, the motion of atoms and molecules is constantly being interfered with by outside forces, because they are perpetually striking one another. These collisions cause changes in direction, and may lead to transfers of energy from one particle to another.

ELECTROMAGNETIC FORCE IN ATOMS. The behavior of molecules cannot be explained in terms of gravitational force. This force, and the motions associated with it, were identified by Sir Isaac Newton (1642-1727), and Newton's model of the universe seemed to answer most physical questions. Then in the late nineteenth century, Maxwell discovered a second kind of force, electromagnetism. (There are two other known varieties of force, strong and weak nuclear, which are exhibited at the subatomic level.) Electromagnetic force, rather than gravitation, explains the attraction between atoms.

Several times up to this point, the subatomic particles have been mentioned but not explained in terms of their electrical charge, which is principal among their defining characteristics. Protons have a positive electrical charge, while neutrons exert no charge. These two types of particles, which make up the vast majority of the atom's mass, are clustered at the center, or nucleus. Orbiting around this nucleus are electrons, much smaller particles which exert a negative charge.

Chemical elements are identified by the number of protons they possess. Hydrogen, first element listed on the periodic table of elements, has one proton and is thus identified as 1; carbon, or element 6, has six protons, and so on.

An atom usually has a neutral charge, meaning that it is composed of an equal number of protons or electrons. In certain situations, however, it may lose one or more electrons and thus acquire a net charge. Such an atom is called an ion. But electrical charge, like energy, is conserved, and the electrons are not "lost" when an atom becomes an ion: they simply go elsewhere.

MOLECULAR BEHAVIOR AND STATES OF MATTER. Positive and negative charges interact at the molecular level in a way that can be compared to the behavior of poles in a pair of magnets. Just as two north poles or two south poles repel one another, so like charges—two positives, or two negatives—repel. Conversely, positive and negative charges exert an attractive force on one another similar to that of a north pole and south pole in contact.

In discussing phases of matter, the attraction between molecules provides a key to distinguishing between states of matter. This is not to say one particular phase of matter is a particularly good conductor of electrical current, however.

For instance, certain solids—particularly metals such as copper—are extremely good conductors. But wood is a solid, too, and conducts electrical current poorly.

The properties of various forms of matter, viewed from the larger electromagnetic picture, are a subject far beyond the scope of this essay. In any case, the electromagnetic properties of concern in the present instance are not the ones demonstrated at a macroscopic level—that is, in view of "the big picture." Rather, the subject of the attractive force operating at the atomic or molecular levels has been introduced to show that certain types of material have a greater intermolecular attraction.

As previously stated, all matter is in motion. The relative speed of that motion, however, is a function of the attraction between molecules, which in turn defines a material according to one of the phases of matter. When the molecules in a material exert a strong attraction toward one another, they move slowly, and the material is called a solid. Molecules of liquid, by contrast, exert a moderate attraction and move at moderate speeds. A material substance whose molecules exert little or no attraction, and therefore, move at high speeds, is known as a gas.

These comparisons of molecular speed and attraction, obviously, are relative. Certainly, it is easy enough in most cases to distinguish between one phase of matter and another, but there are some instances in which they overlap. Examples of these will follow, but first it is necessary to discuss the phases of matter in the context of their behavior in everyday situations.

REAL-LIFE APPLICATIONS

FROM SOLID TO LIQUID

The attractions between particles have a number of consequences in defining the phases of matter. The strong attractive forces in solids cause its particles to be positioned close together. This means that particles of solids resist attempts to compress them, or push them together. Because of their close proximity, solid particles are fixed in an orderly and definite pattern. As a result, a solid usually has a definite volume and shape.

A crystal is a type of solid in which the constituent parts are arranged in a simple, definite geometric arrangement that is repeated in all directions. Metals, for instance, are crystalline solids. Other solids are said to be amorphous, meaning that they possess no definite shape. Amorphous solids—clay, for example—either possess very tiny crystals, or consist of several varieties of crystal mixed randomly. Still other solids, among them glass, do not contain crystals.

VIBRATIONS AND FREEZING. Because of their strong attractions to one another, solid particles move slowly, but like all particles of matter, they do move. Whereas the particles in a liquid or gas move fast enough to be in relative motion with regard to one another, however, solid particles merely vibrate from a fixed position.

This can be shown by the example of a singer hitting a certain note and shattering a glass. Contrary to popular belief, the note does not have to be particularly high: rather, the note should be on the same wavelength as the vibration of the glass. When this occurs, sound energy is transferred directly to the glass, which shatters because of the sudden net intake of energy.

As noted earlier, the attraction and motion of particles in matter has a direct effect on heat and temperature. The cooler the solid, the slower and weaker the vibrations, and the closer the particles are to one another. Thus, most types of matter contract when freezing, and their density increases. Absolute zero, or 0K on the Kelvin scale of temperature—equal to -459.67°F (-273°C)—is the point at which vibration virtually stops. Note that the vibration virtually stops, but does not stop entirely. In any event, the lowest temperature actually achieved, at a Finnish nuclear laboratory in 1993, is $2.8 \cdot 10^{-10}$K, or 0.00000000028K—still above absolute zero.

UNUSUAL CHARACTERISTICS OF ICE. The behavior of water at the freezing/melting point is interesting and exceptional. Above 39.2°F (4°C) water, like most substances, expands when heated. But between 32°F (0°C) and that temperature, however, it actually contracts. And whereas most substances become much denser with lowered temperatures, the density of water reaches its maximum at 39.2°F. Below that point, it starts to decrease again.

Not only does the density of ice begin decreasing just before freezing, but its volume increases. This is the reason ice floats: its weight is less than that of the water it has displaced, and

therefore, it is buoyant. Additionally, the buoyant qualities of ice atop very cold water explain why the top of a lake may freeze, but lakes rarely freeze solid—even in the coldest of inhabited regions.

Instead of freezing from the bottom up, as it would if ice were less buoyant than the water, the lake freezes from the top down. Furthermore, ice is a poorer conductor of heat than water, and, thus, little of the heat from the water below escapes. Therefore, the lake does not freeze completely—only a layer at the top—and this helps preserve animal and plant life in the body of water. On the other hand, the increased volume of frozen water is not always good for humans: when water in pipes freezes, it may increase in volume to the point where the pipe bursts.

MELTING. When heated, particles begin to vibrate more and more, and, therefore, move further apart. If a solid is heated enough, it loses its rigid structure and becomes a liquid. The temperature at which a solid turns into a liquid is called the melting point, and melting points are different for different substances. For the most part, however, solids composed of heavier particles require more energy—and, hence, higher temperatures—to induce the vibrations necessary for freezing. Nitrogen melts at -346°F (-210°C), ice at 32°F (0°C), and copper at 1,985°F (1,085°C). The melting point of a substance, incidentally, is the same as its freezing point: the difference is a matter of orientation—that is, whether the process is one of a solid melting to become a liquid, or of a liquid freezing to become a solid.

The energy required to change a solid to a liquid is called the heat of fusion. In melting, all the heat energy in a solid (energy that exists due to the motion of its particles) is used in breaking up the arrangement of crystals, called a lattice. This is why the water resulting from melted ice does not feel any warmer than when it was frozen: the thermal energy has been expended, with none left over for heating the water. Once all the ice is melted, however, the absorbed energy from the particles—now moving at much greater speeds than when the ice was in a solid state—causes the temperature to rise.

FROM LIQUID TO GAS

The particles of a liquid, as compared to those of a solid, have more energy, more motion, and less attraction to one another. The attraction, however, is still fairly strong: thus, liquid particles are in close enough proximity that the liquid resists compression.

On the other hand, their arrangement is loose enough that the particles tend to move around one another rather than merely vibrating in place, as solid particles do. A liquid is therefore not definite in shape. Both liquids and gases tend to flow, and to conform to the shape of their container; for this reason, they are together classified as fluids.

Owing to the fact that the particles in a liquid are not as close in proximity as those of a solid, liquids tend to be less dense than solids. The liquid phase of substance is thus inclined to be larger in volume than its equivalent in solid form. Again, however, water is exceptional in this regard: liquid water actually takes up less space than an equal mass of frozen water.

BOILING. When a liquid experiences an increase in temperature, its particles take on energy and begin to move faster and faster. They collide with one another, and at some point the particles nearest the surface of the liquid acquire enough energy to break away from their neighbors. It is at this point that the liquid becomes a gas or vapor.

As heating continues, particles throughout the liquid begin to gain energy and move faster, but they do not immediately transform into gas. The reason is that the pressure of the liquid, combined with the pressure of the atmosphere above the liquid, tends to keep particles in place. Those particles below the surface, therefore, remain where they are until they acquire enough energy to rise to the surface.

The heated particle moves upward, leaving behind it a hollow space—a bubble. A bubble is not an empty space: it contains smaller trapped particles, but its small weight relative to that of the liquid it disperses makes it buoyant. Therefore, a bubble floats to the top, releasing its trapped particles as gas or vapor. At that point, the liquid is said to be boiling.

THE EFFECT OF ATMOSPHERIC PRESSURE. As they rise, the particles thus have to overcome atmospheric pressure, and this means that the boiling point for any liquid depends in part on the pressure of the surrounding air. This is why cooking instructions often vary with altitude: the greater the distance from

sea level, the less the air pressure, and the shorter the required cooking time.

Atop Mt. Everest, Earth's highest peak at about 29,000 ft (8,839 m) above sea level, the pressure is approximately one-third normal atmospheric pressure. This means the air is one-third as dense as it is as sea level, which explains why mountain-climbers on Everest and other tall peaks must wear oxygen masks to stay alive. It also means that water boils at a much lower temperature on Everest than it does elsewhere. At sea level, the boiling point of water is 212°F (100°C), but at 29,000 ft it is reduced by one-quarter, to 158°F (70°C).

Of course, no one lives on the top of Mt. Everest—but people do live in Denver, Colorado, where the altitude is 5,577 ft (1,700 m) and the boiling point of water is 203°F (95°C). Given the lower boiling point, one might assume that food would cook faster in Denver than in New York, Los Angeles, or some other city close to sea level. In fact, the opposite is true: because heated particles escape the water so much faster at high altitudes, they do not have time to acquire the energy needed to raise the temperature of the water. It is for this reason that a recipe may contain a statement such as "at altitudes above XX feet, add XX minutes to cooking time."

If lowered atmospheric pressure means a lowered boiling point, what happens in outer space, where there is no atmospheric pressure? Liquids boil at very, very low temperatures. This is one of the reasons why astronauts have to wear pressurized suits: if they did not, their blood would boil—even though space itself is incredibly cold.

LIQUID TO GAS AND BACK AGAIN. Note that the process of a liquid changing to a gas is similar to what occurs when a solid changes to a liquid: particles gain heat and therefore energy, begin to move faster, break free from one another, and pass a certain threshold into a new phase of matter. And just as the freezing and melting point for a given substance are the same temperature, the only difference being one of orientation, the boiling point of a liquid transforming into a gas is the same as the condensation point for a gas turning into a liquid.

The behavior of water in boiling and condensation makes possible distillation, one of the principal methods for purifying seawater in various parts of the world. First, the water is boiled, then, it is allowed to cool and condense, thus forming water again. In the process, the water separates from the salt, leaving it behind in the form of brine. A similar separation takes place when salt water freezes: because salt, like most solids, has a much lower freezing point than water, very little of it remains joined to the water in ice. Instead, the salt takes the form of a briny slush.

GAS AND ITS LAWS. Having reached the gaseous state, a substance takes on characteristics quite different from those of a solid, and somewhat different from those of a liquid. Whereas liquid particles exert a moderate attraction to one another, particles in a gas exert little to no attraction. They are thus free to move, and to move quickly. The shape and arrangement of gas is therefore random and indefinite—and, more importantly, the motion of gas particles give it much greater kinetic energy than the other forms of matter found on Earth.

The constant, fast, and random motion of gas particles means that they are always colliding and thereby transferring kinetic energy back and forth without any net loss. These collisions also have the overall effect of producing uniform pressure in a gas. At the same time, the characteristics and behavior of gas particles indicate that they will tend not to remain in an open container. Therefore, in order to maintain any pressure on a gas—other than the normal atmospheric pressure exerted on the surface of the gas by the atmosphere (which, of course, is also a gas)—it is necessary to keep it in a closed container.

There are a number of gas laws (examined in another essay in this book) describing the response of gases to changes in pressure, temperature, and volume. Among these is Boyle's law, which holds that when the temperature of a gas is constant, there is an inverse relationship between volume and pressure: in other words, the greater the pressure, the less the volume, and vice versa. According to a second gas law, Charles's law, for gases in conditions of constant pressure, the ratio between volume and temperature is constant—that is, the greater the temperature, the greater the volume, and vice versa.

In addition, Gay-Lussac's law shows that the pressure of a gas is directly related to its absolute temperature on the Kelvin scale: the higher the temperature, the higher the pressure, and vice

versa. Gay-Lussac's law is combined, along with Boyle's and Charles's and other gas laws, in the ideal gas law, which makes it possible to find the value of any one variable—pressure, volume, number of moles, or temperature—for a gas, as long as one knows the value of the other three.

OTHER STATES OF MATTER

PLASMA. Principal among states of matter other than solid, liquid, and gas is plasma, which is similar to gas. (The term "plasma," when referring to the state of matter, has nothing to do with the word as it is often used, in reference to blood plasma.) As with gas, plasma particles collide at high speeds—but in plasma, the speeds are even greater, and the kinetic energy levels even higher.

The speed and energy of these collisions is directly related to the underlying property that distinguishes plasma from gas. So violent are the collisions between plasma particles that electrons are knocked away from their atoms. As a result, plasma does not have the atomic structure typical of a gas; rather, it is composed of positive ions and electrons. Plasma particles are thus electrically charged, and, therefore, greatly influenced by electrical and magnetic fields.

Formed at very high temperatures, plasma is found in stars and comets' tails; furthermore, the reaction between plasma and atomic particles in the upper atmosphere is responsible for the aurora borealis or "northern lights." Though not found on Earth, plasma—ubiquitous in other parts of the universe—may be the most plentiful among the four principal states of matter.

QUASI-STATES. Among the quasi-states of matter discussed by physicists are several terms that describe the structure in which particles are joined, rather than the attraction and relative movement of those particles. "Crystalline," "amorphous," and "glassy" are all terms to describe what may be individual states of matter; so too is "colloidal."

A colloid is a structure intermediate in size between a molecule and a visible particle, and it has a tendency to be dispersed in another medium—as smoke, for instance, is dispersed in air. Brownian motion describes the behavior of most colloidal particles. When one sees dust floating in a ray of sunshine through a window, the light reflects off colloids in the dust, which are driven back and forth by motion in the air otherwise imperceptible to the human senses.

DARK MATTER. The number of states or phases of matter is clearly not fixed, and it is quite possible that more will be discovered in outer space, if not on Earth. One intriguing candidate is called dark matter, so described because it neither reflects nor emits light, and is therefore invisible. In fact, luminous or visible matter may very well make up only a small fraction of the mass in the universe, with the rest being taken up by dark matter.

If dark matter is invisible, how do astronomers and physicists know it exists? By analyzing the gravitational force exerted on visible objects when there seems to be no visible object to account for that force. An example is the center of our galaxy, the Milky Way, which appears to be nothing more than a dark "halo." In order to cause the entire galaxy to revolve around it in the same way that planets revolve around the Sun, the Milky Way must contain a staggering quantity of invisible mass.

Dark matter may be the substance at the heart of a black hole, a collapsed star whose mass is so great that its gravitational field prevents light from escaping. It is possible, also, that dark matter is made up of neutrinos, subatomic particles thought to be massless. Perhaps, the theory goes, neutrinos actually possess tiny quantities of mass, and therefore in huge groups—a mole times a mole times a mole—they might possess appreciable mass.

In addition, dark matter may be the deciding factor as to whether the universe is infinite. The more mass the universe possesses, the greater its overall gravity, and if the mass of the universe is above a certain point, it will eventually begin to contract. This, of course, would mean that it is finite; on the other hand, if the mass is below this threshold, it will continue to expand indefinitely. The known mass of the universe is nowhere near that threshold—but, because the nature of dark matter is still largely unknown, it is not possible yet to say what effect its mass may have on the total equation.

A "NEW" FORM OF MATTER? Physicists at the Joint Institute of Laboratory Astrophysics in Boulder, Colorado, in 1995 revealed a highly interesting aspect of atomic behavior at temperatures approaching absolute zero. Some 70 years before, Einstein had predict-ed that, at extremely low temperatures, atoms would fuse to form one large "superatom." This hypothesized structure was dubbed the Bose-Einstein Condensate (BEC) after Einstein and Satyendranath Bose (1894-1974), an Indian physicist whose statistical methods contributed to the development of quantum theory.

Cooling about 2,000 atoms of the element rubidium to a temperature just 170 billionths of a degree Celsius above absolute zero, the physicists succeeded in creating an atom 100 micrometers across—still incredibly small, but vast in comparison to an ordinary atom. The superatom, which lasted for about 15 seconds, cooled down all the way to just 20 billionths of a degree above absolute zero. The Colorado physicists won the Nobel Prize in physics in 1997 for their work.

In 1999, researchers in a lab at Harvard University also created a superatom of BEC, and used it to slow light to just 38 MPH (60.8 km/h)—about 0.02% of its ordinary speed. Dubbed a "new" form of matter, the BEC may lead to a greater understanding of quantum mechanics, and may aid in the design of smaller, more powerful computer chips.

STATES AND PHASES AND IN BETWEEN

At places throughout this essay, references have been made variously to "phases" and "states" of matter. This is not intended to confuse, but rather to emphasize a particular point. Solids, liquids, and gases are referred to as "phases," because substances on Earth—water, for instance—regularly move from one phase to another. This change, a function of temperature, is called (aptly enough) "change of phase."

There is absolutely nothing incorrect in referring to "states of matter." But "phases of matter" is used in the present context as a means of emphasizing the fact that most substances, at the appropriate temperature and pressure, can be solid, liquid, or gas. In fact, a substance may even be solid, liquid, and gas.

AN ANALOGY TO HUMAN LIFE

The phases of matter can be likened to the phases of a person's life: infancy, babyhood, childhood, adolescence, adulthood, old age. The transition between these stages is indefinite, yet it is

THE RESPONSE OF LIQUID CRYSTALS TO LIGHT MAKES THEM USEFUL IN THE DISPLAYS USED ON LAPTOP COMPUT-
ERS. *(AFP/Corbis. Reproduced by permission.)*

easy enough to say when a person is at a certain stage.

At the transition point between adolescence and adulthood—say, at seventeen years old—a young person may say that she is an adult, but her parents may insist that she is still an adolescent or a child. And indeed, she might qualify as either. On the other hand, when she is thirty, it would be ridiculous to assert that she is anything other than an adult.

At the same time, a person at a certain age may exhibit behaviors typically associated with another age. A child, for instance, may behave like an adult, or an adult like a baby. One interesting example of this is the relationship between age two and late adolescence. In both cases, the person is in the process of individualizing, developing an identity separate from that of his or her parents—yet clearly, there are also plenty of differences between a two-year-old and a seventeen-year-old.

As with the transitional phases in human life, in the borderline pressure levels and temperatures for phases of matter it is sometimes difficult to say, for instance, if a substance is fully a liquid or fully a gas. On the other hand, at a certain temperature and pressure level, a substance clearly is what it is: water at very low temperature and pressure, for instance, is indisputably ice—just as an average thirty-year-old is obviously an adult. As for the second observation, that a person at one stage in life may reflect characteristics of another stage, this too is reflected in the behavior of matter.

LIQUID CRYSTALS. A liquid crystal is a substance that, over a specific range of temperature, displays properties both of a liquid and a solid. Below this temperature range, it is unquestionably a solid, and above this range it is just as obviously a liquid. In between, however, liquid crystals exhibit a strange solid-liquid behavior: like a liquid, their particles flow, but like a solid, their molecules maintain specific crystalline arrangements.

Long, wide, and placed alongside one another, liquid crystal molecules exhibit interesting properties in response to light waves. The speed of light through a liquid crystal actually varies, depending on whether the light is traveling along the short or long sides of the molecules. These differences in light speed may lead to a change in the direction of polarization, or the vibration of light waves.

ATOM: The smallest particle of a chemical element. An atom can exist either alone or in combination with other atoms in a molecule. Atoms are made up of protons, neutrons, and electrons. In most cases, the electrical charges in atoms cancel out one another; but when an atom loses one or more electrons, and thus has a net charge, it becomes an ion.

CHEMICAL COMPOUND: A substance made up of atoms of more than one chemical element. These atoms are usually joined in molecules.

CHEMICAL ELEMENT: A substance made up of only one kind of atom.

CONSERVATION OF ENERGY: A law of physics which holds that within a system isolated from all other outside factors, the total amount of energy remains the same, though transformations of energy from one form to another take place.

CONSERVATION OF MASS: A physical principle which states that total mass is constant, and is unaffected by factors such as position, velocity, or temperature, in any system that does not exchange any matter with its environment. Unlike the other conservation laws, however, conservation of mass is not universally applicable, but applies only at speeds significant lower than that of light—186,000 mi (297,600 km) per second. Close to the speed of light, mass begins converting to energy.

CONSERVE: In physics, "to conserve" something means "to result in no net loss of" that particular component. It is possible that within a given system, the component may change form or position, but as long as the net value of the component remains the same, it has been conserved.

ELECTRON: Negatively charged particles in an atom. Electrons, which spin around the nucleus of protons and neutrons, constitute a very small portion of the atom's mass. In most atoms, the number of electrons and protons is the same, thus canceling out one another. When an atom loses one or more electrons, however— thus becoming an ion—it acquires a net electrical charge.

FRICTION: The force that resists motion when the surface of one object comes into contact with the surface of another.

FLUID: Any substance, whether gas or liquid, that tends to flow, and that conforms to the shape of its container. Unlike solids, fluids are typically uniform in molecular structure for instance, one molecule of water is the same as another water molecule.

GAS: A phase of matter in which molecules exert little or no attraction toward one another, and therefore move at high speeds.

ION: An atom that has lost or gained one or more electrons, and thus has a net electrical charge.

LIQUID: A phase of matter in which molecules exert moderate attractions toward one another, and therefore move at moderate speeds.

MATTER: Physical substance that has mass; occupies space; is composed of atoms; and is ultimately (at speeds approaching that of light) convertible to energy. There are several phases of matter, including solids, liquids, and gases.

MOLE: A unit equal to 6.022137×10^{23} (more than 600 billion trillion) molecules. Their size makes it impossible to weigh molecules in relatively small quantities; hence the mole facilitates comparisons of mass between substances.

MOLECULE: A group of atoms, usually of more than one chemical element, joined in a structure.

NEUTRON: A subatomic particle that has no electrical charge. Neutrons are found at the nucleus of an atom, alongside protons.

PHASES OF MATTER: The various forms of material substance (matter), which are defined primarily in terms of the behavior exhibited by their atomic or molecular structures. On Earth, three principal phases of matter exist, namely solid, liquid, and gas. Other forms of matter include plasma.

PLASMA: One of the phases of matter, closely related to gas. Plasma apparently does not exist on Earth, but is found, for instance, in stars and comets' tails. Containing neither atoms nor molecules, plasma is made up of electrons and positive ions.

PROTON: A positively charged particle in an atom. Protons and neutrons, which together form the nucleus around which electrons orbit, have approximately the same mass—a mass that is many times greater than that of an electron.

SOLID: A phase of matter in which molecules exert strong attractions toward one another, and therefore move slowly.

SYSTEM: In physics, the term "system" usually refers to any set of physical interactions isolated from the rest of the universe. Anything outside of the system, including all factors and forces irrelevant to a discussion of that system, is known as the environment.

The cholesteric class of liquid crystals is so named because the spiral patterns of light through the crystal are similar to those which appear in cholesterols. Depending on the physical properties of a cholesteric liquid crystal, only certain colors may be reflected. The response of liquid crystals to light makes them useful in liquid crystal displays (LCDs) found on laptop computer screens, camcorder views, and in other applications.

In some cholesteric liquid crystals, high temperatures lead to a reflection of shorter visible light waves, and lower temperatures to a display of longer visible waves. Liquid crystal thermometers thus show red when cool, and blue as they are warmed. This may seem a bit unusual to someone who does not understand why the thermometer displays those colors, since people typically associate red with heat and blue with cold.

THE TRIPLE POINT. A liquid crystal exhibits aspects of both liquid and solid, and thus, at certain temperatures may be classified within the crystalline quasi-state of matter. On the other hand, the phenomenon known as the triple point shows how an ordinary substance, such as water or carbon dioxide, can actually be a liquid, solid, and vapor—all at once.

Again, water—the basis of all life on Earth—is an unusual substance in many regards. For instance, most people associate water as a gas or vapor (that is, steam) with very high temperatures. Yet, at a level far below normal atmospheric pressure, water can be a vapor at temperatures as low as -4°F (-20 °C). (All of the pressure values in the discussion of water at or near the triple point are far below atmospheric norms: the pressure at which water would turn into a vapor at -4°F, for instance, is about 1/1000 normal atmospheric pressure.)

As everyone knows, at relatively low temperatures, water is a solid—ice. But if the pressure of ice falls below a very low threshold, it will turn straight into a gas (a process known as sublimation) without passing through the liquid stage. On the other hand, by applying enough pressure, it is possible to melt ice, and thereby transform it from a solid to a liquid, at temperatures below its normal freezing point.

The phases and changes of phase for a given substance at specific temperatures and pressure levels can be plotted on a graph called a phase diagram, which typically shows temperature on the x-axis and pressure on the y-axis. The phase diagram of water shows a line between the solid and liquid states that is almost, but not quite, exactly perpendicular to the x-axis: it slopes slightly upward to the left, reflecting the fact that solid ice turns into water with an increase of pressure.

Whereas the line between solid and liquid water is more or less straight, the division between these two states and water vapor is curved. And where the solid-liquid line intersects the vaporization curve, there is a place called the triple point. Just below freezing, in conditions

equivalent to about 0.7% of normal atmospheric pressure, water is a solid, liquid, and vapor all at once.

WHERE TO LEARN MORE

Biel, Timothy L. *Atom: Building Blocks of Matter.* San Diego, CA: Lucent Books, 1990.

Feynman, Richard. *Six Easy Pieces: Essentials of Physics Explained by Its Most Brilliant Teacher.* New introduction by Paul Davies. Cambridge, MA: Perseus Books, 1995.

Hewitt, Sally. *Solid, Liquid, or Gas?* New York: Children's Press, 1998.

"High School Chemistry Table of Contents—Solids and Liquids" Homeworkhelp.com (Web site). <http://www.homeworkhelp.com/homeworkhelp/freemember/text/chem/hig h/topic09.html> (April 10, 2001).

"Matter: Solids, Liquids, Gases." Studyweb (Web site). <http://www.studyweb.com/links/4880.html> (April 10, 2001).

"The Molecular Circus" (Web site). <http://www.cpo.com/Weblabs/circus.html> (April 10, 2001).

Paul, Richard. *A Handbook to the Universe: Explorations of Matter, Energy, Space, and Time for Beginning Scientific Thinkers.* Chicago: Chicago Review Press, 1993.

"Phases of Matter" (Web site). <http://pc65.frontier.osrhe.edu/hs/science/pphase.html> (April 10, 2001).

Royston, Angela. *Solids, Liquids, and Gasses.* Chicago: Heinemann Library, 2001.

Wheeler, Jill C. *The Stuff Life's Made Of: A Book About Matter.* Minneapolis, MN: Abdo & Daughters Publishing, 1996.

THERMODYNAMICS

CONCEPT

Thermodynamics is the study of the relationships between heat, work, and energy. Though rooted in physics, it has a clear application to chemistry, biology, and other sciences: in a sense, physical life itself can be described as a continual thermodynamic cycle of transformations between heat and energy. But these transformations are never perfectly efficient, as the second law of thermodynamics shows. Nor is it possible to get "something for nothing," as the first law of thermodynamics demonstrates: the work output of a system can never be greater than the net energy input. These laws disappointed hopeful industrialists of the early nineteenth century, many of whom believed it might be possible to create a perpetual motion machine. Yet the laws of thermodynamics did make possible such highly useful creations as the internal combustion engine and the refrigerator.

HOW IT WORKS

Historical Context

Machines were, by definition, the focal point of the Industrial Revolution, which began in England during the late eighteenth and early nineteenth centuries. One of the central preoccupations of both scientists and industrialists thus became the efficiency of those machines: the ratio of output to input. The more output that could be produced with a given input, the greater the production, and the greater the economic advantage to the industrialists and (presumably) society as a whole.

At that time, scientists and captains of industry still believed in the possibility of a perpetual motion machine: a device that, upon receiving an initial input of energy, would continue to operate indefinitely without further input. As it emerged that work could be converted into heat, a form of energy, it began to seem possible that heat could be converted directly back into work, thus making possible the operation of a perfectly reversible perpetual motion machine. Unfortunately, the laws of thermodynamics dashed all those dreams.

SNOW'S EXPLANATION. Some texts identify two laws of thermodynamics, while others add a third. For these laws, which will be discussed in detail below, British writer and scientist C. P. Snow (1905-1980) offered a witty, nontechnical explanation. In a 1959 lecture published as *The Two Cultures and the Scientific Revolution,* Snow compared the effort to transform heat into energy, and energy back into heat again, as a sort of game.

The first law of thermodynamics, in Snow's version, teaches that the game is impossible to win. Because energy is conserved, and thus, its quantities throughout the universe are always the same, one cannot get "something for nothing" by extracting more energy than one put into a machine.

The second law, as Snow explained it, offers an even more gloomy prognosis: not only is it impossible to win in the game of energy-work exchanges, one cannot so much as break even. Though energy is conserved, that does not mean the energy is conserved within the machine where it is used: mechanical systems tend toward increasing disorder, and therefore, it is impossi-

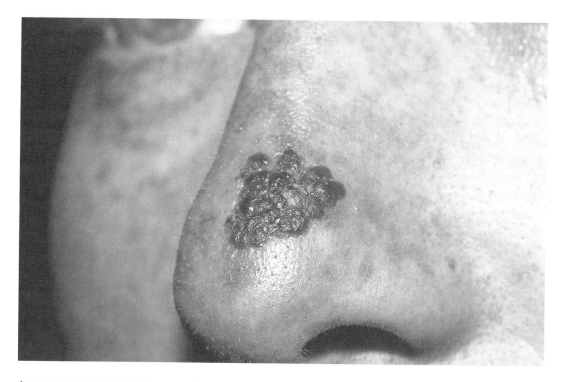

A WOMAN WITH A SUNBURNED NOSE. SUNBURNS ARE CAUSED BY THE SUN'S ULTRAVIOLET RAYS. *(Photograph by Lester V. Bergman/Corbis. Reproduced by permission.)*

ble for the machine even to return to the original level of energy.

The third law, discovered in 1905, seems to offer a possibility of escape from the conditions imposed in the second law: at the temperature of absolute zero, this tendency toward breakdown drops to a level of zero as well. But the third law only proves that absolute zero cannot be attained: hence, Snow's third observation, that it is impossible to step outside the boundaries of this unwinnable heat-energy transformation game.

WORK AND ENERGY

Work and energy, discussed at length elsewhere in this volume, are closely related. Work is the exertion of force over a given distance to displace or move an object. It is thus the product of force and distance exerted in the same direction. Energy is the ability to accomplish work.

There are many manifestations of energy, including one of principal concern in the present context: thermal or heat energy. Other manifestations include electromagnetic (sometimes divided into electrical and magnetic), sound, chemical, and nuclear energy. All these, however, can be described in terms of mechanical energy, which is the sum of potential energy—the energy that an object has due to its position—and kinetic energy, or the energy an object possesses by virtue of its motion.

MECHANICAL ENERGY. Kinetic energy relates to heat more clearly than does potential energy, discussed below; however, it is hard to discuss the one without the other. To use a simple example—one involving mechanical energy in a gravitational field—when a stone is held over the edge of a cliff, it has potential energy. Its potential energy is equal to its weight (mass times the acceleration due to gravity) multiplied by its height above the bottom of the canyon below. Once it is dropped, it acquires kinetic energy, which is the same as one-half its mass multiplied by the square of its velocity.

Just before it hits bottom, the stone's kinetic energy will be at a maximum, and its potential energy will be at a minimum. At no point can the value of its kinetic energy exceed the value of the potential energy it possessed before it fell: the mechanical energy, or the sum of kinetic and potential energy, will always be the same, though the relative values of kinetic and potential energy may change.

CONSERVATION OF ENERGY.
What mechanical energy does the stone possess after it comes to rest at the bottom of the canyon? In terms of the system of the stone dropping from the cliffside to the bottom, none. Or, to put it another way, the stone has just as much mechanical energy as it did at the very beginning. Before it was picked up and held over the side of the cliff, thus giving it potential energy, it was presumably sitting on the ground away from the edge of the cliff. Therefore, it lacked potential energy, inasmuch as it could not be "dropped" from the ground.

If the stone's mechanical energy—at least in relation to the system of height between the cliff and the bottom—has dropped to zero, where did it go? A number of places. When it hit, the stone transferred energy to the ground, manifested as heat. It also made a sound when it landed, and this also used up some of its energy. The stone itself lost energy, but the total energy in the universe was unaffected: the energy simply left the stone and went to other places. This is an example of the conservation of energy, which is closely tied to the first law of thermodynamics.

But does the stone possess any energy at the bottom of the canyon? Absolutely. For one thing, its mass gives it an energy, known as mass or rest energy, that dwarfs the mechanical energy in the system of the stone dropping off the cliff. (Mass energy is the other major form of energy, aside from kinetic and potential, but at speeds well below that of light, it is released in quantities that are virtually negligible.) The stone may have electromagnetic potential energy as well; and of course, if someone picks it up again, it will have gravitational potential energy. Most important to the present discussion, however, is its internal kinetic energy, the result of vibration among the molecules inside the stone.

HEAT AND TEMPERATURE

Thermal energy, or the energy of heat, is really a form of kinetic energy between particles at the atomic or molecular level: the greater the movement of these particles, the greater the thermal energy. Heat itself is internal thermal energy that flows from one body of matter to another. It is not the same as the energy contained in a system—that is, the internal thermal energy of the system. Rather than being "energy-in-residence," heat is "energy-in-transit."

This may be a little hard to comprehend, but it can be explained in terms of the stone-and-cliff kinetic energy illustration used above. Just as a system can have no kinetic energy unless something is moving within it, heat exists only when energy is being transferred. In the above illustration of mechanical energy, when the stone was sitting on the ground at the top of the cliff, it was analogous to a particle of internal energy in body *A*. When, at the end, it was again on the ground—only this time at the bottom of the canyon—it was the same as a particle of internal energy that has transferred to body *B*. In between, however, as it was falling from one to the other, it was equivalent to a unit of heat.

TEMPERATURE. In everyday life, people think they know what temperature is: a measure of heat and cold. This is wrong for two reasons: first, as discussed below, there is no such thing as "cold"—only an absence of heat. So, then, is temperature a measure of heat? Wrong again.

Imagine two objects, one of mass *M* and the other with a mass twice as great, or 2*M*. Both have a certain temperature, and the question is, how much heat will be required to raise their temperature by equal amounts? The answer is that the object of mass 2*M* requires twice as much heat to raise its temperature the same amount. Therefore, temperature cannot possibly be a measure of heat.

What temperature does indicate is the direction of internal energy flow between bodies, and the average molecular kinetic energy in transit between those bodies. More simply, though a bit less precisely, it can be defined as a measure of heat differences. (As for the means by which a thermometer indicates temperature, that is beyond the parameters of the subject at hand; it is discussed elsewhere in this volume, in the context of thermal expansion.)

MEASURING TEMPERATURE AND HEAT. Temperature, of course, can be measured either by the Fahrenheit or Centigrade scales familiar in everyday life. Another temperature scale of relevance to the present discussion is the Kelvin scale, established by William Thomson, Lord Kelvin (1824-1907).

Drawing on the discovery made by French physicist and chemist J. A. C. Charles (1746-

1823), that gas at 0°C (32°F) regularly contracts by about 1/273 of its volume for every Celsius degree drop in temperature, Thomson derived the value of absolute zero (discussed below) as -273.15°C (-459.67°F). The Kelvin and Celsius scales are thus directly related: Celsius temperatures can be converted to Kelvins (for which neither the word nor the symbol for "degree" are used) by adding 273.15.

MEASURING HEAT AND HEAT CAPACITY. Heat, on the other hand, is measured not by degrees (discussed along with the thermometer in the context of thermal expansion), but by the same units as work. Since energy is the ability to perform work, heat or work units are also units of energy. The principal unit of energy in the SI or metric system is the joule (J), equal to 1 newton-meter (N • m), and the primary unit in the British or English system is the foot-pound (ft • lb). One foot-pound is equal to 1.356 J, and 1 joule is equal to 0.7376 ft • lb.

Two other units are frequently used for heat as well. In the British system, there is the Btu, or British thermal unit, equal to 778 ft • lb. or 1,054 J. Btus are often used in reference, for instance, to the capacity of an air conditioner. An SI unit that is also used in the United States—where British measures typically still prevail—is the kilocalorie. This is equal to the heat that must be added to or removed from 1 kilogram of water to change its temperature by 1°C. As its name suggests, a kilocalorie is 1,000 calories. A calorie is the heat required to change the temperature in 1 gram of water by 1°C—but the dietary Calorie (capital C), with which most people are familiar is the same as the kilocalorie.

A kilocalorie is identical to the heat capacity for one kilogram of water. Heat capacity (sometimes called specific heat capacity or specific heat) is the amount of heat that must be added to, or removed from, a unit of mass for a given substance to change its temperature by 1°C. this is measured in units of J/kg • °C (joules per kilogram-degree Centigrade), though for the sake of convenience it is typically rendered in terms of kilojoules (1,000 joules): kJ/kg • °c. Expressed thus, the specific heat of water 4.185—which is fitting, since a kilocalorie is equal to 4.185 kJ. Water is unique in many aspects, with regard to specific heat, in that it requires far more heat to raise the temperature of water than that of mercury or iron.

HOT AND "COLD"

Earlier, it was stated that there is no such thing as "cold"—a statement hard to believe for someone who happens to be in Buffalo, New York, or International Falls, Minnesota, during a February blizzard. Certainly, cold is real as a sensory experience, but in physical terms, cold is not a "thing"—it is simply the absence of heat.

People will say, for instance, that they put an ice cube in a cup of coffee to cool it, but in terms of physics, this description is backward: what actually happens is that heat flows from the coffee to the ice, thus raising its temperature. The resulting temperature is somewhere between that of the ice cube and the coffee, but one cannot obtain the value simply by averaging the two temperatures at the beginning of the transfer.

For one thing, the volume of the water in the ice cube is presumably less than that of the water in the coffee, not to mention the fact that their differing chemical properties may have some minor effect on the interaction. Most important, however, is the fact that the coffee did not simply merge with the ice: in transferring heat to the ice cube, the molecules in the coffee expended some of their internal kinetic energy, losing further heat in the process.

COOLING MACHINES. Even cooling machines, such as refrigerators and air conditioners, actually use heat, simply reversing the usual process by which particles are heated. The refrigerator pulls heat from its inner compartment—the area where food and other perishables are stored—and transfers it to the region outside. This is why the back of a refrigerator is warm.

Inside the refrigerator is an evaporator, into which heat from the refrigerated compartment flows. The evaporator contains a refrigerant—a gas, such as ammonia or Freon 12, that readily liquifies. This gas is released into a pipe from the evaporator at a low pressure, and as a result, it evaporates, a process that cools it. The pipe takes the refrigerant to the compressor, which pumps it into the condenser at a high pressure. Located at the back of the refrigerator, the condenser is a long series of pipes in which pressure turns the gas into liquid. As it moves through the condenser, the gas heats, and this heat is released into the air around the refrigerator.

An air conditioner works in a similar manner. Hot air from the room flows into the evaporator, and a compressor circulates refrigerant from the evaporator to a condenser. Behind the evaporator is a fan, which draws in hot air from the room, and another fan pushes heat from the condenser to the outside. As with a refrigerator, the back of an air conditioner is hot because it is moving heat from the area to be cooled.

Thus, cooling machines do not defy the principles of heat discussed above; nor do they defy the laws of thermodynamics that will be discussed at the conclusion of this essay. In accordance with the second law, in order to move heat in the reverse of its usual direction, external energy is required. Thus, a refrigerator takes in energy from a electric power supply (that is, the outlet it is plugged into), and extracts heat. Nonetheless, it manages to do so efficiently, removing two or three times as much heat from its inner compartment as the amount of energy required to run the refrigerator.

TRANSFERS OF HEAT

It is appropriate now to discuss how heat is transferred. One must remember, again, that in order for heat to be transferred from one point to another, there must be a difference of temperature between those two points. If an object or system has a uniform level of internal thermal energy—no matter how "hot" it may be in ordinary terms—no heat transfer is taking place.

Heat is transferred by one of three methods: conduction, which involves successive molecular collisions; convection, which requires the motion of hot fluid from one place to another; or radiation, which involves electromagnetic waves and requires no physical medium for the transfer.

CONDUCTION. Conduction takes place best in solids and particularly in metals, whose molecules are packed in relatively close proximity. Thus, when one end of an iron rod is heated, eventually the other end will acquire heat due to conduction. Molecules of liquid or nonmetallic solids vary in their ability to conduct heat, but gas—due to the loose attractions between its molecules—is a poor conductor.

When conduction takes place, it is as though a long line of people are standing shoulder to shoulder, passing a secret down the line. In this case, however, the "secret" is kinetic thermal energy. And just as the original phrasing of the secret will almost inevitably become garbled by the time it gets to the tenth or hundredth person, some energy is lost in the transfer from molecule to molecule. Thus, if one end of the iron rod is sitting in a fire and one end is surrounded by air at room temperature, it is unlikely that the end in the air will ever get as hot as the end in the fire.

Incidentally, the qualities that make metallic solids good conductors of heat also make them good conductors of electricity. In the first instance, kinetic energy is being passed from molecule to molecule, whereas in an electrical field, electrons—freed from the atoms of which they are normally a part—are able to move along the line of molecules. Because plastic is much less conductive than metal, an electrician will use a screwdriver with a plastic handle. Similarly, a metal pan typically has a handle of wood or plastic.

CONVECTION. There is a term, "convection oven," that is actually a redundancy: all ovens heat through convection, the principal means of transferring heat through a fluid. In physics, "fluid" refers both to liquids and gases—anything that tends to flow. Instead of simply moving heat, as in conduction, convection involves the movement of heated material—that is, fluid. When air is heated, it displaces cold (that is, unheated) air in its path, setting up a convection current.

Convection takes place naturally, as for instance when hot air rises from the land on a warm day. This heated air has a lower density than that of the less heated air in the atmosphere above it, and, therefore, is buoyant. As it rises, however, it loses energy and cools. This cooled air, now more dense than the air around it, sinks again, creating a repeating cycle.

The preceding example illustrates natural convection; the heat of an oven, on the other hand, is an example of forced convection—a situation in which some sort of pump or mechanism moves heated fluid. So, too, is the cooling work of a refrigerator, though the refrigerator moves heat in the opposite direction.

Forced convection can also take place within a natural system. The human heart is a pump, and blood carries excess heat generated by the body to the skin. The heat passes through the skin by means of conduction, and at the surface of the skin, it is removed from the body in a number of ways, primarily by the cooling evaporation of moisture—that is, perspiration.

RADIATION. If the Sun is hot—hot enough to severely burn the skin of a person who spends too much time exposed to its rays—then why is it cold in the upper atmosphere? After all, the upper atmosphere is closer to the Sun. And why is it colder still in the empty space above the atmosphere, which is still closer to the Sun? The reason is that in outer space there is no medium for convection, and in the upper atmosphere, where the air molecules are very far apart, there is hardly any medium. How, then, does heat come to the Earth from the Sun? By radiation, which is radically different from conduction or convection. The other two involve ordinary thermal energy, but radiation involves electromagnetic energy.

A great deal of "stuff" travels through the electromagnetic spectrum, discussed in another essay in this book: radio waves, microwaves for television and radar, infrared light, visible light, x rays, gamma rays. Though the relatively narrow band of visible-light wavelengths is the only part of the spectrum of which people are aware in everyday life, other parts—particularly the infrared and ultraviolet bands—are involved in the heat one feels from the Sun. (Ultraviolet rays, in fact, cause sunburns.)

Heat by means of radiation is not as "otherworldly" as it might seem: in fact, one does not have to point to the Sun for examples of it. Any time an object glows as a result of heat—as for example, in the case of firelight—that is an example of radiation. Some radiation is emitted in the form of visible light, but the heat component is in infrared rays. This also occurs in an incandescent light bulb. In an incandescent bulb, incidentally, much of the energy is lost to the heat of infrared rays, and the efficiency of a fluorescent bulb lies in the fact that it converts what would otherwise be heat into usable light.

THE LAWS OF THERMODYNAMICS

Having explored the behavior of heat, both at the molecular level and at levels more easily perceived by the senses, it is possible to discuss the laws of thermodynamics alluded to throughout this essay. These laws illustrate the relationships between heat and energy examined earlier, and

BENJAMIN THOMPSON, COUNT RUMFORD. *(Illustration by H. Humphrey. UPI/Corbis-Bettmann. Reproduced by permission.)*

show, for instance, why a refrigerator or air conditioner must have an external source of energy to move heat in a direction opposite to its normal flow.

The story of how these laws came to be discovered is a saga unto itself, involving the contributions of numerous men in various places over a period of more than a century. In 1791, Swiss physicist Pierre Prevost (1751-1839) put forth his theory of exchanges, stating correctly that all bodies radiate heat. Hence, as noted earlier, there is no such thing as "cold": when one holds snow in one's hand, cold does not flow from the snow into the hand; rather, heat flows from the hand to the snow.

Seven years later, an American-British physicist named Benjamin Thompson, Count Rumford (1753) was boring a cannon with a blunt drill when he noticed that this action generated a great deal of heat. This led him to question the prevailing wisdom, which maintained that heat was a fluid form of matter; instead, Thompson began to suspect that heat must arise from some form of motion.

CARNOT'S ENGINE. The next major contribution came from the French physicist and engineer Sadi Carnot (1796-1832).

Though he published only one scientific work, *Reflections on the Motive Power of Fire* (1824), this treatise caused a great stir in the European scientific community. In it, Carnot made the first attempt at a scientific definition of work, describing it as "weight lifted through a height." Even more important was his proposal for a highly efficient steam engine.

A steam engine, like a modern-day internal combustion engine, is an example of a larger class of machine called heat engine. A heat engine absorbs heat at a high temperature, performs mechanical work, and, as a result, gives off heat a lower temperature. (The reason why that temperature must be lower is established in the second law of thermodynamics.)

For its era, the steam engine was what the computer is today: representing the cutting edge in technology, it was the central preoccupation of those interested in finding new ways to accomplish old tasks. Carnot, too, was fascinated by the steam engine, and was determined to help overcome its disgraceful inefficiency: in operation, a steam engine typically lost as much as 95% of its heat energy.

In his *Reflections,* Carnot proposed that the maximum efficiency of any heat engine was equal to $(T_H-T_L)/T_H$, where T_H is the highest operating temperature of the machine, and T_L the lowest. In order to maximize this value, T_L has to be absolute zero, which is impossible to reach, as was later illustrated by the third law of thermodynamics.

In attempting to devise a law for a perfectly efficient machine, Carnot inadvertently proved that such a machine is impossible. Yet his work influenced improvements in steam engine design, leading to levels of up to 80% efficiency. In addition, Carnot's studies influenced Kelvin—who actually coined the term "thermodynamics"—and others.

THE FIRST LAW OF THERMO-DYNAMICS. During the 1840s, Julius Robert Mayer (1814-1878), a German physicist, published several papers in which he expounded the principles known today as the conservation of energy and the first law of thermodynamics. As discussed earlier, the conservation of energy shows that within a system isolated from all outside factors, the total amount of energy remains the same, though transformations of energy from one form to another take place.

The first law of thermodynamics states this fact in a somewhat different manner. As with the other laws, there is no definitive phrasing; instead, there are various versions, all of which say the same thing. One way to express the law is as follows: Because the amount of energy in a system remains constant, it is impossible to perform work that results in an energy output greater than the energy input. For a heat engine, this means that the work output of the engine, combined with its change in internal energy, is equal to its heat input. Most heat engines, however, operate in a cycle, so there is no net change in internal energy.

Earlier, it was stated that a refrigerator extracts two or three times as much heat from its inner compartment as the amount of energy required to run it. On the surface, this seems to contradict the first law: isn't the refrigerator putting out more energy than it received? But the heat it extracts is only part of the picture, and not the most important part from the perspective of the first law.

A regular heat engine, such as a steam or internal-combustion engine, pulls heat from a high-temperature reservoir to a low-temperature reservoir, and, in the process, work is accomplished. Thus, the hot steam from the high-temperature reservoir makes possible the accomplishment of work, and when the energy is extracted from the steam, it condenses in the low-temperature reservoir as relatively cool water.

A refrigerator, on the other hand, reverses this process, taking heat from a low-temperature reservoir (the evaporator inside the cooling compartment) and pumping it to a high-temperature reservoir outside the refrigerator. Instead of producing a work output, as a steam engine does, it requires a work input—the energy supplied via the wall outlet. Of course, a refrigerator does produce an "output," by cooling the food inside, but the work it performs in doing so is equal to the energy supplied for that purpose.

THE SECOND LAW OF THERMODYNAMICS. Just a few years after Mayer's exposition of the first law, another German physicist, Rudolph Julius Emanuel Clausius (1822-1888) published an early version of the second law of thermodynamics. In an 1850 paper, Clausius stated that "Heat cannot, of itself, pass from a colder to a hotter body." He refined this 15 years later, introducing the concept of entropy—the tendency of natural systems toward breakdown, and specifically, the tendency for the energy in a system to be dissipated.

The second law of thermodynamics begins from the fact that the natural flow of heat is always from a high-temperature reservoir to a low-temperature reservoir. As a result, no engine can be constructed that simply takes heat from a source and performs an equivalent amount of work: some of the heat will always be lost. In other words, it is impossible to build a perfectly efficient engine.

Though its relation to the first law is obvious, inasmuch as it further defines the limitations of machine output, the second law of thermodynamics is not derived from the first. Elsewhere in this volume, the first law of thermodynamics—stated as the conservation of energy law—is discussed in depth, and, in that context, it is in fact necessary to explain how the behavior of machines in the real world does not contradict the conservation law.

Even though they mean the same thing, the first law of thermodynamics and the conservation of energy law are expressed in different ways. The first law of thermodynamics states that "the glass is half empty," whereas the conservation of energy law shows that "the glass is half full." The thermodynamics law emphasizes the bad news: that one can never get more energy out of a machine than the energy put into it. Thus, all hopes of a perpetual motion machine were dashed. The conservation of energy, on the other hand, stresses the good news: that energy is never lost.

In this context, the second law of thermodynamics delivers another dose of bad news: though it is true that energy is never lost, the energy available for work output will never be as great as the energy put into a system. A car engine, for instance, cannot transform all of its energy input into usable horsepower; some of the energy will be used up in the form of heat and sound. Though energy is conserved, usable energy is not.

Indeed, the concept of entropy goes far beyond machines as people normally understand them. Entropy explains why it is easier to break something than to build it—and why, for each person, the machine called the human body will inevitably break down and die, or cease to function, someday.

THE THIRD LAW OF THERMO-DYNAMICS. The subject of entropy leads directly to the third law of thermodynamics, formulated by German chemist Hermann Walter Nernst (1864-1941) in 1905. The third law states that at the temperature of absolute zero, entropy also approaches zero. From this statement, Nernst deduced that absolute zero is therefore impossible to reach.

All matter is in motion at the molecular level, which helps define the three major phases of matter found on Earth. At one extreme is a gas, whose molecules exert little attraction toward one another, and are therefore in constant motion at a high rate of speed. At the other end of the phase continuum (with liquids somewhere in the middle) are solids. Because they are close together, solid particles move very little, and instead of moving in relation to one another, they merely vibrate in place. But they do move.

Absolute zero, or 0K on the Kelvin scale of temperature, is the point at which all molecular motion stops entirely—or at least, it virtually stops. (In fact, absolute zero is defined as the temperature at which the motion of the average atom or molecule is zero.) As stated earlier, Carnot's engine achieves perfect efficiency if its lowest temperature is the same as absolute zero; but the second law of thermodynamics shows that a perfectly efficient machine is impossible. This means that absolute zero is an unreachable extreme, rather like matter exceeding the speed of light, also an impossibility.

This does not mean that scientists do not attempt to come as close as possible to absolute zero, and indeed they have come very close. In 1993, physicists at the Helsinki University of Technology Low Temperature Laboratory in Finland used a nuclear demagnetization device to achieve a temperature of $2.8 \cdot 10^{-10}$ K, or 0.00000000028K. This means that a fragment equal to only 28 parts in 100 billion separated this temperature from absolute zero—but it was still above 0K. Such extreme low-temperature research has a number of applications, most notably with superconductors, materials that exhibit virtually no resistance to electrical current at very low temperatures.

KEY TERMS

ABSOLUTE ZERO: The temperature, defined as 0K on the Kelvin scale, at which the motion of molecules in a solid virtually ceases. The third law of thermodynamics establishes the impossibility of actually reaching absolute zero.

BTU (BRITISH THERMAL UNIT): A measure of energy or heat in the British system, often used in reference to the capacity of an air conditioner. A Btu is equal to 778 foot-pounds, or 1,054 joules.

CALORIE: A measure of heat or energy in the SI or metric system, equal to the heat that must be added to or removed from 1 gram of water to change its temperature by 33.8°F (1°C). The dietary Calorie (capital C) with which most people are familiar is the same as the kilocalorie.

CONDUCTION: The transfer of heat by successive molecular collisions. Conduction is the principal means of heat transfer in solids, particularly metals.

CONSERVATION OF ENERGY: A law of physics which holds that within a system isolated from all other outside factors, the total amount of energy remains the same, though transformations of energy from one form to another take place. The first law of thermodynamics is the same as the conservation of energy.

CONSERVE: In physics, "to conserve" something means "to result in no net loss of" that particular component. It is possible that within a given system, the component may change form or position, but as long as the net value of the component remains the same, it has been conserved.

CONVECTION: The transfer of heat through the motion of hot fluid from one place to another. In physics, a "fluid" can be either a gas or a liquid, and convection is the principal means of heat transfer, for instance, in air and water.

ENERGY: The ability to accomplish work.

ENTROPY: The tendency of natural systems toward breakdown, and specifically, the tendency for the energy in a system to be dissipated. Entropy is closely related to the second law of thermodynamics.

FIRST LAW OF THERMODYNAMICS: A law which states the amount of energy in a system remains constant, and therefore it is impossible to perform work that results in an energy output greater than the energy input. This is the same as the conservation of energy.

FOOT-POUND: The principal unit of energy—and thus of heat—in the British or English system. The metric or SI unit is the joule. A foot-pound (ft • lb) is equal to 1.356 J.

HEAT: Internal thermal energy that flows from one body of matter to another. Heat is transferred by three methods conduction, convection, and radiation.

HEAT CAPACITY: The amount of heat that must be added to, or removed from, a unit of mass of a given substance to change its temperature by 33.8°F (1°C). Heat capacity is sometimes called specific heat capacity or specific heat. A kilocalorie is the heat capacity of 1 gram of water.

HEAT ENGINE: A machine that absorbs heat at a high temperature, performs mechanical work, and as a result gives off heat at a lower temperature.

KINETIC ENERGY: The energy that an object possesses by virtue of its motion.

JOULE: The principal unit of energy—and thus of heat—in the SI or metric system, corresponding to 1 newton-meter (N • m). A joule (J) is equal to 0.7376 foot-pounds.

KELVIN SCALE: Established by William Thomson, Lord Kelvin (1824-1907), the Kelvin scale measures temperature in relation to absolute zero, or 0K. (Units in the Kelvin system, known as Kelvins, do not include the word or symbol for degree.) The Kelvin and Celsius scales are directly related; hence Celsius temperatures can be converted to Kelvins by adding 273.15.

KILOCALORIE: A measure of heat or energy in the SI or metric system, equal to the heat that must be added to or removed from 1 kilogram of water to change its temperature by 33.8°F (1°C). As its name suggests, a kilocalorie is 1,000 calories. The dietary Calorie (capital C) with which most people are familiar is the same as the kilocalorie.

MECHANICAL ENERGY: The sum of potential energy and kinetic energy in a given system.

POTENTIAL ENERGY: The energy that an object possesses due to its position.

RADIATION: The transfer of heat by means of electromagnetic waves, which require no physical medium (e.g., water or air) for the transfer. Earth receives the Sun's heat by means of radiation.

SECOND LAW OF THERMODYNAMICS: A law of thermodynamics which states that no engine can be constructed that simply takes heat from a source and performs an equivalent amount of work. Some of the heat will always be lost, and

therefore it is impossible to build a perfectly efficient engine. This is a result of the fact that the natural flow of heat is always from a high-temperature reservoir to a low-temperature reservoir—a fact expressed in the concept of entropy. The second law is sometimes referred to as "the law of entropy."

SYSTEM: In physics, the term "system" usually refers to any set of physical interactions isolated from the rest of the universe. Anything outside of the system, including all factors and forces irrelevant to a discussion of that system, is known as the environment.

TEMPERATURE: The direction of internal energy flow between bodies when heat is being transferred. Temperature measures the average molecular kinetic energy in transit between those bodies.

THERMAL ENERGY: Heat energy, a form of kinetic energy produced by the movement of atomic or molecular particles. The greater the movement of these particles, the greater the thermal energy.

THERMODYNAMICS: The study of the relationships between heat, work, and energy.

THIRD LAW OF THERMODYNAMICS: A law of thermodynamics which states that at the temperature of absolute zero, entropy also approaches zero. Zero entropy would contradict the second law of thermodynamics, meaning that absolute zero is therefore impossible to reach.

WORK: The exertion of force over a given distance to displace or move an object. Work is thus the product of force and distance exerted in the same direction.

WHERE TO LEARN MORE

Beiser, Arthur. *Physics,* 5th ed. Reading, MA: Addison-Wesley, 1991.

Brown, Warren. *Alternative Sources of Energy.* Introduction by Russell E. Train. New York: Chelsea House, 1994.

Encyclopedia of Thermodynamics (Web site). <http://therion.minpet.unibas.ch/minpet/groups/thermodict/> (April 12, 2001).

Entropy and the Second Law of Thermodynamics (Web site). <http://www.2ndlaw.com> (April 12, 2001).

Fleisher, Paul. *Matter and Energy: Principles of Matter and Thermodynamics.* Minneapolis, MN: Lerner Publications, 2002.

Macaulay, David. *The New Way Things Work.* Boston: Houghton Mifflin, 1998.

Moran, Jeffrey B. *How Do We Know the Laws of Thermodynamics?* New York: Rosen Publishing Group, 2001.

Santrey, Laurence. *Heat.* Illustrated by Lloyd Birmingham. Mahwah, N.J.: Troll Associates, 1985.

Suplee, Curt. *Everyday Science Explained.* Washington, D.C.: National Geographic Society, 1996.

"Temperature and Thermodynamics" PhysLINK.com (Web site). <http://www.physlink.com/ae_thermo.cfm> (April 12, 2001).

HEAT

CONCEPT

Heat is a form of energy—specifically, the energy that flows between two bodies because of differences in temperature. Therefore, the scientific definition of heat is different from, and more precise than, the everyday meaning. Physicists working in the area of thermodynamics study heat from a number of perspectives, including specific heat, or the amount of energy required to change the temperature of a substance, and calorimetry, the measurement of changes in heat as a result of physical or chemical changes. Thermodynamics helps us to understand such phenomena as the operation of engines and the gradual breakdown of complexity in physical systems—a phenomenon known as entropy.

HOW IT WORKS

Heat, Work, and Energy

Thermodynamics is the study of the relationships between heat, work, and energy. Work is the exertion of force over a given distance to displace or move an object, and is, thus, the product of force and distance exerted in the same direction. Energy, the ability to accomplish work, appears in numerous manifestations—including thermal energy, or the energy associated with heat.

Thermal and other types of energy, including electromagnetic, sound, chemical, and nuclear energy, can be described in terms of two extremes: kinetic energy, or the energy associated with movement, and potential energy, or the energy associated with position. If a spring is pulled back to its maximum point of tension, its potential energy is also at a maximum; once it is released and begins springing through the air to return to its original position, it begins gaining kinetic energy and losing potential energy.

All manifestations of energy appear in both kinetic and potential forms, somewhat like the way football teams are organized to play both offense or defense. Just as a football team takes an offensive role when it has the ball, and a defensive role when the other team has it, a physical system typically undergoes regular transformations between kinetic and potential energy, and may have more of one or the other, depending on what is taking place in the system.

What Heat Is and Is Not

Thermal energy is actually a form of kinetic energy generated by the movement of particles at the atomic or molecular level: the greater the movement of these particles, the greater the thermal energy. Heat is internal thermal energy that flows from one body of matter to another—or, more specifically, from a system at a higher temperature to one at a lower temperature. Thus, temperature, like heat, requires a scientific definition quite different from its common meaning: temperature measures the average molecular kinetic energy of a system, and governs the direction of internal energy flow between them.

Two systems at the same temperature are said to be in a state of thermal equilibrium. When this occurs, there is no exchange of heat. Though in common usage, "heat" is an expression of relative warmth or coldness, in physical terms, heat exists only in transfer between two systems. What people really mean by "heat" is the internal energy of a system—energy that is a property of that system rather than a property of transferred internal energy.

IF YOU HOLD A SNOWBALL IN YOUR HAND, AS VANNA WHITE AND HER SON ARE DOING IN THIS PICTURE, HEAT WILL MOVE FROM YOUR HAND TO THE SNOWBALL. YOUR HAND EXPERIENCES THIS AS A SENSATION OF COLDNESS. (*Reuters NewMedia Inc./Corbis. Reproduced by permission.*)

NO SUCH THING AS "COLD." Though the term "cold" has plenty of meaning in the everyday world, in physics terminology, it does not. Cold and heat are analogous to darkness and light: again, darkness means something in our daily experience, but in physical terms, darkness is simply the absence of light. To speak of cold or darkness as entities unto themselves is rather like saying, after spending 20 dollars, "I have 20 non-dollars in my pocket."

If you grasp a snowball in your hand, of course, your hand gets cold. The human mind perceives this as a transfer of cold from the snowball, but, in fact, exactly the opposite happens: heat moves from your hand to the snow, and if enough heat enters the snowball, it will melt. At the same time, the departure of heat from your hand results in a loss of internal energy near the surface of your hand, which you experience as a sensation of coldness.

TRANSFERS OF HEAT

In holding the snowball, heat passes from the surface of the hand by one means, conduction, then passes through the snowball by another means, convection. In fact, there are three methods heat is transferred: conduction, involving successive molecular collisions and the transfer of heat between two bodies in contact; convection, which requires the motion of fluid from one place to another; or radiation, which takes place through electromagnetic waves and requires no physical medium, such as water or air, for the transfer.

CONDUCTION. Solids, particularly metals, whose molecules are packed relatively close together, are the best materials for conduction. Molecules of liquid or non-metallic solids vary in their ability to conduct heat, but gas is a poor conductor, because of the loose attractions between its molecules.

The qualities that make metallic solids good conductors of heat, as a matter of fact, also make them good conductors of electricity. In the conduction of heat, kinetic energy is passed from molecule to molecule, like a long line of people standing shoulder to shoulder, passing a secret. (And, just as the original phrasing of the secret becomes garbled, some kinetic energy is inevitably lost in the series of transfers.)

As for electrical conduction, which takes place in a field of electric potential, electrons are freed from their atoms; as a result, they are able to move along the line of molecules. Because plastic is much less conductive than metal, an electrician uses a screwdriver with a plastic handle; similarly, a metal cooking pan typically has a wooden or plastic handle.

CONVECTION. Wherever fluids are involved—and in physics, "fluid" refers both to liquids and gases—convection is a common form of heat transfer. Convection involves the movement of heated material—whether it is air, water, or some other fluid.

Convection is of two types: natural convection and forced convection, in which a pump or other mechanism moves the heated fluid. When heated air rises, this is an example of natural convection. Hot air has a lower density than that of the cooler air in the atmosphere above it, and, therefore, is buoyant; as it rises, however, it loses energy and cools. This cooled air, now denser than the air around it, sinks again, creating a repeating cycle that generates wind.

Examples of forced convection include some types of ovens and even a refrigerator or air conditioner. These two machines both move warm air from an interior to an exterior place. Thus, the refrigerator pulls hot air from the compartment and expels it to the surrounding room, while an air conditioner pulls heat from a building and releases it to the outside.

But forced convection does not necessarily involve humanmade machines: the human heart is a pump, and blood carries excess heat generated by the body to the skin. The heat passes through the skin by means of conduction, and at the surface of the skin, it is removed from the body in a number of ways, primarily by the cooling evaporation of perspiration.

RADIATION. Outer space, of course, is cold, yet the Sun's rays warm the Earth, an apparent paradox. Because there is no atmosphere in space, convection is impossible. In fact, heat from the Sun is not dependant on any fluid medium for its transfer: it comes to Earth by means of radiation. This is a form of heat transfer significantly different from the other two, because it involves electromagnetic energy, instead of ordinary thermal energy generated by the action of molecules. Heat from the Sun comes through a relatively narrow area of the light spectrum, including infrared, visible light, and ultraviolet rays.

Every form of matter emits electromagnetic waves, though their presence may not be readily perceived. Thus, when a metal rod is heated, it experiences conduction, but part of its heat is radiated, manifested by its glow—visible light. Even when the heat in an object is not visible, however, it may be radiating electromagnetic energy, for instance, in the form of infrared light. And, of course, different types of matter radiate better than others: in general, the better an object is at receiving radiation, the better it is at emitting it.

Measuring Heat

The measurement of temperature by degrees in the Fahrenheit or Celsius scales is a part of everyday life, but measurements of heat are not as familiar to the average person. Because heat is a form of energy, and energy is the ability to perform work, heat is, therefore, measured by the same units as work.

The principal unit of work or energy in the metric system (known within the scientific community as SI, or the SI system) is the joule.

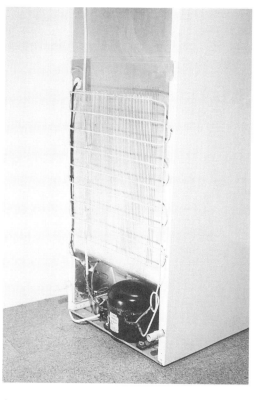

A REFRIGERATOR IS A TYPE OF REVERSE HEAT ENGINE THAT USES A COMPRESSOR, LIKE THE ONE SHOWN AT THE BACK OF THIS REFRIGERATOR, TO COOL THE REFRIGERATOR'S INTERIOR. *(Ecoscene/Corbis. Reproduced by permission.)*

Abbreviated "J," a joule is equal to 1 newton-meter (N • m). The newton is the SI unit of force, and since work is equal to force multiplied by distance, measures of work can also be separated into these components. For instance, the British measure of work brings together a unit of distance, the foot, and a unit of force, the pound. A foot-pound (ft • lb) is equal to 1.356 J, and 1 joule is equal to 0.7376 ft • lb.

In the British system, Btu, or British thermal unit, is another measure of energy used for machines such as air conditioners. One Btu is equal to 778 ft • lb or 1,054 J. The kilocalorie in addition to the joule, is an important SI measure of heat. The amount of energy required to change the temperature of 1 gram of water by 1°C is called a calorie, and a kilocalorie is equal to 1,000 calories. Somewhat confusing is the fact that the dietary Calorie (capital C), with which most people are familiar, is not the same as a calorie (lowercase C)—rather, a dietary Calorie is the equivalent of a kilocalorie.

REAL-LIFE APPLICATIONS

SPECIFIC HEAT

Specific heat is the amount of heat that must be added to, or removed from, a unit of mass for a given substance to change its temperature by 1°C. Thus, a kilocalorie, because it measures the amount of heat necessary to effect that change precisely for a kilogram of water, is identical to the specific heat for that particular substance in that particular unit of mass.

The higher the specific heat, the more resistant the substance is to changes in temperature. Many metals, in fact, have a low specific heat, making them easy to heat up and cool down. This contributes to the tendency of metals to expand when heated (a phenomenon also discussed in the Thermal Expansion essay), and, thus, to their malleability.

MEASURING AND CALCULATING SPECIFIC HEAT. The specific heat of any object is a function of its mass, its composition, and the desired change in temperature. The values of the initial and final temperature are not important—only the difference between them, which is the temperature change.

The components of specific heat are related to one another in the formula $Q = mc\delta T$. Here Q is the quantity of heat, measured in joules, which must be added. The mass of the object is designated by m, and the specific heat of the particular substance in question is represented with c. The Greek letter delta (δ) designates change, and δT stands for "change in temperature."

Specific heat is measured in units of J/kg · °C (joules per kilogram-degree Centigrade), though for the sake of convenience, this is usually rendered in terms of kilojoules (kJ), or 1,000 joules—that is, kJ/kg · °C. The specific heat of water is easily derived from the value of a kilocalorie: it is 4.185, the same number of joules required to equal a kilocalorie.

CALORIMETRY

The measurement of heat gain or loss as a result of physical or chemical change is called calorimetry (pronounced kal-IM-uh-tree). Like the word "calorie," the term is derived from a Latin root meaning "heat."

The foundations of calorimetry go back to the mid-nineteenth century, but the field owes much to scientists' work that took place over a period of about 75 years prior to that time. In 1780, French chemist Antoine Lavoisier (1743-1794) and French astronomer and mathematician Pierre Simon Laplace (1749-1827) had used a rudimentary ice calorimeter for measuring the heats in formations of compounds. Around the same time, Scottish chemist Joseph Black (1728-1799) became the first scientist to make a clear distinction between heat and temperature.

By the mid-1800s, a number of thinkers had come to the realization that—contrary to prevailing theories of the day—heat was a form of energy, not a type of material substance. Among these were American-British physicist Benjamin Thompson, Count Rumford (1753-1814) and English chemist James Joule (1818-1889)—for whom, of course, the joule is named.

Calorimetry as a scientific field of study actually had its beginnings with the work of French chemist Pierre-Eugene Marcelin Berthelot (1827-1907). During the mid-1860s, Berthelot became intrigued with the idea of measuring heat, and by 1880, he had constructed the first real calorimeter.

CALORIMETERS. Essential to calorimetry is the calorimeter, which can be any device for accurately measuring the temperature of a substance before and after a change occurs. A calorimeter can be as simple as a styrofoam cup. Its quality as an insulator, which makes styrofoam ideal for holding in the warmth of coffee and protecting the hand from scalding as well, also makes styrofoam an excellent material for calorimetric testing. With a styrofoam calorimeter, the temperature of the substance inside the cup is measured, a reaction is allowed to take place, and afterward, the temperature is measured a second time.

The most common type of calorimeter used is the bomb calorimeter, designed to measure the heat of combustion. Typically, a bomb calorimeter consists of a large container filled with water, into which is placed a smaller container, the combustion crucible. The crucible is made of metal, having thick walls with an opening through which oxygen can be introduced. In addition, the combustion crucible is designed to be connected to a source of electricity.

In conducting a calorimetric test using a bomb calorimeter, the substance or object to be studied is placed inside the combustion crucible and ignited. The resulting reaction usually occurs so quickly that it resembles the explosion of a bomb—hence, the name "bomb calorimeter." Once the "bomb" goes off, the resulting transfer of heat creates a temperature change in the water, which can be readily gauged with a thermometer.

To study heat changes at temperatures higher than the boiling point of water (212°F or 100°C), physicists use substances with higher boiling points. For experiments involving extremely large temperature ranges, an aneroid (without liquid) calorimeter may be used. In this case, the lining of the combustion crucible must be of a metal, such as copper, with a high coefficient or factor of thermal conductivity.

HEAT ENGINES

The bomb calorimeter that Berthelot designed in 1880 measured the caloric value of fuels, and was applied to determining the thermal efficiency of a heat engine. A heat engine is a machine that absorbs heat at a high temperature, performs mechanical work, and as a result, gives off heat at a lower temperature.

The desire to create efficient heat engines spurred scientists to a greater understanding of thermodynamics, and this resulted in the laws of thermodynamics, discussed at the conclusion of this essay. Their efforts were intimately connected with one of the greatest heat engines ever created, a machine that literally powered the industrialized world during the nineteenth century: the steam engine.

HOW A STEAM ENGINE WORKS. Like all heat engines (except reverse heat engines such as the refrigerator, discussed below), a steam engine pulls heat from a high-temperature reservoir to a low-temperature reservoir, and in the process, work is accomplished. The hot steam from the high-temperature reservoir makes possible the accomplishment of work, and when the energy is extracted from the steam, the steam condenses in the low-temperature reservoir, becoming relatively cool water.

A steam engine is an external-combustion engine, as opposed to the internal-combustion engine that took its place at the forefront of industrial technology at the beginning of the twentieth century. Unlike an internal-combustion engine, a steam engine burns its fuel outside the engine. That fuel may be simply firewood, which is used to heat water and create steam. The thermal energy of the steam is then used to power a piston moving inside a cylinder, thus, converting thermal energy to mechanical energy for purposes such as moving a train.

EVOLUTION OF STEAM POWER. As with a number of advanced concepts in science and technology, the historical roots of the steam engine can be traced to the Greeks, who—just as they did with ideas such as the atom or the Sun-centered model of the universe—thought about it, but failed to develop it. The great inventor Hero of Alexandria (c. 65-125) actually created several steam-powered devices, but he perceived these as mere novelties, hardly worthy of scientific attention. Though Europeans adopted water power, as, for instance, in waterwheels, during the late ancient and medieval periods, further progress in steam power did not occur for some 1,500 years.

Following the work of French physicist Denis Papin (1647-1712), who invented the pressure cooker and conducted the first experiments with the use of steam to move a piston, English engineer Thomas Savery (c. 1650-1715) built the first steam engine. Savery had abandoned the use of the piston in his machine, but another English engineer, Thomas Newcomen (1663-1729), reintroduced the piston for his own steam-engine design.

Then in 1763, a young Scottish engineer named James Watt (1736-1819) was repairing a Newcomen engine and became convinced he could build a more efficient model. His steam engine, introduced in 1769, kept the heating and cooling processes separate, eliminating the need for the engine to pause in order to reheat. These and other innovations that followed—including the introduction of a high-pressure steam engine by English inventor Richard Trevithick (1771-1833)—transformed the world.

CARNOT PROVIDES THEORETICAL UNDERSTANDING. The men who developed the steam engine were mostly practical-minded figures who wanted only to build a better machine; they were not particularly concerned with the theoretical explanation for its workings. Then in 1824, a French physicist and engineer by the name of Sadi Carnot (1796-1832) published his sole work, the highly influential *Reflections on the Motive Power of Fire* (1824), in which he discussed heat engines scientifically.

In *Reflections*, Carnot offered the first definition of work in terms of physics, describing it as "weight lifted through a height." Analyzing Watt's steam engine, he also conducted groundbreaking studies in the nascent science of thermodynamics. Every heat engine, he explained, has a theoretical limit of efficiency related to the temperature difference in the engine: the greater the difference between the lowest and highest temperature, the more efficient the engine.

Carnot's work influenced the development of more efficient steam engines, and also had an impact on the studies of other physicists investigating the relationship between work, heat, and energy. Among these was William Thomson, Lord Kelvin (1824-1907). In addition to coining the term "thermodynamics," Kelvin developed the Kelvin scale of absolute temperature and established the value of absolute zero, equal to -273.15°C or -459.67°F.

According to Carnot's theory, maximum effectiveness was achieved by a machine that could reach absolute zero. However, later developments in the understanding of thermodynamics, as discussed below, proved that both maximum efficiency and absolute zero are impossible to attain.

REVERSE HEAT ENGINES. It is easy to understand that a steam engine is a heat engine: after all, it produces heat. But how is it that a refrigerator, an air conditioner, and other cooling machines are also heat engines? Moreover, given the fact that cold is the absence of heat and heat is energy, one might ask how a refrigerator or air conditioner can possibly use energy to produce cold, which is the same as the absence of energy. In fact, cooling machines simply reverse the usual process by which heat engines operate, and for this reason, they are called "reverse heat engines." Furthermore, they use energy to extract heat.

A steam engine takes heat from a high-temperature reservoir—the place where the water is turned into steam—and uses that energy to produce work. In the process, energy is lost and the heat moves to a low-temperature reservoir, where it condenses to form relatively cool water. A refrigerator, on the other hand, pulls heat from a low-temperature reservoir called the evaporator,

into which flows heat from the refrigerated compartment—the place where food and other perishables are kept. The coolant from the evaporator take this heat to the condenser, a high-temperature reservoir at the back of the refrigerator, and in the process it becomes a gas. Heat is released into the surrounding air; this is why the back of a refrigerator is hot.

Instead of producing a work output, as a steam engine does, a refrigerator requires a work input—the energy supplied via the wall outlet. The principles of thermodynamics show that heat always flows from a high-temperature to a low-temperature reservoir, and reverse heat engines do not defy these laws. Rather, they require an external power source in order to effect the transfer of heat from a low-temperature reservoir, through the gases in the evaporator, to a high-temperature reservoir.

THE LAWS OF THERMODYNAMICS

THE FIRST LAW OF THERMODYNAMICS. There are three laws of thermodynamics, which provide parameters as to the operation of thermal systems in general, and heat engines in particular. The history behind the derivation of these laws is discussed in the essay on Thermodynamics; here, the laws themselves will be examined in brief form.

The physical law known as conservation of energy shows that within a system isolated from all outside factors, the total amount of energy remains the same, though transformations of energy from one form to another take place. The first law of thermodynamics states the same fact in a somewhat different manner.

According to the first law of thermodynamics, because the amount of energy in a system remains constant, it is impossible to perform work that results in an energy output greater than the energy input. Thus, it could be said that the conservation of energy law shows that "the glass is half full": energy is never lost. On the hand, the first law of thermodynamics shows that "the glass is half empty": no machine can ever produce more energy than was put into it. Hence, a perpetual motion machine is impossible, because in order to keep a machine running continually, there must be a continual input of energy.

THE SECOND LAW OF THERMODYNAMICS. The second law of ther-

KEY TERMS

ABSOLUTE ZERO: The temperature, defined as 0K on the Kelvin scale, at which the motion of molecules in a solid virtually ceases. The third law of thermodynamics establishes the impossibility of actually reaching absolute zero.

BTU (BRITISH THERMAL UNIT): A measure of energy or heat in the British system, often used in reference to the capacity of an air conditioner. A Btu is equal to 778 foot-pounds, or 1,054 joules.

CALORIE: A measure of heat or energy in the SI or metric system, equal to the heat that must be added to or removed from 1 gram of water to change its temperature by 1°C. The dietary Calorie (capital C) with which most people are familiar is the same as the kilocalorie.

CALORIMETRY: The measurement of heat gain or loss as a result of physical or chemical change.

CONDUCTION: The transfer of heat by successive molecular collisions. Conduction is the principal means of heat transfer in solids, particularly metals.

CONSERVATION OF ENERGY: A law of physics stating that within a system isolated from all other outside factors, the total amount of energy remains the same, though transformations of energy from one form to another take place. The first law of thermodynamics is the same as the conservation of energy.

CONVECTION: The transfer of heat through the motion of hot fluid from one place to another. In physics, a "fluid" can be either a gas or a liquid, and convection is the principal means of heat transfer, for instance, in air and water.

ENERGY: The ability to accomplish work.

ENTROPY: The tendency of natural systems toward breakdown, and specifically, the tendency for the energy in a system to be dissipated. Entropy is closely related to the second law of thermodynamics.

FIRST LAW OF THERMODYNAMICS: A law stating that the amount of energy in a system remains constant, and therefore, it is impossible to perform work that results in an energy, output greater than the energy input. This is the same as the conservation of energy.

FOOT-POUND: The principal unit of energy—and, thus, of heat—in the British or English system. The metric or SI unit is the joule. A foot-pound (ft • lb) is equal to 1.356 J.

HEAT: Internal thermal energy that flows from one body of matter to another. Heat is transferred by three methods conduction, convection, and radiation.

HEAT ENGINE: A machine that absorbs heat at a high temperature, performs mechanical work, and, as a result, gives off heat at a lower temperature.

JOULE: The principal unit of energy—and, thus, of heat—in the SI or metric system, corresponding to 1 newton-meter (N • m). A joule (J) is equal to 0.7376 foot-pounds.

KELVIN SCALE: Established by William Thomson, Lord Kelvin (1824-1907), the Kelvin scale measures temperature in relation to absolute zero, or 0K. (Units in the Kelvin system, known as Kelvins, do not include the word or symbol for degree.) The Kelvin and Celsius scales are directly related; hence, Celsius temperatures can be converted to Kelvins by adding 273.15.

KILOCALORIE: A measure of heat or energy in the SI or metric system, equal to the heat that must be added to or removed from 1 kilogram of water to change its temperature by 1°C. As its name suggests, a kilocalorie is 1,000 calories. The dietary Calorie (capital C) with which most people are familiar, is the same as the kilocalorie.

KINETIC ENERGY: The energy that an object possesses by virtue of its motion.

POTENTIAL ENERGY: The energy that an object possesses due to its position.

RADIATION: The transfer of heat by means of electromagnetic waves, which require no physical medium (for example, water or air) for the transfer. Earth receives the Sun's heat by means of radiation.

SECOND LAW OF THERMODYNAMICS: A law of thermodynamics stating that no engine can be constructed that simply takes heat from a source and performs an equivalent amount of work. Some of the heat will always be lost, and, therefore, it is impossible to build a perfectly efficient engine. This is a result of the fact that the natural flow of heat is always from a high-temperature reservoir to a low-temperature reservoir—a fact expressed in the concept of entropy. The second law is sometimes referred to as "the law of entropy."

modynamics begins from the fact that the natural flow of heat is always from a high-temperature to a low-temperature reservoir. As a result, no engine can be constructed that simply takes heat from a source and performs an equivalent amount of work: some of the heat will always be lost. In other words, it is impossible to build a perfectly efficient engine.

In effect, the second law of thermodynamics compounds the "bad news" delivered by the first law with some even worse news: though it is true that energy is never lost, the energy available for work output will never be as great as the energy put into a system. Linked to the second law is the concept of entropy, the tendency of natural systems toward breakdown, and specifically, the tendency for the energy in a system to be dissipated. "Dissipated" in this context means that the high- and low-temperature reservoirs approach equal

temperatures, and as this occurs, entropy increases.

THE THIRD LAW OF THERMODYNAMICS. Entropy also plays a part in the third law of thermodynamics, which states that at the temperature of absolute zero, entropy also approaches zero. This might seem to counteract the "worse news" of the second law, but in fact, what the third law shows is that absolute zero is impossible to reach.

As stated earlier, Carnot's engine would achieve perfect efficiency if its lowest temperature were the same as absolute zero; but the second law of thermodynamics shows that a perfectly efficient machine is impossible. Relativity theory (which first appeared in 1905, the same year as the third law of thermodynamics) showed that matter can never exceed the speed of light. In the same way, the collective effect of the second and third laws is to prove that absolute

KEY TERMS CONTINUED

SPECIFIC HEAT: The amount of heat that must be added to, or removed from, a unit of mass of a given substance to change its temperature by 1°C. A kilocalorie is the specific heat of 1 gram of water.

SYSTEM: In physics, the term "system" usually refers to any set of physical interactions isolated from the rest of the universe. Anything outside of the system, including all factors and forces irrelevant to a discussion of that system, is known as the environment.

TEMPERATURE: The direction of internal energy flow between two systems when heat is being transferred. Temperature measures the average molecular kinetic energy in transit between those systems.

THERMAL ENERGY: Heat energy, a form of kinetic energy produced by the movement of atomic or molecular parti-

cles. The greater the movement of these particles, the greater the thermal energy.

THERMAL EQUILIBRIUM: The state that exists when two systems have the same temperature. As a result, there is no exchange of heat between them.

THERMODYNAMICS: The study of the relationships between heat, work, and energy.

THIRD LAW OF THERMODYNAMICS: A law of thermodynamics which states that at the temperature of absolute zero, entropy also approaches zero. Zero entropy would contradict the second law of thermodynamics, meaning that absolute zero is, therefore, impossible to reach.

WORK: The exertion of force over a given distance to displace or move an object. Work is, thus, the product of force and distance exerted in the same direction.

---

zero—the temperature at which molecular motion in all forms of matter theoretically ceases—can never be reached.

WHERE TO LEARN MORE

Beiser, Arthur. *Physics,* 5th ed. Reading, MA: Addison-Wesley, 1991.

Bonnet, Robert L and Dan Keen. *Science Fair Projects: Physics.* Illustrated by Frances Zweifel. New York: Sterling, 1999.

Encyclopedia of Thermodynamics (Web site). <http://therion.minpet.unibas.ch/minpet/groups/thermodict/> (April 12, 2001).

Friedhoffer, Robert. *Physics Lab in the Home.* Illustrated by Joe Hosking. New York: Franklin Watts, 1997.

Manning, Mick and Brita Granström. *Science School.* New York: Kingfisher, 1998.

Macaulay, David. *The New Way Things Work.* Boston: Houghton Mifflin, 1998.

Moran, Jeffrey B. *How Do We Know the Laws of Thermodynamics?* New York: Rosen Publishing Group, 2001.

Santrey, Laurence. *Heat.* Illustrated by Lloyd Birmingham. Mahwah, NJ: Troll Associates, 1985.

Suplee, Curt. *Everyday Science Explained.* Washington, D.C.: National Geographic Society, 1996.

"Temperature and Thermodynamics" PhysLINK.com (Web site). <http://www.physlink.com/ae_thermo.cfm> (April 12, 2001).

TEMPERATURE

CONCEPT

Temperature is one of those aspects of the everyday world that seems rather abstract when viewed from the standpoint of physics. In scientific terms, it is not simply a measure of hot and cold, but an indicator of molecular motion and energy flow. Thermometers measure temperature by a number of means, including the expansion that takes place in a medium such as mercury or alcohol. These measuring devices are gauged in several different ways, with scales based on the freezing and boiling points of water—as well as, in the case of the absolute temperature scale, the point at which all molecular motion virtually ceases.

HOW IT WORKS

Heat

Energy appears in many forms, including thermal energy, or the energy associated with heat. Heat is internal thermal energy that flows from one body of matter to another—or, more specifically, from a system at a higher temperature to one at a lower temperature.

Two systems at the same temperature are said to be in a state of thermal equilibrium. When this occurs, there is no exchange of heat. Though people ordinarily speak of "heat" as an expression of relative warmth or coldness, in physical terms, heat only exists in transfer between two systems. It is never something inherently part of a system; thus, unless there is a transfer of internal energy, there is no heat, scientifically speaking.

HEAT: ENERGY IN TRANSIT. Thus, heat cannot be said to exist unless there is one system in contact with another system of differing temperature. This can be illustrated by way of the old philosophical question: "If a tree falls in the woods when there is no one to hear it, does it make a sound?" From a physicist's point of view, of course, sound waves are emitted whether or not there is an ear to receive their vibrations; but, consider this same scenario in terms of heat. First, replace the falling tree with a hypothetical object possessing a certain amount of internal energy; then replace sound waves with heat. In this case, if this object is not in contact with something else that has a different temperature, it "does not make a sound"—in other words, it transfers no internal energy, and, thus, there is no heat from the standpoint of physics.

This could even be true of two incredibly "hot" objects placed next to one another inside a vacuum—an area devoid of matter, including air. If both have the same temperature, there is no heat, only two objects with high levels of internal energy. Note that a vacuum was specified: assuming there was air around them, and that the air was of a lower temperature, both objects would then be transferring heat to the air.

RELATIVE MOTION BETWEEN MOLECULES. If heat is internal thermal energy in transfer, from whence does this energy originate? From the movement of molecules. Every type of matter is composed of molecules, and those molecules are in motion relative to one another. The greater the amount of relative motion between molecules, the greater the kinetic energy, or the energy of movement, which is manifested as thermal energy. Thus, "heat"—to use the everyday term for what physicists describe as thermal energy—is really nothing more than the result of relative molecular motion. Thus, thermal energy is sometimes identified as molecular translational energy.

Note that the molecules are in relative motion, meaning that if one were "standing" on a molecule, one would see the other molecules moving. This is not the same as movement on the part of a large object composed of molecules; in this case, molecules themselves are not directly involved in relative motion.

Put another way, the movement of Earth through space is an entirely different type of movement from the relative motion of objects on Earth—people, animals, natural forms such as clouds, manmade forms of transportation, and so forth. In this example, Earth is analogous to a "large" item of matter, such as a baseball, a stream of water, or a cloud of gas.

The smaller objects on Earth are analogous to molecules, and, in both cases, the motion of the larger object has little direct impact on the motion of smaller objects. Hence, as discussed in the Frame of Reference essay, it is impossible to perceive with one's senses the fact that Earth is actually hurling through space at incredible speeds.

MOLECULAR MOTION AND PHASES OF MATTER. The relative motion of molecules determines phase of matter—that is, whether something is a solid, liquid, or gas. When molecules move quickly in relation to one another, they exert a small electromagnetic attraction toward one another, and the larger material of which they are a part is called a gas. A liquid, on the other hand, is a type of matter in which molecules move at moderate speeds in relation to one another, and therefore exert a moderate intermolecular attraction.

The kinetic theory of gases relates molecular motion to energy in gaseous substances. It does not work as well in relation to liquids and solids; nonetheless, it is safe to say that—generally speaking—a gas has more energy than a liquid, and a liquid more energy than a solid. In a solid, the molecules undergo very little relative motion: instead of bumping into each other, like gas molecules and (to a lesser extent) liquid molecules, solid molecules merely vibrate in place.

WILLIAM THOMSON, (BETTER KNOWN AS LORD KELVIN) ESTABLISHED WHAT IS NOW KNOWN AS THE KELVIN SCALE.

UNDERSTANDING TEMPERATURE

As with heat, temperature requires a scientific definition quite different from its common meaning. Temperature may be defined as a measure of the average molecular translational energy in a system—that is, in any material body.

Because it is an average, the mass or other characteristics of the body do not matter. A large quantity of one substance, because it has more molecules, possesses more thermal energy than a smaller quantity of that same substance. Since it has more thermal energy, it transfers more heat to any body or system with which it is in contact. Yet, assuming that the substance is exactly the same, the temperature, as a measure of average energy, will be the same as well.

Temperature determines the direction of internal energy flow between two systems when heat is being transferred. This can be illustrated through an experience familiar to everyone: having one's temperature taken with a thermometer. If one has a fever, one's mouth will be warmer than the thermometer, and therefore heat will be transferred to the thermometer from the mouth until the two objects have the same temperature.

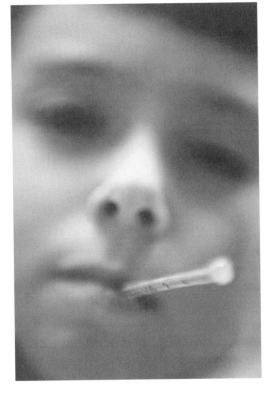

A THERMOMETER WORKS BY MEASURING THE LEVEL OF THERMAL EXPANSION EXPERIENCED BY A MATERIAL WITHIN THE THERMOMETER. MERCURY HAS BEEN A COMMON THERMOMETER MATERIAL SINCE THE 1700S. *(Photograph by Michael Prince/Corbis. Reproduced by permission.)*

At that point of thermal equilibrium, a temperature reading can be taken from the thermometer.

TEMPERATURE AND HEAT FLOW. The principles of thermodynamics—the study of the relationships between heat, work, and energy, show that heat always flows from an area of higher temperature to an area of lower temperature. The opposite simply cannot happen, because coldness, though it is very real in terms of sensory experience, is not an independent phenomenon. There is not, strictly speaking, such a thing as "cold"—only the absence of heat, which produces the sensation of coldness.

One might pour a kettle of boiling water into a cold bathtub to heat it up; or put an ice cube in a hot cup of coffee "to cool it down." These seem like two very different events, but from the standpoint of thermodynamics, they are exactly the same. In both cases, a body of high temperature is placed in contact with a body of low temperature, and in both cases, heat passes from the high-temperature body to the low-temperature one.

The boiling water warms the tub of cool water, and due to the high ratio of cool water to boiling water in the bathtub, the boiling water expends all its energy raising the temperature in the bathtub as a whole. The greater the ratio of very hot water to cool water, on the other hand, the warmer the bathtub will be in the end. But even after the bath water is heated, it will continue to lose heat, assuming the air in the room is not warmer than the water in the tub. If the water in the tub is warmer than the air, it immediately begins transferring thermal energy to the low-temperature air until their temperatures are equalized.

As for the coffee and the ice cube, what happens is quite different from, indeed, opposite to, the common understanding of the process. In other words, the ice does not "cool down" the coffee: the coffee warms up the ice and presumably melts it. Once again, however, it expends at least some of its thermal energy in doing so, and as a result, the coffee becomes cooler than it was.

If the coffee is placed inside a freezer, there is a large temperature difference between it and the surrounding environment—so much so that if it is left for hours, the once-hot coffee will freeze. But again, the freezer does not cool down the coffee; the molecules in the coffee respond to the temperature difference by working to warm up the freezer. In this case, they have to "work overtime," and since the freezer has a constant supply of electrical energy, the heated molecules of the coffee continue to expend themselves in a futile effort to warm the freezer. Eventually, the coffee loses so much energy that it is frozen solid; meanwhile, the heat from the coffee has been transferred outside the freezer to the atmosphere in the surrounding room.

THERMAL EXPANSION AND EQUILIBRIUM. Temperature is related to the concept of thermal equilibrium, and has an effect on thermal expansion. As discussed below, as well as within the context of thermal expansion, a thermometer provides a gauge of temperature by measuring the level of thermal expansion experienced by a material (for example, mercury) within the thermometer.

In the examples used earlier—the thermometer in the mouth, the hot water in the cool bathtub, and the ice cube in the cup of coffee—the systems in question eventually reach thermal equilibrium. This is rather like averaging their temperatures, though, in fact, the equation involved is more complicated than a simple arithmetic average.

In the case of an ordinary mercury thermometer, the need to achieve thermal equilibrium explains why one cannot get an instantaneous temperature reading: first, the mouth transfers heat to the thermometer, and once both mouth and thermometer reach the same temperature, they are in thermal equilibrium. At that point, it is possible to gauge the temperature of the mouth by reading the thermometer.

REAL-LIFE APPLICATIONS

DEVELOPMENT OF THE THERMOMETER

A thermometer can be defined scientifically as a device that gauges temperature by measuring a temperature-dependent property, such as the expansion of a liquid in a sealed tube. As with many aspects of scientific or technological knowledge, the idea of the thermometer appeared in ancient times, but was never developed. Again, like so many other intellectual phenomena, it lay dormant during the medieval period, only to be resurrected at the beginning of the modern era.

The Greco-Roman physician Galen (c. 129-216) was among the first thinkers to envision a scale for measuring temperature. Of course, what he conceived of as "temperature" was closer to the everyday meaning of that term, not its more precise scientific definition: the ideas of molecular motion, heat, and temperature discussed in this essay emerged only in the period beginning about 1750. In any case, Galen proposed that equal amounts of boiling water and ice be combined to establish a "neutral" temperature, with four units of warmth above it and four degrees of cold below.

THE THERMOSCOPE. The great physicist Galileo Galilei (1564-1642) is sometimes credited with creating the first practical temperature measuring device, called a thermoscope. Certainly Galileo—whether or not he was the first—did build a thermoscope, which consisted of a long glass tube planted in a container of liquid. Prior to inserting the tube into the liquid—which was usually colored water, though Galileo's thermoscope used wine—as much air as possible was removed from the tube. This created a vacuum, and as a result of pressure differences between the liquid and the interior of the thermoscope tube, some of the liquid went into the tube.

But the liquid was not the thermometric medium—that is, the substance whose temperature-dependent property changes the thermoscope measured. (Mercury, for instance, is the thermometric medium in most thermometers today.) Instead, the air was the medium whose changes the thermoscope measured: when it was warm, the air expanded, pushing down on the liquid; and when the air cooled, it contracted, allowing the liquid to rise.

It is interesting to note the similarity in design between the thermoscope and the barometer, a device for measuring atmospheric pressure invented by Italian physicist Evangelista Torricelli (1608-1647) around the same time. Neither were sealed, but by the mid-seventeenth century, scientists had begun using sealed tubes containing liquid instead of air. These were the first true thermometers.

EARLY THERMOMETERS. Ferdinand II, Grand Duke of Tuscany (1610-1670), is credited with developing the first thermometer in 1641. Ferdinand's thermometer used alcohol sealed in glass, which was marked with a temperature scale containing 50 units. It did not, however, designate a value for zero.

English physicist Robert Hooke (1635-1703) created a thermometer using alcohol dyed red. Hooke's scale was divided into units equal to about 1/500 of the volume of the thermometric medium, and for the zero point, he chose the temperature at which water freezes. Thus, Hooke established a standard still used today; likewise, his thermometer itself set a standard. Built in 1664, it remained in use by the Royal Society—the foremost organization for the advancement of science in England during the early modern period—until 1709.

Olaus Roemer (1644-1710), a Danish astronomer, introduced another important standard. In 1702, he built a thermometer based not on one but two fixed points, which he designated as the temperature of snow or crushed ice, and the boiling point of water. As with Hooke's use of the freezing point, Roemer's idea of the freezing and boiling points of water as the two parameters for temperature measurements has remained in use ever since.

TEMPERATURE SCALES

FAHRENHEIT. Not only did he develop the Fahrenheit scale, oldest of the temperature scales still used in Western nations today, but German physicist Daniel Fahrenheit (1686-1736) also built the first thermometer to contain mercury as a thermometric medium. Alcohol has a low boiling point, whereas mercury remains fluid at a wide range of temperatures. In addition, it expands and contracts at a very constant rate, and tends not to stick to glass. Furthermore, its silvery color makes a mercury thermometer easy to read.

Fahrenheit also conceived the idea of using "degrees" to measure temperature in his thermometer, which he introduced in 1714. It is no mistake that the same word refers to portions of a circle, or that exactly 180 degrees—half the number in a circle—separate the freezing and boiling points for water on Fahrenheit's thermometer. Ancient astronomers attempting to denote movement in the skies used a circle with 360 degrees as a close approximation of the ratio between days and years. The number 360 is also useful for computations, because it has a large quantity of divisors, as does 180—a total of 16 whole-number divisors other than 1 and itself.

Though it might seem obvious that 0 should denote the freezing point and 180 the boiling point on Fahrenheit's scale, such an idea was far from obvious in the early eighteenth century. Fahrenheit considered the idea not only of a 0-to-180 scale, but also of a 180-to-360 scale. In the end, he chose neither—or rather, he chose not to equate the freezing point of water with zero on his scale. For zero, he chose the coldest possible temperature he could create in his laboratory, using what he described as "a mixture of sal ammoniac or sea salt, ice, and water." Salt lowers the melting point of ice (which is why it is used in the northern United States to melt snow and ice from the streets on cold winter days), and, thus, the mixture of salt and ice produced an extremely cold liquid water whose temperature he equated to zero.

With Fahrenheit's scale, the ordinary freezing point of water was established at 32°, and the boiling point exactly 180° above it, at 212°. Just a few years after he introduced his scale, in 1730, a French naturalist and physicist named Rene Antoine Ferchault de Reaumur (1683-1757) presented a scale for which 0° represented the freezing point of water and 80° the boiling point. Although the Reaumur scale never caught on to the same extent as Fahrenheit's, it did include one valuable addition: the specification that temperature values be determined at standard sea-level atmospheric pressure.

CELSIUS. With its 32-degree freezing point and its 212-degree boiling point, the Fahrenheit system is rather ungainly, lacking the neat orderliness of a decimal or base-10 scale. The latter quality became particularly important when, 10 years after the French Revolution of 1789, France adopted the metric system for measuring length, mass, and other physical phenomena. The metric system eventually spread to virtually the entire world, with the exception of English-speaking countries, where the more cumbersome British system still prevails. But even in the United States and Great Britain, scientists use the metric system. The metric temperature measure is the Celsius scale, created in 1742 by Swedish astronomer Anders Celsius (1701-1744).

Like Fahrenheit, Celsius chose the freezing and boiling points of water as his two reference points, but he determined to set them 100, rather than 180, degrees apart. Interestingly, he planned to equate 0° with the boiling point, and 100° with the freezing point—proving that even the most apparently obvious aspects of a temperature scale were once open to question. Only in 1750 did fellow Swedish physicist Martin Strömer change the orientation of the Celsius scale.

Celsius's scale was based not simply on the boiling and freezing points of water, but, specifically, those points at normal sea-level atmospheric pressure. The latter, itself a unit of measure known as an atmosphere (atm), is equal to 14.7 lb/in², or 101,325 pascals in the metric system. A Celsius degree is equal to 1/100 of the difference between the freezing and boiling temperatures of water at 1 atm.

The Celsius scale is sometimes called the centigrade scale, because it is divided into 100 degrees, *cent* being a Latin root meaning "hundred." By international convention, its values were refined in 1948, when the scale was redefined in terms of temperature change for an ideal gas, as well as the triple point of water. (Triple point is the temperature and pressure at which a substance is at once a solid, liquid, and vapor.) As a result of these refinements, the boiling point of water on the Celsius scale is actually 99.975°. This represents a difference equal to about 1 part in 4,000—hardly significant in daily life, though a significant change from the standpoint of the precise measurements made in scientific laboratories.

KELVIN. In about 1787, French physicist and chemist J. A. C. Charles (1746-1823) made an interesting discovery: that at 0°C, the volume of gas at constant pressure drops by 1/273 for every Celsius degree drop in temperature. This seemed to suggest that the gas would simply disappear if cooled to -273°C, which, of course, made no sense. In any case, the gas would most likely become first a liquid, and then a solid, long before it reached that temperature.

The man who solved the quandary raised by Charles's discovery was born a year after Charles—who also formulated Charles's law—died. He was William Thomson, Lord Kelvin (1824-1907), and in 1848, he put forward the suggestion that it was molecular translational energy, and not volume, that would become zero at -273°C. He went on to establish what came to be known as the Kelvin scale.

Sometimes known as the absolute temperature scale, the Kelvin scale is based not on the freezing point of water, but on absolute zero—the temperature at which molecular motion comes to a virtual stop. This is -273.15°C (-459.67°F), which in the Kelvin scale is designated as 0K. (Kelvin measures do not use the term or symbol for "degree.")

Though scientists normally use metric or SI measures, they prefer the Kelvin scale to Celsius, because the absolute temperature scale is directly related to average molecular translational energy. Thus, if the Kelvin temperature of an object is doubled, this means that its average molecular translational energy has doubled as well. The same cannot be said if the temperature were doubled from, say, 10°C to 20°C, or from 40°C to 80°F, since neither the Celsius nor the Fahrenheit scale is based on absolute zero.

CONVERSIONS. The Kelvin scale is, however, closely related to the Celsius scale, in that a difference of 1 degree measures the same amount of temperature in both. Therefore, Celsius temperatures can be converted to Kelvins by adding 273.15. There is also an absolute temperature scale that uses Fahrenheit degrees. This is the Rankine scale, created by Scottish engineer William Rankine (1820-1872), but it is seldom used today: scientists and others who desire absolute temperature measures prefer the precision and simplicity of the Celsius-based Kelvin scale.

Conversion between Celsius and Fahrenheit figures is a bit more challenging. To convert a temperature from Celsius to Fahrenheit, multiply by 9/5 and add 32. It is important to perform the steps in that order, because reversing them will produce a wrong answer. Thus, 100°C multiplied by 9/5 or 1.8 equals 180, which, when added to 32 equals 212°F. Obviously, this is correct, since 100°C and 212°F each represent the boiling point of water. But, if one adds 32 to 100°, then multiplies it by 9/5, the result is 237.6°F—an incorrect answer.

For converting Fahrenheit temperatures to Celsius, there are also two steps, involving multiplication and subtraction, but the order is reversed. Here, the subtraction step is performed before the multiplication step: thus, 32 is subtracted from the Fahrenheit temperature, then the result is multiplied by 5/9. Beginning with 212°F, if 32 is subtracted, this equals 180. Multiplied by 5/9, the result is 100°C—the correct answer.

One reason the conversion formulae use fractions instead of decimal fractions (what most people simply call "decimals") is that 5/9 is a repeating decimal fraction (0.55555....) Furthermore, the symmetry of 5/9 and 9/5 makes memorization easy. One way to remember the formula is that *Fahrenheit* is multiplied by a *fraction*—since 5/9 is a real fraction, whereas 9/5 is actually a whole number plus a fraction.

THERMOMETERS

As discussed earlier, with regard to the early history of the thermometer, it is important that the glass tube be kept sealed; otherwise, atmospheric pressure contributes to inaccurate readings, because it influences the movement of the thermometric medium. Also important is the choice of the thermometric medium itself.

Water quickly proved unreliable, due to its unusual properties: it does not expand uniformly with a rise in temperature, or contract uniformly with a lowered temperature. Rather, it

KEY TERMS

ABSOLUTE ZERO: The temperature, defined as 0K on the Kelvin scale, at which the motion of molecules in a solid virtually ceases.

CELSIUS SCALE: A scale of temperature, sometimes known as the centigrade scale, created in 1742 by Swedish astronomer Anders Celsius (1701-1744). The Celsius scale establishes the freezing and boiling points of water at 0° and 100°, respectively. To convert a temperature from the Celsius to the Fahrenheit scale, multiply by 9/5 and add 32. The Celsius scale is part of the metric system used by most non-English speaking countries today. Though the worldwide scientific community uses the metric or SI system for most measurements, scientists prefer the related Kelvin scale.

FAHRENHEIT SCALE: The oldest of the temperature scales still used in Western nations today, created in 1714 by German physicist Daniel Fahrenheit (1686-1736).

The Fahrenheit scale establishes the freezing and boiling points of water at 32° and 212° respectively. To convert a temperature from the Fahrenheit to the Celsius scale, subtract 32 and multiply by 5/9. Most English-speaking countries use the Fahrenheit scale.

HEAT: Internal thermal energy that flows from one body of matter to another.

KELVIN SCALE: Established by William Thomson, Lord Kelvin (1824-1907), the Kelvin scale measures temperature in relation to absolute zero, or 0K. (Units in the Kelvin system, known as Kelvins, do not include the word or symbol for degree.) The Kelvin and Celsius scales are directly related; hence, Celsius temperatures can be converted to Kelvins by adding 273.15. The Kelvin scale is used almost exclusively by scientists.

KINETIC ENERGY: The energy that an object possesses by virtue of its motion.

reaches its maximum density at 39.2°F (4°C), and is less dense both above and below that temperature. Therefore, alcohol, which responds in a much more uniform fashion to changes in temperature, took its place.

MERCURY THERMOMETERS. Alcohol is still used in thermometers today, but the preferred thermometric medium is mercury. As noted earlier, its advantages include a much higher boiling point, a tendency not to stick to glass, and a silvery color that makes its levels easy to gauge visually. Like alcohol, mercury expands at a uniform rate with an increase in temperature: hence, the higher the temperature, the higher the mercury stands in the thermometer.

In a typical mercury thermometer, mercury is placed in a long, narrow sealed tube called a capillary. The capillary is inscribed with figures for a calibrated scale, usually in such a way as to allow easy conversions between Fahrenheit and Celsius. A thermometer is calibrated by measuring the difference in height between mercury at the freezing point of water, and mercury at the boiling point of water. The interval between these two points is then divided into equal increments—180, as we have seen, for the Fahrenheit scale, and 100 for the Celsius scale.

ELECTRIC THERMOMETERS. Faster temperature measures can be obtained by thermometers using electricity. All matter displays a certain resistance to electrical current, a resistance that changes with temperature. Therefore, a resistance thermometer uses a fine wire wrapped around an insulator, and when a change in temperature occurs, the resistance in the wire changes as well. This makes possible much quick-

KEY TERMS CONTINUED

MOLECULAR TRANSLATIONAL ENERGY: The kinetic energy in a system produced by the movement of molecules in relation to one another.

SYSTEM: In physics, the term "system" usually refers to any set of physical interactions, or any material body, isolated from the rest of the universe. Anything outside of the system, including all factors and forces irrelevant to a discussion of that system, is known as the environment.

TEMPERATURE: A measure of the average kinetic energy—or molecular translational energy in a system. Differences in temperature determine the direction of internal energy flow between two systems when heat is being transferred.

THERMAL ENERGY: Heat energy, a form of kinetic energy produced by the movement of atomic or molecular particles. The greater the movement of these particles, the greater the thermal energy.

THERMAL EQUILIBRIUM: The state that exists when two systems have the same temperature. As a result, there is no exchange of heat between them.

THERMODYNAMICS: The study of the relationships between heat, work, and energy.

THERMOMETRIC MEDIUM: A substance whose properties change with temperature. A mercury or alcohol thermometer measures such changes.

THERMOMETER: A device that gauges temperature by measuring a temperature-dependent property, such as the expansion of a liquid in a sealed tube, or resistance to electric current.

TRIPLE POINT: The temperature and pressure at which a substance is at once a solid, liquid, and vapor.

VACUUM: Space entirely devoid of matter, including air.

---

er temperature readings than those offered by a thermometer containing a traditional thermometric medium.

Resistance thermometers are highly reliable, but expensive, and are used primarily for very precise measurements. More practical for everyday use is a thermistor, which also uses the principle of electric resistance, but is much simpler and less expensive. Thermistors are used for providing measurements of the internal temperature of food, for instance, and for measuring human body temperature.

Another electric temperature-measurement device is a thermocouple. When wires of two different materials are connected, this creates a small level of voltage that varies as a function of temperature. A typical thermocouple uses two junctions: a reference junction, kept at some con-

stant temperature, and a measurement junction. The measurement junction is applied to the item whose temperature is to be measured, and any temperature difference between it and the reference junction registers as a voltage change, which is measured with a meter connected to the system.

OTHER TYPES OF THERMOMETER. A pyrometer also uses electromagnetic properties, but of a very different kind. Rather than responding to changes in current or voltage, the pyrometer is a gauge that responds to visible and infrared radiation. Temperature and color are closely related: thus, it is no accident that greens, blues, and purples, at one end of the visible light spectrum, are associated with coolness, while reds, oranges, and yellows at the other end are associated with heat. As with the thermocouple, a pyrometer has both a reference element and a measurement element, which compares light readings between the reference filament and the object whose temperature is being measured.

Still other thermometers, such as those in an oven that tell the user its internal temperature, are based on the expansion of metals with heat. In fact, there are a wide variety of thermometers, each suited to a specific purpose. A pyrometer, for instance, is good for measuring the temperature of an object that the thermometer itself does not touch.

WHERE TO LEARN MORE

About Temperature (Web site). <http://www.unidata.ucar.edu/staff/blynds/tmp.html> (April 18, 2001).

About Temperature Sensors (Web site). <http://www.temperatures.com> (April 18, 2001).

Gardner, Robert. *Science Projects About Methods of Measuring.* Berkeley Heights, N.J.: Enslow Publishers, 2000.

Maestro, Betsy and Giulio Maestro. *Temperature and You.* New York: Macmillan/McGraw-Hill School Publishing, 1990.

Megaconverter (Web site). <http://www.megaconverter.com> (April 18, 2001).

NPL: National Physics Laboratory: Thermal Stuff: Beginners' Guides (Web site). <http://www.npl.co.uk/npl/cbtm/thermal/stuff/guides.html> (April 18, 2001).

Royston, Angela. *Hot and Cold.* Chicago: Heinemann Library, 2001.

Santrey, Laurence. *Heat.* Illustrated by Lloyd Birmingham. Mahwah, N.J.: Troll Associates, 1985.

Suplee, Curt. *Everyday Science Explained.* Washington, D.C.: National Geographic Society, 1996.

Walpole, Brenda. *Temperature.* Illustrated by Chris Fairclough and Dennis Tinkler. Milwaukee, WI: Gareth Stevens Publishing, 1995.

THERMAL EXPANSION

CONCEPT

Most materials are subject to thermal expansion: a tendency to expand when heated, and to contract when cooled. For this reason, bridges are built with metal expansion joints, so that they can expand and contract without causing faults in the overall structure of the bridge. Other machines and structures likewise have built-in protection against the hazards of thermal expansion. But thermal expansion can also be advantageous, making possible the workings of thermometers and thermostats.

HOW IT WORKS

MOLECULAR TRANSLATIONAL ENERGY

In scientific terms, heat is internal energy that flows from a system of relatively high temperature to one at a relatively low temperature. The internal energy itself, identified as thermal energy, is what people commonly mean when they say "heat." A form of kinetic energy due to the movement of molecules, thermal energy is sometimes called molecular translational energy.

Temperature is defined as a measure of the average molecular translational energy in a system, and the greater the temperature change for most materials, as we shall see, the greater the amount of thermal expansion. Thus, all these aspects of "heat"—heat itself (in the scientific sense), as well as thermal energy, temperature, and thermal expansion—are ultimately affected by the motion of molecules in relation to one another.

MOLECULAR MOTION AND NEWTONIAN PHYSICS. In general, the kinetic energy created by molecular motion can be understood within the framework of classical physics—that is, the paradigm associated with Sir Isaac Newton (1642-1727) and his laws of motion. Newton was the first to understand the physical force known as gravity, and he explained the behavior of objects within the context of gravitational force. Among the concepts essential to an understanding of Newtonian physics are the mass of an object, its rate of motion (whether in terms of velocity or acceleration), and the distance between objects. These, in turn, are all components central to an understanding of how molecules in relative motion generate thermal energy.

The greater the momentum of an object—that is, the product of its mass multiplied by its rate of velocity—the greater the impact it has on another object with which it collides. The greater, also, is its kinetic energy, which is equal to one-half its mass multiplied by the square of its velocity. The mass of a molecule, of course, is very small, yet if all the molecules within an object are in relative motion—many of them colliding and, thus, transferring kinetic energy—this is bound to lead to a relatively large amount of thermal energy on the part of the larger object.

MOLECULAR ATTRACTION AND PHASES OF MATTER. Yet, precisely because molecular mass is so small, gravitational force alone cannot explain the attraction between molecules. That attraction instead must be understood in terms of a second type of force—electromagnetism—discovered by Scottish physicist James Clerk Maxwell (1831-1879). The details of electromagnetic force are not

BECAUSE STEEL HAS A RELATIVELY HIGH COEFFICIENT OF THERMAL EXPANSION, STANDARD RAILROAD TRACKS ARE
CONSTRUCTED SO THAT THEY CAN SAFELY EXPAND ON A HOT DAY WITHOUT DERAILING THE TRAINS TRAVELING OVER
THEM. *(Milepost 92 1/2/Corbis. Reproduced by permission.)*

important here; it is necessary only to know that all molecules possess some component of electrical charge. Since like charges repel and opposite charges attract, there is constant electromagnetic interaction between molecules, and this produces differing degrees of attraction.

The greater the relative motion between molecules, generally speaking, the less their attraction toward one another. Indeed, these two aspects of a material—relative attraction and motion at the molecular level—determine whether that material can be classified as a solid, liquid, or gas. When molecules move slowly in relation to one another, they exert a strong attraction, and the material of which they are a part is usually classified as a solid. Molecules of liquid, on the other hand, move at moderate speeds, and therefore exert a moderate attraction. When molecules move at high speeds, they exert little or no attraction, and the material is known as a gas.

PREDICTING THERMAL EXPANSION

COEFFICIENT OF LINEAR EX-PANSION. A coefficient is a number that serves as a measure for some characteristic or property. It may also be a factor against which other values are multiplied to provide a desired result. For any type of material, it is possible to

A MAN ICE FISHING IN MONTANA. BECAUSE OF THE UNIQUE THERMAL EXPANSION PROPERTIES OF WATER, ICE FORMS
AT THE TOP OF A LAKE RATHER THAN THE BOTTOM, THUS ALLOWING MARINE LIFE TO CONTINUE LIVING BELOW ITS
SURFACE DURING THE WINTER. *(Corbis. Reproduced by permission.)*

calculate the degree to which that material will expand or contract when exposed to changes in temperature. This is known, in general terms, as its coefficient of expansion, though, in fact, there are two varieties of expansion coefficient.

The coefficient of linear expansion is a constant that governs the degree to which the length of a solid will change as a result of an alteration in temperature For any given substance, the coefficient of linear expansion is typically a number expressed in terms of $10^{-5}/°C$. In other words, the value of a particular solid's linear expansion coefficient is multiplied by 0.00001 per °C. (The °C in the denominator, shown in the equation below, simply "drops out" when the coefficient of linear expansion is multiplied by the change in temperature.)

For quartz, the coefficient of linear expansion is 0.05. By contrast, iron, with a coefficient of 1.2, is 24 times more likely to expand or contract as a result of changes in temperature. (Steel has the same value as iron.) The coefficient for aluminum is 2.4, twice that of iron or steel. This means that an equal temperature change will produce twice as much change in the length of a bar of aluminum as for a bar of iron. Lead is among the most expansive solid materials, with a coefficient equal to 3.0.

CALCULATING LINEAR EXPANSION. The linear expansion of a given solid can be calculated according to the formula $\delta L = aL_O\Delta T$. The Greek letter delta (d) means "a change in"; hence, the first figure represents change in length, while the last figure in the equation stands for change in temperature. The letter a is the coefficient of linear expansion, and L_O is the original length.

Suppose a bar of lead 5 meters long experiences a temperature change of 10°C; what will its change in length be? To answer this, a (3.0 · 10^{-5}/°C) must be multiplied by L_O (5 m) and δT (10°C). The answer should be 150 & 10^{-5} m, or 1.5 mm. Note that this is simply a change in length related to a change in temperature: if the temperature is raised, the length will increase, and if the temperature is lowered by 10°C, the length will decrease by 1.5 mm.

VOLUME EXPANSION. Obviously, linear equations can only be applied to solids. Liquids and gases, classified together as fluids, conform to the shape of their container; hence, the "length" of any given fluid sample is the same as that of the solid that contains it. Fluids are, however, subject to volume expansion—that is, a change in volume as a result of a change in temperature.

To calculate change in volume, the formula is very much the same as for change in length; only a few particulars are different. In the formula $\delta V = bV_O\delta T$, the last term, again, means change in temperature, while δV means change in volume and V_O is the original volume. The letter b refers to the coefficient of volume expansion. The latter is expressed in terms of 10^{-4}/°C, or 0.0001 per °C.

Glass has a very low coefficient of volume expansion, 0.2, and that of Pyrex glass is extremely low—only 0.09. For this reason, items made of Pyrex are ideally suited for cooking. Significantly higher is the coefficient of volume expansion for glycerin, an oily substance associated with soap, which expands proportionally to a factor of 5.1. Even higher is ethyl alcohol, with a volume expansion coefficient of 7.5.

REAL-LIFE APPLICATIONS

LIQUIDS

Most liquids follow a fairly predictable pattern of gradual volume increase, as a response to an increase in temperature, and volume decrease, in response to a decrease in temperature. Indeed, the coefficient of volume expansion for a liquid generally tends to be higher than for a solid, and—with one notable exception discussed below— a liquid will contract when frozen.

The behavior of gasoline pumped on a hot day provides an example of liquid thermal expansion in response to an increase in temperature. When it comes from its underground tank at the gas station, the gasoline is relatively cool, but it will warm when sitting in the tank of an already warm car. If the car's tank is filled and the vehicle left to sit in the sun—in other words, if the car is not driven after the tank is filled—the gasoline might very well expand in volume faster than the fuel tank, overflowing onto the pavement.

ENGINE COOLANT. Another example of thermal expansion on the part of a liquid can be found inside the car's radiator. If the radiator is "topped off" with coolant on a cold day, an increase in temperature could very well cause the coolant to expand until it overflows. In the past, this produced a problem for car owners, because car engines released the excess volume of coolant onto the ground, requiring periodic replacement of the fluid.

Later-model cars, however, have an overflow container to collect fluid released as a result of volume expansion. As the engine cools down again, the container returns the excess fluid to the radiator, thus, "recycling" it. This means that newer cars are much less prone to overheating as older cars. Combined with improvements in radiator fluid mixtures, which act as antifreeze in cold weather and coolant in hot, the "recycling" process has led to a significant decrease in breakdowns related to thermal expansion.

WATER. One good reason not to use pure water in one's radiator is that water has a far higher coefficient of volume expansion than a typical engine coolant. This can be particularly hazardous in cold weather, because frozen water in a radiator could expand enough to crack the engine block.

In general, water—whose volume expansion coefficient in the liquid state is 2.1, and 0.5 in the solid state—exhibits a number of interesting characteristics where thermal expansion is concerned. If water is reduced from its boiling point—212°F (100°C) to 39.2°F (4°C) it will steadily contract, like any other substance responding to a drop in temperature. Normally, however, a substance continues to become denser as it turns from liquid to solid; but this does not occur with water.

At 32.9°F, water reaches it maximum density, meaning that its volume, for a given unit of mass, is at a minimum. Below that temperature, it "should" (if it were like most types of matter) continue to decrease in volume per unit of mass, but, in fact, it steadily begins to expand. Thus, it is less dense, with a greater volume per unit of mass, when it reaches the freezing point. It is for this reason that when pipes freeze in winter, they often burst—explaining why a radiator filled with water could be a serious problem in very cold weather.

In addition, this unusual behavior with regard to thermal expansion and contraction explains why ice floats: solid water is less dense than the liquid water below it. As a result, frozen water stays at the top of a lake in winter; since ice is a poor conductor of heat, energy cannot escape from the water below it in sufficient amounts to freeze the rest of the lake water. Thus, the water below the ice stays liquid, preserving plant and animal life.

GASES

THE GAS LAWS. As discussed, liquids expand by larger factors than solids do. Given the increasing amount of molecular kinetic energy for a liquid as compared to a solid, and for a gas as compared to a liquid, it should not be surprising, then, to learn that gases respond to changes in temperature with a volume change even greater than that of liquids. Of course, where a gas is concerned, "volume" is more difficult to measure, because a gas simply expands to fill its container. In order for the term to have any meaning, pressure and temperature must be specified as well.

A number of the gas laws describe the three parameters for gases: volume, temperature, and pressure. Boyle's law, for example, holds that in conditions of constant temperature, an inverse relationship exists between the volume and pressure of a gas: the greater the pressure, the less the volume, and vice versa. Even more relevant to the subject of thermal expansion is Charles's law.

Charles's law states that when pressure is kept constant, there is a direct relationship

between volume and temperature. As a gas heats up, its volume increases, and when it cools down, its volume reduces accordingly. Thus, if an air mattress is filled in an air-conditioned room, and the mattress is then taken to the beach on a hot day, the air inside will expand. Depending on how much its volume increases, the expansion of the hot air could cause the mattress to "pop."

VOLUME GAS THERMOMETERS. Whereas liquids and solids vary significantly with regard to their expansion coefficients, most gases follow more or less the same pattern of expansion in response to increases in temperature. The predictable behavior of gases in these situations led to the development of the constant gas thermometer, a highly reliable instrument against which other thermometers—including those containing mercury (see below)—are often gauged.

In a volume gas thermometer, an empty container is attached to a glass tube containing mercury. As gas is released into the empty container, this causes the column of mercury to move upward. The difference between the former position of the mercury and its position after the introduction of the gas shows the difference between normal atmospheric pressure and the pressure of the gas in the container. It is, then, possible to use the changes in volume on the part of the gas as a measure of temperature. The response of most gases, under conditions of low pressure, to changes in temperature is so uniform that volume gas thermometers are often used to calibrate other types of thermometers.

SOLIDS

Many solids are made up of crystals, regular shapes composed of molecules joined to one another as though on springs. A spring that is pulled back, just before it is released, is an example of potential energy, or the energy that an object possesses by virtue of its position. For a crystalline solid at room temperature, potential energy and spacing between molecules are relatively low. But as temperature increases and the solid expands, the space between molecules increases—as does the potential energy in the solid.

In fact, the responses of solids to changes in temperature tend to be more dramatic, at least when they are seen in daily life, than are the behaviors of liquids or gases under conditions of thermal expansion. Of course, solids actually respond less to changes in temperature than fluids do; but since they are solids, people expect their contours to be immovable. Thus, when the volume of a solid changes as a result of an increase in thermal energy, the outcome is more noteworthy.

JAR LIDS AND POWER LINES. An everyday example of thermal expansion can be seen in the kitchen. Almost everyone has had the experience of trying unsuccessfully to budge a tight metal lid on a glass container, and after running hot water over the lid, finding that it gives way and opens at last. The reason for this is that the high-temperature water causes the metal lid to expand. On the other hand, glass—as noted earlier—has a low coefficient of expansion. Otherwise, it would expand with the lid, which would defeat the purpose of running hot water over it. If glass jars had a high coefficient of expansion, they would deform when exposed to relatively low levels of heat.

Another example of thermal expansion in a solid is the sagging of electrical power lines on a hot day. This happens because heat causes them to expand, and, thus, there is a greater length of power line extending from pole to pole than under lower temperature conditions. It is highly unlikely, of course, that the heat of summer could be so great as to pose a danger of power lines breaking; on the other hand, heat can create a serious threat with regard to larger structures.

EXPANSION JOINTS. Most large bridges include expansion joints, which look rather like two metal combs facing one another, their teeth interlocking. When heat causes the bridge to expand during the sunlight hours of a hot day, the two sides of the expansion joint move toward one another; then, as the bridge cools down after dark, they begin gradually to retract. Thus the bridge has a built-in safety zone; otherwise, it would have no room for expansion or contraction in response to temperature changes. As for the use of the comb shape, this staggers the gap between the two sides of the expansion joint, thus minimizing the bump motorists experience as they drive over it.

Expansion joints of a different design can also be found in highways, and on "highways" of rail. Thermal expansion is a particularly serious problem where railroad tracks are concerned, since the tracks on which the trains run are made

of steel. Steel, as noted earlier, expands by a factor of 12 parts in 1 million for every Celsius degree change in temperature, and while this may not seem like much, it can create a serious problem under conditions of high temperature.

Most tracks are built from pieces of steel supported by wooden ties, and laid with a gap between the ends. This gap provides a buffer for thermal expansion, but there is another matter to consider: the tracks are bolted to the wooden ties, and if the steel expands too much, it could pull out these bolts. Hence, instead of being placed in a hole the same size as the bolt, the bolts are fitted in slots, so that there is room for the track to slide in place slowly when the temperature rises.

Such an arrangement works agreeably for trains that run at ordinary speeds: their wheels merely make a noise as they pass over the gaps, which are rarely wider than 0.5 in (0.013 m). A high-speed train, however, cannot travel over irregular track; therefore, tracks for high-speed trains are laid under conditions of relatively high tension. Hydraulic equipment is used to pull sections of the track taut; then, once the track is secured in place along the cross ties, the tension is distributed down the length of the track.

THERMOMETERS AND THERMOSTATS

MERCURY IN THERMOMETERS. A thermometer gauges temperature by measuring a temperature-dependent property. A thermostat, by contrast, is a device for adjusting the temperature of a heating or cooling system. Both use the principle of thermal expansion in their operation. As noted in the example of the metal lid and glass jar above, glass expands little with changes in temperature; therefore, it makes an ideal container for the mercury in a thermometer. As for mercury, it is an ideal thermometric medium—that is, a material used to gauge temperature—for several reasons. Among these is a high boiling point, and a highly predictable, uniform response to changes in temperature.

In a typical mercury thermometer, mercury is placed in a long, narrow sealed tube called a capillary. Because it expands at a much faster rate than the glass capillary, mercury rises and falls with the temperature. A thermometer is calibrated by measuring the difference in height between mercury at the freezing point of water, and mercury at the boiling point of water. The interval

KEY TERMS

COEFFICIENT: A number that serves as a measure for some characteristic or property. A coefficient may also be a factor against which other values are multiplied to provide a desired result.

COEFFICIENT OF LINEAR EXPANSION: A figure, constant for any particular type of solid, used in calculating the amount by which the length of that solid will change as a result of temperature change. For any given substance, the coefficient of linear expansion is typically a number expressed in terms of $10^{-5}/°C$.

COEFFICIENT OF VOLUME EXPANSION: A figure, constant for any particular type of material, used in calculating the amount by which the volume of that material will change as a result of temperature change. For any given substance, the coefficient of volume expansion is typically a number expressed in terms of $10^{-4}/°C$.

HEAT: Internal thermal energy that flows from one body of matter to another.

KINETIC ENERGY: The energy that an object possesses by virtue of its motion.

MOLECULAR TRANSLATIONAL ENERGY: The kinetic energy in a system produced by the movement of molecules in relation to one another.

POTENTIAL ENERGY: The energy that an object possesses by virtue of its position.

SYSTEM: In physics, the term "system" usually refers to any set of physical interactions, or any material body, isolated from the rest of the universe. Anything outside of the system, including all factors and forces irrelevant to a discussion of that system, is known as the environment.

TEMPERATURE: A measure of the average kinetic energy—or molecular translational energy in a system. Differences in temperature determine the direction of internal energy flow between two systems when heat is being transferred.

THERMAL ENERGY: Heat energy, a form of kinetic energy produced by the movement of atomic or molecular particles. The greater the movement of these particles, the greater the thermal energy.

THERMAL EXPANSION: A property in all types of matter that display a tendency to expand when heated, and to contract when cooled.

between these two points is then divided into equal increments in accordance with one of the well-known temperature scales.

THE BIMETALLIC STRIP IN THERMOSTATS. In a thermostat, the central component is a bimetallic strip, consisting of thin strips of two different metals placed back to back. One of these metals is of a kind that possesses a high coefficient of linear expansion, while the other metal has a low coefficient. A temperature increase will cause the side with a higher coefficient to expand more than the side

that is less responsive to temperature changes. As a result, the bimetallic strip will bend to one side.

When the strip bends far enough, it will close an electrical circuit, and, thus, direct the air conditioner to go into action. By adjusting the thermostat, one varies the distance that the bimetallic strip must be bent in order to close the circuit. Once the air in the room reaches the desired temperature, the high-coefficient metal will begin to contract, and the bimetallic strip will straighten. This will cause an opening of the electrical circuit, disengaging the air conditioner.

In cold weather, when the temperature-control system is geared toward heating rather than cooling, the bimetallic strip acts in much the same way—only this time, the high-coefficient metal contracts with cold, engaging the heater. Another type of thermostat uses the expansion of a vapor rather than a solid. In this case, heating of the vapor causes it to expand, pushing on a set of brass bellows and closing the circuit, thus, engaging the air conditioner.

WHERE TO LEARN MORE

Beiser, Arthur. *Physics,* 5th ed. Reading, MA: Addison-Wesley, 1991.

"Comparison of Materials: Coefficient of Thermal Expansion" (Web site). <http://www.handyharmancanada.com/TheBrazingBook/comparis.html> (April 21, 2001).

Encyclopedia of Thermodynamics (Web site). <http://therion.minpet.unibas.ch/minpet/groups/thermodict/> (April 12, 2001).

Fleisher, Paul. *Matter and Energy: Principles of Matter and Thermodynamics.* Minneapolis, MN: Lerner Publications, 2002.

NPL: National Physics Laboratory: Thermal Stuff: Beginners' Guides (Web site). <http://www.npl.co.uk/npl/cbtm/thermal/stuff/guides.html> (April 18, 2001).

Royston, Angela. *Hot and Cold.* Chicago: Heinemann Library, 2001.

Suplee, Curt. *Everyday Science Explained.* Washington, D.C.: National Geographic Society, 1996.

"Thermal Expansion Measurement" (Web site). <http://www.measurementsgroup.com/guide/tn/tn513/513intro.html> (April 21, 2001).

"Thermal Expansion of Solids and Liquids" (Web site). <http://www.physics.mun.ca/~gquirion/P2053/html19b/> (April 21, 2001).

Walpole, Brenda. *Temperature.* Illustrated by Chris Fairclough and Dennis Tinkler. Milwaukee, WI: Gareth Stevens Publishing, 1995.

WAVE MOTION AND OSCILLATION

WAVE MOTION

CONCEPT

Wave motion is activity that carries energy from one place to another without actually moving any matter. Studies of wave motion are most commonly associated with sound or radio transmissions, and, indeed, these are among the most common forms of wave activity experienced in daily life. Then, of course, there are waves on the ocean or the waves produced by an object falling into a pool of still water—two very visual examples of a phenomenon that takes place everywhere in the world around us.

HOW IT WORKS

Related Forms of Motion

In wave motion, energy—the ability to perform work, or to exert force over distance—is transmitted from one place to another without actually moving any matter along the wave. In some types of waves, such as those on the ocean, it might seem as though matter itself has been displaced; that is, it appears that the water has actually moved from its original position. In fact, this is not the case: molecules of water in an ocean wave move up and down, but they do not actually travel with the wave itself. Only the energy is moved.

A wave is an example of a larger class of regular, repeated, and/or back-and-forth types of motion. As with wave motion, these varieties of movement may or may not involve matter, but, in any case, the key component is not matter, but energy. Broadest among these is periodic motion, or motion that is repeated at regular intervals called periods. A period might be the amount of time that it takes an object orbiting another (as, for instance, a satellite going around Earth) to complete one cycle of orbit. With wave motion, a period is the amount of time required to complete one full cycle of the wave, from trough to crest and back to trough.

HARMONIC MOTION. Harmonic motion is the repeated movement of a particle about a position of equilibrium, or balance. In harmonic motion—or, more specifically, simple harmonic motion—the object moves back and forth under the influence of a force directed toward the position of equilibrium, or the place where the object stops if it ceases to be in motion. A familiar example of harmonic motion, to anyone who has seen an old movie with a clichéd depiction of a hypnotist, is the back-and-forth movement of the hypnotist's watch, as he tries to control the mind of his patient.

One variety of harmonic motion is vibration, which wave motion resembles in some respects. Both wave motion and vibration are periodic, involving the regular repetition of a certain form of movement. In both, there is a continual conversion and reconversion between potential energy (the energy of an object due to its position, as for instance with a sled at the top of a hill) and kinetic energy (the energy of an object due to its motion, as with the sled when sliding down the hill.) The principal difference between vibration and wave motion is that, in the first instance, the energy remains in place, whereas waves actually transport energy from one place to another.

OSCILLATION. Oscillation is a type of harmonic motion, typically periodic, in one or more dimensions. Suppose a spring is fixed in

HEINRICH HERTZ. *(Hulton-Deutsch Collection/Corbis. Reproduced by permission.)*

place to a ceiling, such that it hangs downward. At this point, the spring is in a position of equilibrium. Now, consider what happens if the spring is grasped at a certain point and lifted, then let go. It will, of course, fall downward with the force of gravity until it comes to a stop—but it will not stop at the earlier position of equilibrium. Instead, it will continue downward to a point of maximum tension, where it possesses maximum potential energy as well. Then, it will spring upward again, and as it moves, its kinetic energy increases, while potential energy decreases. At the high point of this period of oscillation, the spring will not be as high as it was before it was originally released, but it will be higher than the position of equilibrium.

Once it falls, the spring will again go lower than the position of equilibrium, but not as low as before—and so on. This is an example of oscillation. Now, imagine what happens if another spring is placed beside the first one, and they are connected by a rubber band. If just the first spring is disturbed, as before, the second spring will still move, because the energy created by the movement of the first spring will be transmitted to the second one via the rubber band. The same will happen if a row of springs, all side-by-side, are attached by multiple rubber bands, and the

first spring is once again disturbed: the energy will pass through the rubber bands, from spring to spring, causing the entire row to oscillate. This is similar to what happens in the motion of a wave.

TYPES AND PROPERTIES OF WAVES

There are some types of waves that do not follow regular, repeated patterns; these are discussed below, in the illustration concerning a string, in which a pulse is created and reflected. Of principal concern here, however, is the periodic wave, a series of wave motions, following one after the other in regular succession. Examples of periodic waves include waves on the ocean, sound waves, and electromagnetic waves. The last of these include visible light and radio, among others.

Electromagnetic waves involve only energy; on the other hand, a mechanical wave involves matter as well. Ocean waves are mechanical waves; so, too, are sound waves, as well as the waves produced by pulling a string. It is important to note, again, that the matter itself is not moved from place to place, though it may move in place without leaving its position. For example, water molecules in the crest of an ocean wave rotate in the same direction as the wave, while those in the trough of the wave rotate in a direction opposite to that of the wave, yet there is no net motion of the water: only energy is transmitted along the wave.

FIVE PROPERTIES OF WAVES. There are three notable interrelated characteristics of periodic waves. One of these is wave speed, symbolized by v and typically calculated in meters per second. Another is wavelength, represented as λ (the Greek letter lambda), which is the distance between a crest and the adjacent crest, or a trough and the adjacent trough. The third is frequency, abbreviated as f, which is the number of waves passing through a given point during the interval of 1 second.

Frequency is measured in terms of cycles per second, or Hertz (Hz), named in honor of nineteenth-century German physicist Heinrich Rudolf Hertz (1857-1894). If a wave has a frequency of 100 Hz, this means that 100 waves are passing through a given point during the interval of 1 second. Higher frequencies are expressed in terms of kilohertz (kHz; 10^3 or 1,000 cycles per

TRANSVERSE WAVES PRODUCED BY A WATER DROPLET PENETRATING THE SURFACE OF A BODY OF LIQUID. *(Photograph by Martin Dohrn/Science Photo Library, National Audubon Society Collection/Photo Researchers, Inc. Reproduced with permission.)*

second) or megahertz (MHz; 10^6 or 1 million cycles per second.)

Frequency is clearly related to wave speed, and there is also a relationship—though it is not so immediately grasped—between wavelength and speed. Over the interval of 1 second, a given number of waves pass a certain point (frequency), and each wave occupies a certain distance (wavelength). Multiplied by one another, these two properties equal the speed of the wave. This can be stated as a formula: $v = f\lambda$.

Earlier, the term "period" was defined in terms of wave motion as the amount of time required to complete one full cycle of the wave. Period, symbolized by T, can be expressed in terms of frequency, and, thus, can also be related to the other two properties identified above. It is the inverse of frequency, meaning that $T = 1/f$. Furthermore, period is equal to the ratio of wavelength to wave speed; in other words, $T = \lambda/v$.

A fifth property of waves—one not mathematically related to wavelength, wave speed, frequency, or period, is amplitude. Amplitude can be defined as the maximum displacement of oscillating particles from their normal position. For an ocean wave, amplitude is the distance

from either the crest or the trough to the level that the ocean would maintain if it were perfectly still.

WAVE SHAPES. When most people think of waves, naturally, one of the first images that comes to mind is that of waves on the ocean. These are an example of a transverse wave, or one in which the vibration or motion is perpendicular to the direction the wave is moving. (Actually, ocean waves are simply perceived as transverse waves; in fact, as discussed below, their behavior is rather more complicated.) In a longitudinal wave, on the other hand, the movement of vibration is in the same direction as the wave itself.

Transverse waves are easier to visualize, particularly with regard to the aspects of wave motion—for example, frequency and amplitude—discussed above. Yet, longitudinal waves can be understood in terms of a common example. Sound waves, for instance, are longitudinal: thus, when a stereo is turned up to a high volume, the speakers vibrate in the same direction as the sound itself.

A longitudinal wave may be understood as a series of fluctuations in density. If one were to take a coiled spring (such as the toy known as the "Slinky") and release one end while holding the other, the motion of the springs would produce longitudinal waves. As these waves pass through the spring, they cause some portions of it to be compressed and others extended. The distance between each point of compression is the wavelength.

Now, to return to the qualified statement made above: that ocean waves are an example of transverse waves. We perceive them as transverse waves, but, in fact, they are also longitudinal. In fact, all types of waves on the surface of a liquid are a combination of longitudinal and transverse, and are known as surface waves. Thus, if one drops a stone into a body of still water, waves radiate outward (longitudinal), but these waves also have a component that is perpendicular to the surface of the water, meaning that they are also transverse.

REAL-LIFE APPLICATIONS

Pulses on a String

There is another variety of wave, though it is defined in terms of behavior rather than the direction of disturbance. (In terms of direction, it is simply a variety of transverse wave.) This is a standing wave, produced by causing vibrations on a string or other piece of material whose ends are fixed in place. Standing waves are really a series of pulses that travel down the string and are reflected back to the point of the original disturbance.

Suppose you hold a string in one hand, with the other end attached to a wall. If you give the string a shake, this causes a pulse—an isolated, non-periodic disturbance—to move down it. A pulse is a single wave, and the behavior of this lone wave helps us to understand what happens within the larger framework of wave motion. As with wave motion in general, the movement of the pulse involves both kinetic and potential energy. The tension of the string itself creates potential energy; then, as the movement of the pulse causes the string to oscillate upward and downward, this generates a certain amount of kinetic energy.

TENSION AND REFLECTION. The speed of the pulse is a function of the string and its properties, not of the way that the pulse

was originally delivered. The tighter the string, and the less its mass per unit of length, the faster the pulse travels down it. The greater the mass per unit of length, however, the greater the inertia resisting the movement of the pulse. Furthermore, the more loosely you hold the string, the less it will respond to the movement of the pulse.

In accordance with the third law of motion, there should be an equal and opposite reaction once the pulse comes into contact with the wall. Assuming that you are holding the string tightly, this reaction will be manifested in the form of an inverted wave, or one that is upside-down in relation to the original pulse. In this case, the tension on the end attached to the support is equal and opposite to the tension exerted by your hand. As a result, the pulse comes back in the same shape as before, but inverted.

If, on the other hand, you hold the other end of the string loosely; instead, once it reaches the wall, its kinetic energy will be converted into potential energy, which will cause the end of the string closest to the wall to move downward. This will result in sending back a pulse that is reversed in horizontal direction, but the same in vertical direction.

In both cases, the energy in the string is reflected backward to its source—that is, to the place from which the pulse was originally produced by the action of your hand. If, however, you hold the string so that its level of tension is exactly between perfect rigidity and perfect looseness, then the pulse will not be reflected. In other words, there will be no reflected wave.

TRANSMISSION AND REFLECTION. If two strings are joined end-to-end, and a pulse is produced at one end, the pulse would, of course, be transmitted to the second string. If, however, the second string has a greater mass per unit of length than the first one, the result would be two pulses: a transmitted pulse moving in the "right" direction, and a reflected, inverted pulse, moving toward the original source of energy. If, on the other hand, the first string has a greater mass per unit of length than the second one, the reflected pulse would be erect (right side up), not inverted.

For simplicity's sake, this illustration has been presented in terms of a string attached to a wall, but, in fact, transmission and reflection occur in a number of varieties of wave motion—

not just those involving pulses or standing waves. A striking example occurs when light hits an ordinary window. The majority of the light, of course, is transmitted through the window pane, but a portion is reflected. Thus, as one looks through the window, one also sees one's reflection.

Similarly, sound waves are reflected depending on the medium with which they are in contact. A canyon wall, for instance, will reflect a great deal of sound, and, thus, it is easy to produce an echo in such a situation. On the other hand, there are many instances in which the desire is to "absorb" sound by transmitting it to some other form of material. Thus, for example, the lobby of an upscale hotel will include a number of plants, as well as tapestries and various wall hangings. In addition to adding beauty, these provide a medium into which the sound of voices and other noises can be transmitted and, thus, absorbed.

SOUND WAVES

PRODUCTION. The experience of sound involves production, or the generation of sound waves; transmission, or the movement of those waves from their source; and reception, the principal example of which is hearing. Sound itself is discussed in detail elsewhere. Of primary concern here is the transmission, and to a lesser extent, the production of sound waves.

In terms of production, sound waves are, as noted, longitudinal waves: changes in pressure, or alternations between condensation and rarefaction. Vibration is integral to the generation of sound. When the diaphragm of a loudspeaker pushes outward, it forces nearby air molecules closer together, creating a high-pressure region all around the loudspeaker. The loudspeaker's diaphragm is pushed backward in response, thus freeing up a volume of space for the air molecules. These, then, rush toward the diaphragm, creating a low-pressure region behind the high-pressure one. As a result, the loudspeaker sends out alternating waves of high pressure (condensation) and low pressure (rarefaction).

FREQUENCY AND WAVELENGTH. As sound waves pass through a medium such as air, they create fluctuations between condensation and rarefaction. These result in pressure changes that cause the listener's eardrum to vibrate with the same frequency as the sound wave, a vibration that the ear's inner mechanisms translate and pass on to the brain. The range of audibility for the human ear is from 20 Hz to 20 kHz. The lowest note of the eighty-eight keys on a piano is 27 Hz and the highest 4.186 kHz. This places the middle and upper register of the piano well within the optimal range for audibility, which is between 3 and 4 kHz.

Sound travels at a speed of about 1,088 ft (331 m) per second through air at sea level, and the range of sound audible to human ears includes wavelengths as large as 11 ft (3.3 m) and as small as 1.3 in (3.3 cm). Unlike light waves, which are very small, the wavelengths of audible sound are comparable to the sizes of ordinary objects. This creates an interesting contrast between the behaviors of sound and light when confronted with an obstacle to their transmission.

It is fairly easy to block out light by simply holding up a hand in front of one's eyes. When this happens, the Sun casts a shadow on the other side of one's hand. The same action does not work with one's ears and the source of a sound, however, because the wavelengths of sound are large enough to go right past a relatively small object such as a hand. However, if one were to put up a tall, wide cement wall between oneself and the source of a sound—as is often done in areas where an interstate highway passes right by a residential community—the object would be sufficiently large to block out much of the sound.

RADIO WAVES

Radio waves, like visible light waves, are part of the electromagnetic spectrum. They are characterized by relatively long wavelengths and low frequencies—low, that is, in contrast to the much higher frequencies of both visible and invisible light waves. The frequency range of radio is between 10 KHz and about 2,000 MHz—in other words, from 10,000 Hz to as much as 2 billion Hz—an impressively wide range.

AM radio broadcasts are found between 0.6 and 1.6 MHz, and FM broadcasts between 88 and 108 MHz. Thus, FM is at a much, much higher frequency than AM, with the lowest frequency on the FM dial 55 times as great as the highest on the AM dial. There are other ranges of frequency assigned by the FCC (Federal Communications Commission) to other varieties of radio trans-

AMPLITUDE: The maximum displacement of particles in oscillation from their normal position. For an ocean wave, amplitude is the distance from either the crest or the trough to the level that the ocean would maintain if it were perfectly still.

ENERGY: The ability to perform work, which is the exertion of force over a given distance. Work is the product of force and distance, where force and distance are exerted in the same direction.

FREQUENCY: The number of waves passing through a given point during the interval of one second. The higher the frequency, the shorter the wavelength. Frequency can also be mathematically related to wave speed and period.

HARMONIC MOTION: The repeated movement of a particle about a position of equilibrium, or balance.

HERTZ: A unit for measuring frequency, equal to one cycle per second. If a sound wave has a frequency of 20,000 Hz, this means that 20,000 waves are passing through a given point during the interval of one second. Higher frequencies are expressed in terms of kilohertz (kHz; 10^3 or 1,000 cycles per second) or megahertz (MHz; 10^6 or 1 million cycles per second).

KINETIC ENERGY: The energy that an object possesses due to its motion, as with a sled when sliding down a hill. This is contrasted with potential energy.

LONGITUDINAL WAVE: A wave in which the movement of vibration is in the same direction as the wave itself. This is contrasted to a transverse wave.

MATTER: Physical substance that has mass; occupies space; is composed of atoms; and is ultimately convertible to energy.

MECHANICAL WAVE: A type of wave that involves matter. Ocean waves are mechanical waves; so, too, are the waves produced by pulling a string. The matter itself may move in place, but, as with all

---

mission: for instance, citizens' band (CB) radios are in a region between AM and FM, ranging from 26.985 MHz to 27.405 MHz.

Frequency does not indicate power. The power of a radio station is a function of the wattage available to its transmitter: hence, radio stations often promote themselves with announcements such as "operating with 100,000 watts of power...." Thus, an AM station, though it has a much lower frequency than an FM station, may possess more power, depending on the wattage of the transmitter. Indeed, as we shall see, it is precisely because of its high frequency that an FM station lacks the broadcast range of an AM station.

AMPLITUDE AND FREQUENCY MODULATIONS. What is the difference between AM and FM? Or to put it another way, why is it that an AM station may be heard halfway across the country, yet its sound on a car radio fades out when the car goes under an overpass? The difference relates to how the various radio signals are modulated.

A radio signal is simply a carrier: it may carry Morse code, or it may carry complex sounds, but in order to transmit voices and music, its signal must be modulated. This can be done, for instance, by varying the instantaneous amplitude of the radio wave, which is a function of the radio station's power. These variations in amplitude are called amplitude modulation, or

KEY TERMS CONTINUED

types of wave motion, there is no net movement of matter—only of energy.

OSCILLATION: A type of harmonic motion, typically periodic, in one or more dimensions.

PERIOD: For wave motion, a period is the amount of time required to complete one full cycle of the wave, from trough to crest and back to trough. Period can be mathematically related to frequency, wavelength, and wave speed.

PERIODIC MOTION: Motion that is repeated at regular intervals. These intervals are known as periods.

PERIODIC WAVE: A wave in which a uniform series of crests and troughs follow one after the other in regular succession. By contrast, the wave produced by applying a pulse to a stretched string does not follow regular, repeated patterns.

POTENTIAL ENERGY: The energy that an object possesses due to its position, as for instance with a sled at the top of a hill. This is contrasted with kinetic energy.

PULSE: An isolated, non-periodic disturbance that takes place in wave motion of a type other than that of a periodic wave.

STANDING WAVE: A type of transverse wave produced by causing vibrations on a string or other piece of material whose ends are fixed in place.

SURFACE WAVE: A wave that exhibits the behavior of both a transverse wave and a longitudinal wave.

TRANSVERSE WAVE: A wave in which the vibration or motion is perpendicular to the direction in which the wave is moving. This is contrasted to a longitudinal wave.

WAVELENGTH: The distance between a crest and the adjacent crest, or the trough and an adjacent trough, of a wave. Wavelength, abbreviated λ (the Greek letter lambda) is mathematically related to wave speed, period, and frequency.

WAVE MOTION: Activity that carries energy from one place to another without actually moving any matter.

AM, and this was the first type of commercial radio to appear. Developed in the period before World War I, AM made its debut as a popular phenomenon shortly after the war.

Ironically, FM (frequency modulation) was developed not long after AM, but it did not become commercially viable until well after World War II. As its name suggests, frequency modulation involves variation in the signal's frequency. The amplitude stays the same, and this—combined with the high frequency—produces a nice, even sound for FM radio.

But the high frequency also means that FM signals do not travel as far. If a person is listening to an FM station while moving away from the station's signal, eventually the station will be below the horizon relative to the car, and the car radio will no longer be able to receive the signal. In contrast to the direct, or line-of-sight, transmissions of FM stations, AM signals (with their longer wavelengths) are reflected off of layers in Earth's ionosphere. As a result, a nighttime signal from a "clear channel station" may be heard across much of the continental United States.

WHERE TO LEARN MORE

Ardley, Neil. *Sound Waves to Music.* New York: Gloucester Press, 1990.

Berger, Melvin and Gilda Berger. *What Makes an Ocean Wave?: Questions and Answers About Oceans and Ocean Life.* New York: Scholastic, 2001.

Catherall, Ed. *Exploring Sound.* Austin, TX: Steck-Vaughn Library, 1990.

Glover, David. *Sound and Light.* New York: Kingfisher Books, 1993.

"Longitudinal and Transverse Wave Motion" (Web site). <http://www.kettering.edu/~drussell/Demos/waves/wavemotion.html> (April 22, 2001).

"Multimedia Activities: Wave Motion." ExploreScience.com (Web site). <http://www.explorescience.com/activities/activity_list.cfm?catego ryID=3> (April 22, 2001).

Ruchlis, Hyman. *Bathtub Physics.* Edited by Donald Barr; illustrated by Ray Skibinski. New York: Harcourt, Brace, and World, 1967.

"Wave Motion" (Web site). <http://www.media.uwe.ac.uk/masoud/projects/water/wave.html> (April 22, 2001).

"Wave Motion and Sound." The Physics Web (Web site). <http://www.hcrhs.hunterdon.k12.nj.us/disk2/Physics/wave.html> (April 22, 2001).

"Wave Motion Menu." Carson City-Crystal High School Physics and Chemistry Departments (Web site). <http://members.aol.com/cepeirce/b2.html> (April 22, 2001).

OSCILLATION

CONCEPT

When a particle experiences repeated movement about a position of stable equilibrium, or balance, it is said to be in harmonic motion, and if this motion is repeated at regular intervals, it is called periodic motion. Oscillation is a type of harmonic motion, typically periodic, in one or more dimensions. Among the examples of oscillation in the physical world are the motion of a spring, a pendulum, or even the steady back-and-forth movement of a child on a swing.

HOW IT WORKS

STABLE AND UNSTABLE EQUILIBRIUM

When a state of equilibrium exists, the vector sum of the forces on an object is equal to zero. There are three varieties of equilibrium: stable, unstable, and neutral. Neutral equilibrium, discussed in the essay on Statics and Equilibrium elsewhere in this book, does not play a significant role in oscillation; on the other hand, stable and unstable equilibrium do.

In the example of a playground swing, when the swing is simply hanging downward—either empty or occupied—it is in a position of stable equilibrium. The vector sums are balanced, because the swing hangs downward with a force (its weight) equal to the force of the bars on the swing set that hold it up. If it were disturbed from this position, as, for instance, by someone pushing the swing, it would tend to return to its original position.

If, on the other hand, the swing were raised to a certain height—if, say, a child were swinging

and an adult caught the child at the point of maximum displacement—this would be an example of unstable equilibrium. The swing is in equilibrium because the forces on it are balanced: it is being held upward with a force equal to its weight. Yet, this equilibrium is unstable, because a disturbance (for instance, if the adult lets go of the swing) will cause it to move. Since the swing tends to oscillate, it will move back and forth across the position of stable equilibrium before finally coming to a rest in the stable position.

PROPERTIES OF OSCILLATION

There are two basic models of oscillation to consider, and these can be related to the motion of two well-known everyday objects: a spring and a swing. As noted below, objects not commonly considered "springs," such as rubber bands, display spring-like behavior; likewise one could substitute "pendulum" for swing. In any case, it is easy enough to envision the motion of these two varieties of oscillation: a spring generally oscillates along a straight line, whereas a swing describes an arc.

Either case involves properties common to all objects experiencing oscillation. There is always a position of stable equilibrium, and there is always a cycle of oscillation. In a single cycle, the oscillating particle moves from a certain point in a certain direction, then reverses direction and returns to the original point. The amount of time it takes to complete one cycle is called a period, and the number of cycles that take place during one second is the frequency of the oscillation. Frequency is measured in Hertz (Hz), with 1 Hz—the term is both singular and plural—equal to one cycle per second.

THE BOUNCE PROVIDED BY A TRAMPOLINE IS DUE TO ELASTIC POTENTIAL ENERGY. *(Photograph by Kevin Fleming/Corbis. Reproduced by permission.)*

It is easiest to think of a cycle as the movement from a position of stable equilibrium to one of maximum displacement, or the furthest possible point from stable equilibrium. Because stable equilibrium is directly in the middle of a cycle, there are two points of maximum displacement. For a swing or pendulum, maximum displacement occurs when the object is at its highest point on either side of the stable equilibrium position. For example, maximum displacement in a spring occurs when the spring reaches the furthest point of being either stretched or compressed.

The amplitude of a cycle is the maximum displacement of particles during a single period of oscillation, and the greater the amplitude, the greater the energy of the oscillation. When an object reaches maximum displacement, it reverses direction, and, therefore, it comes to a stop for an instant of time. Thus, the speed of movement is slowest at that position, and fastest as it passes back through the position of stable equilibrium. An increase in amplitude brings with it an increase in speed, but this does not lead to a change in the period: the greater the amplitude, the further the oscillating object has to move, and, therefore, it takes just as long to complete a cycle.

RESTORING FORCE

Imagine a spring hanging vertically from a ceiling, one end attached to the ceiling for support and the other free to hang. It would thus be in a position of stable equilibrium: the spring hangs downward with a force equal to its weight, and the ceiling pulls it upward with an equal and opposite force. Suppose, now, that the spring is pulled downward.

A spring is highly elastic, meaning that it can experience a temporary stress and still rebound to its original position; by contrast, some objects (for instance, a piece of clay) respond to deformation with plastic behavior, permanently assuming the shape into which they were deformed. The force that directs the spring back to a position of stable equilibrium—the force, in other words, which must be overcome when the spring is pulled downward—is called a restoring force.

The more the spring is stretched, the greater the amount of restoring force that must be overcome. The same is true if the spring is compressed: once again, the spring is removed from a position of equilibrium, and, once again, the restoring force tends to pull it outward to its "natural" position. Here, the example is a spring, but restoring force can be understood just as easily in terms of a swing: once again, it is the force that tends to return the swing to a position of stable equilibrium. There is, however, one significant difference: the restoring force on a swing is gravity, whereas, in the spring, it is related to the properties of the spring itself.

ELASTIC POTENTIAL ENERGY

For any solid that has not exceeded the elastic limit—the maximum stress it can endure without experiencing permanent deformation—there is a proportional relationship between force and the distance it can be stretched. This is expressed in the formula $F = ks$, where s is the distance and k is a constant related to the size and composition of the material in question.

The amount of force required to stretch the spring is the same as the force that acts to bring it back to equilibrium— that is, the restoring force. Using the value of force, thus derived, it is possible, by a series of steps, to establish a formula for elastic potential energy. The latter, sometimes called strain potential energy, is the potential energy that a spring or a spring-like object possesses by virtue of its deformation from the state of equilibrium. It is equal to $\frac{1}{2}ks^2$.

POTENTIAL AND KINETIC EN-ERGY. Potential energy, as its name suggests, involves the potential of something to move across a given interval of space—for example, when a sled is perched at the top of a hill. As it begins moving through that interval, the object will gain kinetic energy. Hence, the elastic potential energy of the spring, when the spring is held at a position of the greatest possible displacement from equilibrium, is at a maximum. Once it is released, and the restoring force begins to move it toward the equilibrium position, potential energy drops and kinetic energy increases. But the spring will not just return to equilibrium and stop: its kinetic energy will cause it to keep going.

In the case of the "swing" model of oscillation, elastic potential energy is not a factor. (Unless, of course, the swing itself were suspended on some sort of spring, in which case the object will oscillate in two directions at once.) Nonetheless, all systems of motion involve potential and kinetic energy. When the swing is at a position of maximum displacement, its

A BUNGEE JUMPER HELPS ILLUSTRATE A REAL-WORLD EXAMPLE OF OSCILLATION. (*Eye Ubiquitous/Corbis. Reproduced by permission.*)

potential energy is at a maximum as well. Then, as it moves toward the position of stable equilibrium, it loses potential energy and gains kinetic energy. Upon passing through the stable equilibrium position, kinetic energy again decreases, while potential energy increases. The sum of the two forms of energy is always the same, but the greater the amplitude, the greater the value of this sum.

REAL-LIFE APPLICATIONS

SPRINGS AND DAMPING

Elastic potential energy relates primarily to springs, but springs are a major part of everyday life. They can be found in everything from the shock-absorber assembly of a motor vehicle to the supports of a trampoline fabric, and in both cases, springs blunt the force of impact.

If one were to jump on a piece of trampoline fabric stretched across an ordinary table—one with no springs—the experience would not be much fun, because there would be little bounce. On the other hand, the elastic potential energy of the trampoline's springs ensures that anyone of normal weight who jumps on the trampoline is liable to bounce some distance into the air. As a person's body comes down onto the trampoline fabric, this stretches the fabric (itself highly elastic) and, hence, the springs. Pulled from a position of equilibrium, the springs acquire elastic potential energy, and this energy makes possible the upward bounce.

As a car goes over a bump, the spring in its shock-absorber assembly is compressed, but the elastic potential energy of the spring immediately forces it back to a position of equilibrium, thus ensuring that the bump is not felt throughout the entire vehicle. However, springs alone would make for a bouncy ride; hence, a modern vehicle also has shock absorbers. The shock absorber, a cylinder in which a piston pushes down on a quantity of oil, acts as a damper—that is, an inhibitor of the springs' oscillation.

SIMPLE HARMONIC MOTION AND DAMPING. Simple harmonic motion occurs when a particle or object moves back and forth within a stable equilibrium position under the influence of a restoring force proportional to its displacement. In an ideal situation, where friction played no part, an object would continue to oscillate indefinitely.

Of course, objects in the real world do not experience perpetual oscillation; instead, most oscillating particles are subject to damping, or the dissipation of energy, primarily as a result of friction. In the earlier illustration of the spring suspended from a ceiling, if the string is pulled to a position of maximum displacement and then released, it will, of course, behave dramatically at first. Over time, however, its movements will become slower and slower, because of the damping effect of frictional forces.

HOW DAMPING WORKS. When the spring is first released, most likely it will fly upward with so much kinetic energy that it will, quite literally, bounce off the ceiling. But with each transit within the position of equilibrium, the friction produced by contact between the metal spring and the air, and by contact between molecules within the spring itself, will gradually reduce the energy that gives it movement. In time, it will come to a stop.

If the damping effect is small, the amplitude will gradually decrease, as the object continues to oscillate, until eventually oscillation ceases. On the other hand, the object may be "overdamped," such that it completes only a few cycles before ceasing to oscillate altogether. In the spring illustration, overdamping would occur if one were to grab the spring on a downward cycle, then slowly let it go, such that it no longer bounced.

There is a type of damping less forceful than overdamping, but not so gradual as the slow dissipation of energy due to frictional forces alone. This is called critical damping. In a critically damped oscillator, the oscillating material is made to return to equilibrium as quickly as possible without oscillating. An example of a critically damped oscillator is the shock-absorber assembly described earlier.

Even without its shock absorbers, the springs in a car would be subject to some degree of damping that would eventually bring a halt to their oscillation; but because this damping is of a very gradual nature, their tendency is to continue oscillating more or less evenly. Over time, of course, the friction in the springs would wear down their energy and bring an end to their oscillation, but by then, the car would most likely have hit another bump. Therefore, it makes sense to apply critical damping to the oscillation of the springs by using shock absorbers.

Bungee Cords and Rubber Bands

Many objects in daily life oscillate in a spring-like way, yet people do not commonly associate them with springs. For example, a rubber band, which behaves very much like a spring, possesses high elastic potential energy. It will oscillate when stretched from a position of stable equilibrium.

Rubber is composed of long, thin molecules called polymers, which are arranged side by side. The chemical bonds between the atoms in a polymer are flexible and tend to rotate, producing kinks and loops along the length of the molecule. The super-elastic polymers in rubber are called elastomers, and when a piece of rubber is pulled, the kinks and loops in the elastomers straighten.

The structure of rubber gives it a high degree of elastic potential energy, and in order to stretch rubber to maximum displacement, there is a powerful restoring force that must be overcome. This can be illustrated if a rubber band is attached to a ceiling, like the spring in the earlier example, and allowed to hang downward. If it is pulled down and released, it will behave much as the spring did.

The oscillation of a rubber band will be even more appreciable if a weight is attached to the "free" end—that is, the end hanging downward. This is equivalent, on a small scale, to a bungee jumper attached to a cord. The type of cord used for bungee jumping is highly elastic; otherwise, the sport would be even more dangerous than it already is. Because of the cord's elasticity, when the bungee jumper "reaches the end of his rope," he bounces back up. At a certain point, he begins to fall again, then bounces back up, and so on, oscillating until he reaches the point of stable equilibrium.

The Pendulum

As noted earlier, a pendulum operates in much the same way as a swing; the difference between them is primarily one of purpose. A swing exists to give pleasure to a child, or a certain bittersweet pleasure to an adult reliving a childhood experience. A pendulum, on the other hand, is not for play; it performs the function of providing a reading, or measurement.

One type of pendulum is a metronome, which registers the tempo or speed of music. Housed in a hollow box shaped like a pyramid, a metronome consists of a pendulum attached to a sliding weight, with a fixed weight attached to the bottom end of the pendulum. It includes a number scale indicating the number of oscillations per minute, and by moving the upper weight, one can change the beat to be indicated.

ZHANG HENG'S SEISMO-SCOPE. Metronomes were developed in the early nineteenth century, but, by then, the concept of a pendulum was already old. In the second century A.D., Chinese mathematician and astronomer Zhang Heng (78-139) used a pendulum to develop the world's first seismoscope, an instrument for measuring motion on Earth's surface as a result of earthquakes.

Zhang Heng's seismoscope, which he unveiled in 132 A.D., consisted of a cylinder surrounded by bronze dragons with frogs (also made of bronze) beneath. When the earth shook, a ball would drop from a dragon's mouth into that of a frog, making a noise. The number of balls released, and the direction in which they fell, indicated the magnitude and location of the seismic disruption.

CLOCKS, SCIENTIFIC INSTRU-MENTS, AND "FAX MACHINE". In 718 A.D., during a period of intellectual flowering that attended the early T'ang Dynasty (618-907), a Buddhist monk named I-hsing and a military engineer named Liang Ling-tsan built an astronomical clock using a pendulum. Many clocks today—for example, the stately and imposing "grandfather clock" found in some homes—likewise, use a pendulum to mark time.

Physicists of the early modern era used pendula (the plural of pendulum) for a number of interesting purposes, including calculations regarding gravitational force. Experiments with pendula by Galileo Galilei (1564-1642) led to the creation of the mechanical pendulum clock—the grandfather clock, that is—by distinguished Dutch physicist and astronomer Christiaan Huygens (1629-1695).

In the nineteenth century, A Scottish inventor named Alexander Bain (1810-1877) even used a pendulum to create the first "fax machine." Using matching pendulum transmitters and receivers that sent and received electrical

KEY TERMS

AMPLITUDE: The maximum displacement of particles from their normal position during a single period of oscillation.

CYCLE: One full repetition of oscillation. In a single cycle, the oscillating particle moves from a certain point in a certain direction, then switches direction and moves back to the original point. Typically, this is from the position of stable equilibrium to maximum displacement and back again to the stable equilibrium position.

DAMPING: The dissipation of energy during oscillation, which prevents an object from continuing in simple harmonic motion and will eventually force it to stop oscillating altogether. Damping is usually caused by friction.

ELASTIC POTENTIAL ENERGY: The potential energy that a spring or a spring-like object possesses by virtue of its deformation from the state of equilibrium. Sometimes called strain potential energy, it is equal to $\frac{1}{2}KS^2$, WHERE S is the distance stretched and k is a figure related to the size and composition of the material in question.

EQUILIBRIUM: A state in which the vector sum for all lines of force on an object is equal to zero.

FREQUENCY: For a particle experiencing oscillation, frequency is the number of cycles that take place during one second. Frequency is measured in Hertz.

FRICTION: The force that resists motion when the surface of one object comes into contact with the surface of another.

HARMONIC MOTION: The repeated movement of a particle within a position of equilibrium, or balance.

HERTZ: A unit for measuring frequency. The number of Hertz is the number of cycles per second.

KINETIC ENERGY: The energy that an object possesses due to its motion, as with a sled, when sliding down a hill. This is contrasted with potential energy.

MAXIMUM DISPLACEMENT: For an object in oscillation, maximum displacement is the furthest point from stable equilibrium. Since stable equilibrium is in the middle of a cycle, there are two points of maximum displacement. For a swing or pendulum, this occurs when the object is at its highest point on either side of the stable equilibrium position. Maximum displace-

impulses, he created a crude device that, at the time, seemed to have little practical purpose. In fact, Bain's "fax machine," invented in 1840, was more than a century ahead of its time.

THE FOUCAULT PENDULUM. By far the most important experiments with pendula during the nineteenth century, however, were those of the French physicist Jean Bernard Leon Foucault (1819-1868). Swinging a heavy iron ball from a wire more than 200 ft (61 m) in length, he was able to demonstrate that Earth rotates on its axis.

Foucault conducted his famous demonstration in the Panthéon, a large domed building in Paris named after the ancient Pantheon of Rome. He arranged to have sand placed on the floor of the Panthéon, and placed a pin on the bottom of the iron ball, so that it would mark the sand as the pendulum moved. A pendulum in oscillation maintains its orientation, yet the Foucault pen-

KEY TERMS CONTINUED

ment in a spring occurs when the spring is either stretched or compressed as far as it will go.

OSCILLATION: A type of harmonic motion, typically periodic, in one or more dimensions.

PERIOD: The amount of time required for one cycle in oscillating motion—for instance, from a position of maximum displacement to one of stable equilibrium, and, once again, to maximum displacement.

PERIODIC MOTION: Motion that is repeated at regular intervals. These intervals are known as periods.

POTENTIAL ENERGY: The energy that an object possesses due to its position, as for instance, with a sled at the top of a hill. This is contrasted with kinetic energy.

RESTORING FORCE: A force that directs an object back to a position of stable equilibrium. An example is the resistance of a spring, when it is extended.

SIMPLE HARMONIC MOTION: Harmonic motion, in which a particle moves back and forth about a stable equilibrium position under the influence of a restoring force proportional to its displacement. Simple harmonic motion is, in fact, an ideal situation; most types of oscillation are subject to some form of damping.

STABLE EQUILIBRIUM: A type of equilibrium in which, if an object were disturbed, it would tend to return to its original position. For an object in oscillation, stable equilibrium is in the middle of a cycle, between two points of maximum displacement.

VECTOR: A quantity that possesses both magnitude and direction.

VECTOR SUM: A calculation that yields the net result of all the vectors applied in a particular situation. Because direction is involved, it is necessary when calculating the vector sum of forces on an object (as, for instance, when determining whether or not it is in a state of equilibrium), to assign a positive value to forces in one direction, and a negative value to forces in the opposite direction. If the object is in equilibrium, these forces will cancel one another out.

dulum (as it came to be called) seemed to be shifting continually toward the right, as indicated by the marks in the sand.

The confusion related to reference point: since Earth's rotation is not something that can be perceived with the senses, it was natural to assume that the pendulum itself was changing orientation—or rather, that only the pendulum was moving. In fact, the path of Foucault's pendulum did not vary nearly as much as it seemed.

Earth itself was moving beneath the pendulum, providing an additional force which caused the pendulum's plane of oscillation to rotate.

WHERE TO LEARN MORE

Brynie, Faith Hickman. *Six-Minute Science Experiments.* Illustrated by Kim Whittingham. New York: Sterling Publishing Company, 1996.

Ehrlich, Robert. *Turning the World Inside Out, and 174 Other Simple Physics Demonstrations.* Princeton, N.J.: Princeton University Press, 1990.

"Foucault Pendulum" Smithsonian Institution FAQs (Web site). <http://www.si.edu/resource/faq/nmah/pendulum.html> (April 23, 2001).

Kruszelnicki, Karl S. The Foucault Pendulum (Web site). <http://www.abc.net.au/surf/pendulum/pendulum.html> (April 23, 2001).

Schaefer, Lola M. Back and Forth. Edited by Gail Saunders-Smith; P. W. Hammer, consultant. Mankato, MN: Pebble Books, 2000.

Shirley, Jean. Galileo. Illustrated by Raymond Renard. St. Louis: McGraw-Hill, 1967.

Suplee, Curt. Everyday Science Explained. Washington, D.C.: National Geographic Society, 1996.

Topp, Patricia. This Strange Quantum World and You. Nevada City, CA: Blue Dolphin, 1999.

Zubrowski, Bernie. Making Waves: Finding Out About Rhythmic Motion. Illustrated by Roy Doty. New York: Morrow Junior Books, 1994.

FREQUENCY

CONCEPT

Everywhere in daily life, there are frequencies of sound and electromagnetic waves, constantly changing and creating the features of the visible and audible world familiar to everyone. Some aspects of frequency can only be perceived indirectly, yet people are conscious of them without even thinking about it: a favorite radio station, for instance, may have a frequency of 99.7 MHz, and fans of that station knows that every time they turn the FM dial to that position, the station's signal will be there. Of course, people cannot "hear" radio and television frequencies—part of the electromagnetic spectrum—but the evidence for them is everywhere. Similarly, people are not conscious, in any direct sense, of frequencies in sound and light—yet without differences in frequency, there could be no speech or music, nor would there be any variations of color.

HOW IT WORKS

HARMONIC MOTION AND ENERGY

In order to understand frequency, it is first necessary to comprehend two related varieties of movement: oscillation and wave motion. Both are examples of a broader category, periodic motion: movement that is repeated at regular intervals called periods. Oscillation and wave motion are also examples of harmonic motion, or the repeated movement of a particle about a position of equilibrium, or balance.

KINETIC AND POTENTIAL EN-
ERGY. In harmonic motion, and in some types of periodic motion, there is a continual conversion of energy from one form to another. On the one hand is potential energy, or the energy of an object due to its position and, hence, its potential for movement. On the other hand, there is kinetic energy, the energy of movement itself.

Potential-kinetic conversions take place constantly in daily life: any time an object is at a distance from a position of stable equilibrium, and some force (for instance, gravity) is capable of moving it to that position, it possesses potential energy. Once it begins to move toward that equilibrium position, it loses potential energy and gains kinetic energy. Likewise, a wave at its crest has potential energy, and gains kinetic energy as it moves toward its trough. Similarly, an oscillating object that is as far as possible from the stable-equilibrium position has enormous potential energy, which dissipates as it begins to move toward stable equilibrium.

VIBRATION. Though many examples of periodic and harmonic motion can be found in daily life, the terms themselves are certainly not part of everyday experience. On the other hand, everyone knows what "vibration" means: to move back and forth in place. Oscillation, discussed in more detail below, is simply a more scientific term for vibration; and while waves are not themselves merely vibrations, they involve—and may produce—vibrations. This, in fact, is how the human ear hears: by interpreting vibrations resulting from sound waves.

Indeed, the entire world is in a state of vibration, though people seldom perceive this movement—except, perhaps, in dramatic situations such as earthquakes, when the vibrations of plates beneath Earth's surface become too force-

GRANDFATHER CLOCKS ARE ONE OF THE BEST-KNOWN VARIETIES OF A PENDULUM. *(Photograph by Peter Harholdt/Corbis. Reproduced by permission.)*

ful to ignore. All matter vibrates at the molecular level, and every object possesses what is called a natural frequency, which depends on its size, shape, and composition. This explains how a singer can shatter a glass by hitting a certain note, which does not happen because the singer's voice has reached a particularly high pitch; rather, it is a matter of attaining the natural frequency of the glass. As a result, all the energy in the sound of the singer's voice is transferred to the glass, and it shatters.

OSCILLATION

Oscillation is a type of harmonic motion, typically periodic, in one or more dimensions. There are two basic types of oscillation: that of a swing or pendulum and that of a spring. In each case, an object is disturbed from a position of stable equilibrium, and, as a result, it continues to move back and forth around that stable equilibrium position. If a spring is pulled from stable equilibrium, it will generally oscillate along a straight path; a swing, on the other hand, will oscillate along an arc.

In oscillation, whether the oscillator be spring-like or swing-like, there is always a cycle in which the oscillating particle moves from a certain point in a certain direction, then reverses direction and returns to the original point. Usually a cycle is viewed as the movement from a position of stable equilibrium to one of maximum displacement, or the furthest possible point from stable equilibrium. Because stable equilibrium is directly in the middle of a cycle, there are two points of maximum displacement: on a swing, this occurs when the object is at its highest point on either side of the stable equilibrium position, and on a spring, maximum displacement occurs when the spring is either stretched or compressed as far as it will go.

WAVE MOTION

Wave motion is a type of harmonic motion that carries energy from one place to another without actually moving any matter. While oscillation involves the movement of "an object," whether it be a pendulum, a stretched rubber band, or some other type of matter, a wave may or may not involve matter. Example of a wave made out of matter—that is, a mechanical wave—is a wave on the ocean, or a sound wave, in which energy vibrates through a medium such as air. Even in the case of the mechanical wave, however, the matter does not experience any net displacement from its original position. (Water molecules do rotate as a result of wave motion, but they end up where they began.)

There are waves that do not follow regular, repeated patterns; however, within the context of frequency, our principal concern is with periodic waves, or waves that follow one another in regular succession. Examples of periodic waves include ocean waves, sound waves, and electromagnetic waves.

Periodic waves may be further divided into transverse and longitudinal waves. A transverse wave is the shape that most people imagine when they think of waves: a regular up-and-down pattern (called "sinusoidal" in mathematical terms) in which the vibration or motion is perpendicular to the direction the wave is moving.

A longitudinal wave is one in which the movement of vibration is in the same direction as the wave itself. Though these are a little harder to picture, longitudinal waves can be visualized as a series of concentric circles emanating from a single point. Sound waves are longitudinal: thus when someone speaks, waves of sound vibrations radiate out in all directions.

AMPLITUDE

There are certain properties of waves, such as wavelength, or the distance between waves, that are not properties of oscillation. However, both types of motion can be described in terms of amplitude, period, and frequency. The first of these is not related to frequency in any mathematical sense; nonetheless, where sound waves are concerned, both amplitude and frequency play a significant role in what people hear.

Though waves and oscillators share the properties of amplitude, period, and frequency, the definitions of these differ slightly depending on whether one is discussing wave motion or oscillation. Amplitude, generally speaking, is the value of maximum displacement from an average value or position—or, in simpler terms, amplitude is "size." For an object experiencing oscillation, it is the value of the object's maximum displacement from a position of stable equilibrium during a single period. It is thus the "size" of the oscillation.

In the case of wave motion, amplitude is also the "size" of a wave, but the precise definition varies, depending on whether the wave in question is transverse or longitudinal. In the first instance, amplitude is the distance from either the crest or the trough to the average position between them. For a sound wave, which is longitudinal, amplitude is the maximum value of the pressure change between waves.

PERIOD AND FREQUENCY

Unlike amplitude, period is directly related to frequency. For a transverse wave, a period is the amount of time required to complete one full cycle of the wave, from trough to crest and back to trough. In a longitudinal wave, a period is the interval between waves. With an oscillator, a period is the amount of time it takes to complete one cycle. The value of a period is usually expressed in seconds.

Frequency in oscillation is the number of cycles per second, and in wave motion, it is the number of waves that pass through a given point per second. These cycles per second are called Hertz (Hz) in honor of nineteenth-century Ger-

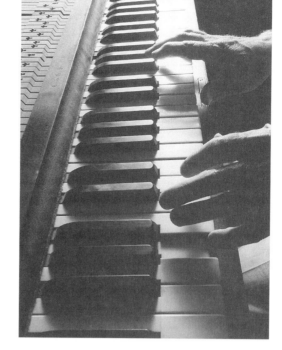

MIDDLE C—WHICH IS AT THE MIDDLE OF A PIANO KEYBOARD—IS THE STARTING POINT OF A BASIC MUSICAL SCALE. IT IS CALLED THE FUNDAMENTAL FREQUENCY, OR THE FIRST HARMONIC. *(Photograph by Francoise Gervais/Corbis. Reproduced by permission.)*

man physicist Heinrich Rudolf Hertz (1857-1894), who greatly advanced understanding of electromagnetic wave behavior during his short career.

If something has a frequency of 100 Hz, this means that 100 waves are passing through a given point during the interval of one second, or that an oscillator is completing 100 cycles in a second. Higher frequencies are expressed in terms of kilohertz (kHz; 10^3 or 1,000 cycles per second); megahertz (MHz; 10^6 or 1 million cycles per second); and gigahertz (GHz; 10^9 or 1 billion cycles per second.).

A clear mathematical relationship exists between period, symbolized by T, and frequency (f): each is the inverse of the other. Hence,

$$T = \frac{1}{f},$$

and

$$f = \frac{1}{T}.$$

If an object in harmonic motion has a frequency of 50 Hz, its period is 1/50 of a second (0.02 sec). Or, if it has a period of 1/20,000 of a second (0.00005 sec), that means it has a frequency of 20,000 Hz.

REAL-LIFE APPLICATIONS

GRANDFATHER CLOCKS AND METRONOMES

One of the best-known varieties of pendulum (plural, pendula) is a grandfather clock. Its invention was an indirect result of experiments with pendula by Galileo Galilei (1564-1642), work that influenced Dutch physicist and astronomer Christiaan Huygens (1629-1695) in the creation of the mechanical pendulum clock—or grandfather clock, as it is commonly known.

The frequency of a pendulum, a swing-like oscillator, is the number of "swings" per minute. Its frequency is proportional to the square root of the downward acceleration due to gravity (32 ft or 9.8 m/sec^2) divided by the length of the pendulum. This means that by adjusting the length of the pendulum on the clock, one can change its frequency: if the pendulum length is shortened, the clock will run faster, and if it is lengthened, the clock will run more slowly.

Another variety of pendulum, this one dating to the early nineteenth century, is a metronome, an instrument that registers the tempo or speed of music. Consisting of a pendulum attached to a sliding weight, with a fixed weight attached to the bottom end of the pendulum, a metronome includes a number scale indicating the frequency—that is, the number of oscillations per minute. By moving the upper weight, one can speed up or slow down the beat.

HARMONICS

As noted earlier, the volume of any sound is related to the amplitude of the sound waves. Frequency, on the other hand, determines the pitch or tone. Though there is no direct correlation between intensity and frequency, in order for a person to hear a very low-frequency sound, it must be above a certain decibel level.

The range of audibility for the human ear is from 20 Hz to 20,000 Hz. The optimal range for hearing, however, is between 3,000 and 4,000 Hz.

This places the piano, whose 88 keys range from 27 Hz to 4,186 Hz, well within the range of human audibility. Many animals have a much wider range: bats, whales, and dolphins can hear sounds at a frequency up to 150,000 Hz. But humans have something that few animals can appreciate: music, a realm in which frequency changes are essential.

Each note has its own frequency: middle C, for instance, is 264 Hz. But in order to produce what people understand as music—that is, pleasing combinations of notes—it is necessary to employ principles of harmonics, which express the relationships between notes. These mathematical relations between musical notes are among the most intriguing aspects of the connection between art and science.

It is no wonder, perhaps, that the great Greek mathematician Pythagoras (c. 580-500 B.C.) believed that there was something spiritual or mystical in the connection between mathematics and music. Pythagoras had no concept of frequency, of course, but he did recognize that there were certain numerical relationships between the lengths of strings, and that the production of harmonious music depended on these ratios.

RATIOS OF FREQUENCY AND PLEASING TONES. Middle C—located,, appropriately enough, in the middle of a piano keyboard—is the starting point of a basic musical scale. It is called the fundamental frequency, or the first harmonic. The second harmonic, one octave above middle C, has a frequency of 528 Hz, exactly twice that of the first harmonic; and the third harmonic (two octaves above middle C) has a frequency of 792 cycles, or three times that of middle C. So it goes, up the scale.

As it turns out, the groups of notes that people consider harmonious just happen to involve specific whole-number ratios. In one of those curious interrelations of music and math that would have delighted Pythagoras, the smaller the numbers involved in the ratios, the more pleasing the tone to the human psyche.

An example of a pleasing interval within an octave is a fifth, so named because it spans five notes that are a whole step apart. The C Major scale is easiest to comprehend in this regard, because it does not require reference to the "black keys," which are a half-step above or below the "white keys." Thus, the major fifth in the C-Major scale is C, D, E, F, G. It so happens that the

KEY TERMS

AMPLITUDE: For an object oscillation, amplitude is the value of the object's maximum displacement from a position of stable equilibrium during a single period. In a transverse wave, amplitude is the distance from either the crest or the trough to the average position between them. For a sound wave, the best-known example of a longitudinal wave, amplitude is the maximum value of the pressure change between waves.

CYCLE: In oscillation, a cycle occurs when the oscillating particle moves from a certain point in a certain direction, then switches direction and moves back to the original point. Typically, this is from the position of stable equilibrium to maximum displacement and back again to the stable equilibrium position.

FREQUENCY: For a particle experiencing oscillation, frequency is the number of cycles that take place during one second. In wave motion, frequency is the number of waves passing through a given point during the interval of one second. In either case, frequency is measured in Hertz. Period (T) is the mathematical inverse of frequency (f) hence $f=1/T$.

HARMONIC MOTION: The repeated movement of a particle about a position of equilibrium, or balance.

HERTZ: A unit for measuring frequency, named after nineteenth-century German physicist Heinrich Rudolf Hertz

(1857-1894). Higher frequencies are expressed in terms of kilohertz (kHz; 10^3 or 1,000 cycles per second); megahertz (MHz; 10^6 or 1 million cycles per second); and gigahertz (GHz; 10^9 or 1 billion cycles per second.)

KINETIC ENERGY: The energy that an object possesses due to its motion, as with a sled when sliding down a hill. This is contrasted with potential energy.

LONGITUDINAL WAVE: A wave in which the movement of vibration is in the same direction as the wave itself. This is contrasted to a transverse wave.

MAXIMUM DISPLACEMENT: For an object in oscillation, maximum displacement is the farthest point from stable equilibrium.

OSCILLATION: A type of harmonic motion, typically periodic, in one or more dimensions.

PERIOD: In oscillation, a period is the amount of time required for one cycle. For a transverse wave, a period is the amount of time required to complete one full cycle of the wave, from trough to crest and back to trough. In a longitudinal wave, a period is the interval between waves. Frequency is the mathematical inverse of period (T): hence, $T=1/f$.

PERIODIC MOTION: Motion that is repeated at regular intervals. These intervals are known as periods.

KEY TERMS CONTINUED

POTENTIAL ENERGY: The energy that an object possesses due to its position, as, for instance, with a sled at the top of a hill. This is contrasted with kinetic energy.

STABLE EQUILIBRIUM: A position in which, if an object were disturbed, it would tend to return to its original position. For an object in oscillation, stable equilibrium is in the middle of a cycle,

between two points of maximum displacement.

TRANSVERSE WAVE: A wave in which the vibration or motion is perpendicular to the direction in which the wave is moving. This is contrasted to a longitudinal wave.

WAVE MOTION: A type of harmonic motion that carries energy from one place to another without actually moving any matter.

---

ratio in frequency between middle C and G (396 Hz) is 2:3.

Less melodious, but still certainly tolerable, is an interval known as a third. Three steps up from middle C is E, with a frequency of 330 Hz, yielding a ratio involving higher numbers than that of a fifth—4:5. Again, the higher the numbers involved in the ratio, the less appealing the sound is to the human ear: the combination E-F, with a ratio of 15:16, sounds positively grating.

THE ELECTROMAGNETIC SPECTRUM

Everyone who has vision is aware of sunlight, but, in fact, the portion of the electromagnetic spectrum that people perceive is only a small part of it. The frequency range of visible light is from $4.3 \cdot 10^{14}$ Hz to $7.5 \cdot 10^{14}$ Hz—in other words, from 430 to 750 trillion Hertz. Two things should be obvious about these numbers: that both the range and the frequencies are extremely high. Yet, the values for visible light are small compared to the higher reaches of the spectrum, and the range is also comparatively small.

Each of the colors has a frequency, and the value grows higher from red to orange, and so on through yellow, green, blue, indigo, and violet. Beyond violet is ultraviolet light, which human eyes cannot see. At an even higher frequency are x rays, which occupy a broad band extending almost to 10^{20} Hz—in other words, 1 followed by 20 zeroes. Higher still is the very broad range

of gamma rays, reaching to frequencies as high as 10^{25}. The latter value is equal to 10 trillion trillion.

Obviously, these ultra-ultra high-frequency waves must be very small, and they are: the higher gamma rays have a wavelength of around 10^{-15} meters (0.000000000000001 m). For frequencies lower than those of visible light, the wavelengths get larger, but for a wide range of the electromagnetic spectrum, the wavelengths are still much too small to be seen, even if they were visible. Such is the case with infrared light, or the relatively lower-frequency millimeter waves.

Only at the low end of the spectrum, with frequencies below about 10^{10} Hz—still an incredibly large number—do wavelengths become the size of everyday objects. The center of the microwave range within the spectrum, for instance, has a wavelength of about 3.28 ft (1 m). At this end of the spectrum—which includes television and radar (both examples of microwaves), short-wave radio, and long-wave radio—there are numerous segments devoted to various types of communication.

RADIO AND MICROWAVE FREQUENCIES. The divisions of these sections of the electromagnetic spectrum are arbitrary and manmade, but in the United States—where they are administered by the Federal Communications Commission (FCC)—they have the force of law. When AM (amplitude modulation) radio first came into widespread use in the early

1920s—Congress assigned AM stations the frequency range that they now occupy: 535 kHz to 1.7 MHz.

A few decades after the establishment of the FCC in 1927, new forms of electronic communication came into being, and these too were assigned frequencies—sometimes in ways that were apparently haphazard. Today, television stations 2-6 are in the 54-88 MHz range, while stations 7-13 occupy the region from 174-220 MHz. In between is the 88 to 108 MHz band, assigned to FM radio. Likewise, short-wave radio (5.9 to 26.1 MHz) and citizens' band or CB radio (26.96 to 27.41 MHz) occupy positions between AM and FM.

In fact, there are a huge variety of frequency ranges accorded to all manner of other communication technologies. Garage-door openers and alarm systems have their place at around 40 MHz. Much, much higher than these—higher, in fact, than TV broadcasts—is the band allotted to deep-space radio communications: 2,290 to 2,300 MHz. Cell phones have their own realm, of course, as do cordless phones; but so too do radio controlled cars (75 MHz) and even baby monitors (49 MHz).

Beiser, Arthur. Physics, 5th ed. Reading, MA: Addison-Wesley, 1991.

Allocation of Radio Spectrum in the United States (Web site). <http://members.aol.com/jneuhaus/fccindex/spectrum.html> (April 25, 2001).

DiSpezio, Michael and Catherine Leary. Awesome Experiments in Light and Sound. New York: Sterling Juvenile, 2001.

Electromagnetic Spectrum (Web site). <http://www.jsc.mil/images/speccht.jpg> (April 25, 2001).

"How the Radio Spectrum Works." How Stuff Works (Web site). <http://www.howstuffworks.com/radio~spectrum.html> (April 25, 2001).

Internet Resources for Sound and Light (Web site). <http://electro.sau.edu/SLResources.html> (April 25, 2001).

"NIST Time and Frequency Division." NIST: National Institute of Standards and Technology (Web site). <http://www.boulder.nist.gov/timefreq/> (April 25, 2001).

Parker, Steve. Light and Sound. Austin, TX: Raintree Steck-Vaughn, 2000.

Physics Tutorial System: Sound Waves Modules (Web site). <http://csgrad.cs.vt.edu/~chin/chin_sound.html> (April 25, 2001).

"Radio Electronics Pages" ePanorama.net (Web site). <http://www.epanorama.net/radio.html> (April 25, 2001).

RESONANCE

CONCEPT

Though people seldom witness it directly, the entire world is in a state of motion, and where solid objects are concerned, this motion is manifested as vibration. When the vibrations produced by one object come into alignment with those of another, this is called resonance. The power of resonance can be as gentle as an adult pushing a child on a swing, or as ferocious as the force that toppled what was once the world's third-longest suspension bridge. Resonance helps to explain all manner of familiar events, from the feedback produced by an electric guitar to the cooking of food in a microwave oven.

HOW IT WORKS

VIBRATION OF MOLECULES

The possibility of resonance always exists wherever there is periodic motion, movement that is repeated at regular intervals called periods, and/or harmonic motion, the repeated movement of a particle about a position of equilibrium or balance. Many examples of resonance involve large objects: a glass, a child on a swing, a bridge. But resonance also takes place at a level invisible to the human eye using even the most powerful optical microscope.

All molecules exert a certain electromagnetic attraction toward each other, and generally speaking, the less the attraction between molecules, the greater their motion relative to one another. This, in turn, helps define the object in relation to its particular phase of matter.

A substance in which molecules move at high speeds, and therefore hardly attract one another at all, is called a gas. Liquids are materials in which the rate of motion, and hence of intermolecular attraction, is moderate. In a solid, on the other hand, there is little relative motion, and therefore molecules exert enormous attractive forces. Instead of moving in relation to one another, the molecules that make up a solid tend to vibrate in place.

Due to the high rate of motion in gas molecules, gases possess enormous internal kinetic energy. The internal energy of solids and liquids is much less than in gases, yet, as we shall see, the use of resonance to transfer energy to these objects can yield powerful results.

OSCILLATION

In colloquial terms, oscillation is the same as vibration, but, in more scientific terms, oscillation can be identified as a type of harmonic motion, typically periodic, in one or more dimensions. All things that oscillate do so either along a more or less straight path, like that of a spring pulled from a position of stable equilibrium; or they oscillate along an arc, like a swing or pendulum.

In the case of the swing or pendulum, stable equilibrium is the point at which the object is hanging straight downward—that is, the position to which gravitation force would take it if no other net forces were acting on the object. For a spring, stable equilibrium lies somewhere between the point at which the spring is stretched to its maximum length and the point at which it is subjected to maximum compression without permanent deformation.

CYCLES AND FREQUENCY. A cycle of oscillation involves movement from a certain point in a certain direction, then a reversal of direction and a return to the original point. It is simplest to treat a cycle as the movement from a position of stable equilibrium to one of maximum displacement, or the furthest possible point from stable equilibrium.

The amount of time it takes to complete one cycle is called a period, and the number of cycles in one second is the frequency of the oscillation. Frequency is measured in Hertz. Named after nineteenth-century German physicist Heinrich Rudolf Hertz (1857-1894), a single Hertz (Hz)—the term is both singular and plural—is equal to one cycle per second.

AMPLITUDE AND ENERGY. The amplitude of a cycle is the maximum displacement of particles during a single period of oscillation. When an oscillator is at maximum displacement, its potential energy is at a maximum as well. From there, it begins moving toward the position of stable equilibrium, and as it does so, it loses potential energy and gains kinetic energy. Once it reaches the stable equilibrium position, kinetic energy is at a maximum and potential energy at a minimum.

As the oscillating object passes through the position of stable equilibrium, kinetic energy begins to decrease and potential energy increases. By the time it has reached maximum displacement again—this time on the other side of the stable equilibrium position—potential energy is once again at a maximum.

OSCILLATION IN WAVE MOTION. The particles in a mechanical wave (a wave that moves through a material medium) have potential energy at the crest and trough, and gain kinetic energy as they move between these points. This is just one of many ways in which wave motion can be compared to oscillation. There is one critical difference between oscillation and wave motion: whereas oscillation involves no net movement, but merely movement in place, the harmonic motion of waves carries energy from one place to another. Nonetheless, the analogies than can be made between waves and oscillations are many, and understandably so: oscillation, after all, is an aspect of wave motion.

A periodic wave is one in which a uniform series of crests and troughs follow one after the

A COMMON EXAMPLE OF RESONANCE: A PARENT PUSH-ES HER CHILD ON A SWING. *(Photograph by Annie Griffiths Belt/Corbis. Reproduced by permission.)*

other in regular succession. Two basic types of periodic waves exist, and these are defined by the relationship between the direction of oscillation and the direction of the wave itself. A transverse wave forms a regular up-and-down pattern, in which the oscillation is perpendicular to the direction in which the wave is moving. On the other hand, in a longitudinal wave (of which a sound wave is the best example), oscillation is in the same direction as the wave itself.

Again, the wave itself experiences net movement, but within the wave—one of its defining characteristics, as a matter of fact—are oscillations, which (also by definition) experience no net movement. In a transverse wave, which is usually easier to visualize than a longitudinal wave, the oscillation is from the crest to the trough and back again. At the crest or trough, potential energy is at a maximum, while kinetic energy reaches a maximum at the point of equilibrium between crest and trough. In a longitudinal wave, oscillation is a matter of density fluctuations: the greater the value of these fluctuations, the greater the energy in the wave.

THE POWER OF RESONANCE CAN DESTROY A BRIDGE. ON NOVEMBER 7, 1940, THE ACCLAIMED TACOMA NARROWS BRIDGE COLLAPSED DUE TO OVERWHELMING RESONANCE. *(UPI/Corbis-Bettmann. Reproduced by permission.)*

PARAMETERS FOR DESCRIBING HARMONIC MOTION

The maximum value of the pressure change between waves is the amplitude of a longitudinal wave. In fact, waves can be described according to many of the same parameters used for oscillation—frequency, period, amplitude, and so on. The definitions of these terms vary somewhat, depending on whether one is discussing oscillation or wave motion; or, where wave motion is concerned, on whether the subject is a transverse wave or a longitudinal wave.

For the present purposes, however, it is necessary to focus on just a few specifics of harmonic motion. First of all, the type of motion with which we will be concerned is oscillation, and though wave motion will be mentioned, our principal concern is the oscillations within the waves, not the waves themselves. Second, the two parameters of importance in understanding resonance are amplitude and frequency.

RESONANCE AND ENERGY TRANSFER

Resonance can be defined as the condition in which force is applied to an oscillator at the point of maximum amplitude. In this way, the motion of the outside force is perfectly matched to that of the oscillator, making possible a transfer of energy.

As its name suggests, resonance is a matter of one object or force "getting in tune with" another object. One literal example of this involves shattering a wine glass by hitting a musical note that is on the same frequency as the natural frequency of the glass. (Natural frequency depends on the size, shape, and composition of the object in question.) Because the frequencies resonate, or are in sync with one another, maximum energy transfer is possible.

The same can be true of soldiers walking across a bridge, or of winds striking the bridge at a resonant frequency—that is, a frequency that matches that of the bridge. In such situations, a large structure may collapse under a force that would not normally destroy it, but the effects of resonance are not always so dramatic. Sometimes resonance can be a simple matter, like pushing a child in a swing in such a way as to ensure that the child gets maximum enjoyment for the effort expended.

REAL-LIFE APPLICATIONS

A Child on a Swing and a Pendulum in a Museum

Suppose a father is pushing his daughter on a swing, so that she glides back and forth through the air. A swing, as noted earlier, is a classic example of an oscillator. When the child gets in the seat, the swing is in a position of stable equilibrium, but as the father pulls her back before releasing her, she is at maximum displacement.

He releases her, and quickly, potential energy becomes kinetic energy as she swings toward the position of stable equilibrium, then up again on the other side. Now the half-cycle is repeated, only in reverse, as she swings backward toward her father. As she reaches the position from which he first pushed her, he again gives her a little push. This push is essential, if she is to keep going. Without friction, she could keep on swinging forever at the same rate at which she begun. But in the real world, the wearing of the swing's chain against the support along the bar above the swing will eventually bring the swing itself to a halt.

TIMING THE PUSH. Therefore, the father pushes her—but in order for his push to be effective, he must apply force at just the right moment. That right moment is the point of greatest amplitude—the point, that is, at which the father's pushing motion and the motion of the swing are in perfect resonance.

If the father waits until she is already on the downswing before he pushes her, not all the energy of his push will actually be applied to keeping her moving. He will have failed to efficiently add energy to his daughter's movement on the swing. On the other hand, if he pushes her too soon—that is, while she is on the upswing—he will actually take energy away from her movement.

If his purpose were to bring the swing to a stop, then it would make good sense to push her on the upswing, because this would produce a cycle of smaller amplitude and hence less energy. But if the father's purpose is to help his daughter keep swinging, then the time to apply energy is at the position of maximum displacement.

It so happens that this is also the position at which the swing's speed is the slowest. Once it reaches maximum displacement, the swing is about to reverse direction, and, therefore, it stops for a split-second. Once it starts moving again, now in a new direction, both kinetic energy and speed increase until the swing passes through the position of stable equilibrium, where it reaches its highest rate.

THE FOUCAULT PENDULUM. Hanging from a ceiling in Washington, D.C.'s Smithsonian Institution is a pendulum 52 ft (15.85 m) long, at the end of which is an iron ball weighing 240 lb (109 kg). Back and forth it swings, and if one sits and watches it long enough, the pendulum appears to move gradually toward the right. Over the course of 24 hours, in fact, it seems to complete a full circuit, moving back to its original orientation.

There is just one thing wrong with this picture: though the pendulum is shifting direction, this does not nearly account for the total change in orientation. At the same time the pendulum is moving, Earth is rotating beneath it, and it is the viewer's frame of reference that creates the mistaken impression that only the pendulum is rotating. In fact it is oscillating, swinging back and forth from the Smithsonian ceiling, but though it shifts orientation somewhat, the greater component of this shift comes from the movement of the Earth itself.

This particular type of oscillator is known as a Foucault pendulum, after French physicist Jean Bernard Leon Foucault (1819-1868), who in 1851 used just such an instrument to prove that Earth is rotating. Visitors to the Smithsonian, after they get over their initial bewilderment at the fact that the pendulum is not actually rotating, may well have another question: how exactly does the pendulum keep moving?

As indicated earlier, in an ideal situation, a pendulum continues oscillating. But situations on Earth are not ideal: with each swing, the Foucault pendulum loses energy, due to friction from the air through which it moves. In addition, the cable suspending it from the ceiling is also oscillating slightly, and this, too, contributes to energy loss. Therefore, it is necessary to add energy to the pendulum's swing.

Surrounding the cable where it attaches to the ceiling is an electromagnet shaped like a donut, and on either side, near the top of the cable, are two iron collars. An electronic device senses when the pendulum reaches maximum amplitude, switching on the electromagnet, which causes the appropriate collar to give the cable a slight jolt. Because the jolt is delivered at the right moment, the resonance is perfect, and energy is restored to the pendulum.

RESONANCE IN ELECTRICITY AND ELECTROMAGNETIC WAVES

Resonance is a factor in electromagnetism, and in electromagnetic waves, such as those of light or radio. Though much about electricity tends to be rather abstract, the idea of current is fairly easy to understand, because it is more or less analogous to a water current: hence, the less impedance to flow, the stronger the current. Minimal impedance is achieved when the impressed voltage has a certain resonant frequency.

NUCLEAR MAGNETIC RESONANCE. The term "nuclear magnetic resonance" (NMR) is hardly a household world, but thanks to its usefulness in medicine, MRI—short for magnetic resonance imagining—is certainly a well-known term. In fact, MRI is simply the medical application of NMR. The latter is a process in which a rotating magnetic field is produced, causing the nuclei of certain atoms to absorb energy from the field. It is used in a range of areas, from making nuclear measurements to

medical imaging, or MRI. In the NMR process, the nucleus of an atom is forced to wobble like a top, and this speed of wobbling is increased by applying a magnetic force that resonates with the frequency of the wobble.

The principles of NMR were first developed in the late 1930s, and by the early 1970s they had been applied to medicine. Thanks to MRI, physicians can make diagnoses without the patient having to undergo either surgery or x rays. When a patient undergoes MRI, he or she is made to lie down inside a large tube-like chamber. A technician then activates a powerful magnetic field that, depending on its position, resonates with the frequencies of specific body tissues. It is thus possible to isolate specific cells and analyze them independently, a process that would be virtually impossible otherwise without employing highly invasive procedures.

LIGHT AND RADIO WAVES. One example of resonance involving visible and invisible light in the electromagnetic spectrum is resonance fluorescence. Fluorescence itself is a process whereby a material absorbs electromagnetic radiation from one source, then re-emits that radiation on a wavelength longer than that of the illuminating radiation. Among its many applications are the fluorescent lights found in many homes and public buildings. Sometimes the emitted radiation has the same wavelength as the absorbed radiation, and this is called resonance fluorescence. Resonance fluorescence is used in laboratories for analyzing phenomena such as the flow of gases in a wind tunnel.

Though most people do not realize that radio waves are part of the electromagnetic spectrum, radio itself is certainly a part of daily life, and, here again, resonance plays a part. Radio waves are relatively large compared to visible light waves, and still larger in comparison to higher-frequency waves, such as those in ultraviolet light or x rays. Because the wavelength of a radio signal is as large as objects in ordinary experience, there can sometimes be conflict if the size of an antenna does not match properly with a radio wave. When the sizes are compatible, this, too, is an example of resonance.

MICROWAVES. Microwaves occupy a part of the electromagnetic spectrum with higher frequencies than those of radio waves. Examples of microwaves include television signals, radar—and of course the microwave oven, which cooks food without applying external heat. Like many other useful products, the microwave oven ultimately arose from military-industrial research, in this case, during World War II. Introduced for home use in 1955, its popularity grew slowly for the first few decades, but in the 1970s and 1980s, microwave use increased dramatically. Today, most American homes have microwaves ovens.

Of course there will always be types of food that cook better in a conventional oven, but the beauty of a microwave is that it makes possible the quick heating and cooking of foods—all without the drying effect of conventional baking. The basis for the microwave oven is the fact that the molecules in all forms of matter are vibrating. By achieving resonant frequency, the oven adds energy—heat—to food. The oven is not equipped in such a way as to detect the frequency of molecular vibration in all possible substances, however; instead, the microwaves resonant with the frequency of a single item found in nearly all types of food: water.

Emitted from a small antenna, the microwaves are directed into the cooking compartment of the oven, and, as they enter, they pass a set of turning metal fan blades. This is the stirrer, which disperses the microwaves uniformly over the surface of the food to be heated. As a microwave strikes a water molecule, resonance causes the molecule to align with the direction of the wave. An oscillating magnetron, a tube that generates radio waves, causes the microwaves to oscillate as well, and this, in turn, compels the water molecules to do the same. Thus, the water molecules are shifting in position several million times a second, and this vibration generates energy that heats the water.

Microwave ovens do not heat food from the inside out: like a conventional oven, they can only cook from the outside in. But so much energy is transferred to the water molecules that conduction does the rest, ensuring relatively uniform heating of the food. Incidentally, the resonance between microwaves and water molecules explains why many materials used in cooking dishes—materials that do not contain water—can be placed in a microwave oven without being melted or burned. Yet metal, though it also contains no water, is unsafe.

Metals have free electrons, which makes them good electrical conductors, and the pres-

ence of these free electrons means that the microwaves produce electric currents in the surfaces of metal objects placed in the oven. Depending on the shape of the object, these currents can jump, or arc, between points on the surface, thus producing sparks. On the other hand, the interior of the microwave oven itself is in fact metal, and this is so precisely because microwaves do bounce back and forth off of metal. Because the walls are flat and painted, however, currents do not arc between them.

RESONANCE OF SOUND WAVES

A highly trained singer can hit a note that causes a wine glass to shatter, but what causes this to happen is not the frequency of the note, per se. In other words, the shattering is not necessarily because of the fact that the note is extremely high; rather, it is due to the phenomenon of resonance. The natural, or resonant, frequency in the wine glass, as with all objects, is determined by its shape and composition. If the singer's voice (or a note from an instrument) hits the resonant frequency, there will be a transfer of energy, as with the father pushing his daughter on the swing. In this case, however, a full transfer of energy from the voice or musical instrument can overload the glass, causing it to shatter.

Another example of resonance and sound waves is feedback, popularized in the 1960s by rock guitarists such as Jimi Hendrix and Pete Townsend of the Who. When a musician strikes a note on an electric guitar string, the string oscillates, and an electromagnetic device in the guitar converts this oscillation into an electrical pulse that it sends to an amplifier. The amplifier passes this oscillation on to the speaker, but if the frequency of the speaker is the same as that of the vibrations in the guitar, the result is feedback.

Both in scientific terms and in the view of a music fan, feedback adds energy. The feedback from the speaker adds energy to the guitar body, which, in turn, increases the energy in the vibration of the guitar strings and, ultimately, the power of the electrical signal is passed on to the amp. The result is increasing volume, and the feedback thus creates a loop that continues to repeat until the volume drowns out all other notes.

KEY TERMS

AMPLITUDE: The maximum displacement of particles from their normal position during a single period of oscillation.

CYCLE: One full repetition of oscillation.

FREQUENCY: For a particle experiencing oscillation, frequency is the number of cycles that take place during one second. Frequency is measured in Hertz.

HARMONIC MOTION: The repeated movement of a particle about a position of equilibrium, or balance.

HERTZ: A unit for measuring frequency, named after nineteenth-century German physicist Heinrich Rudolf Hertz (1857-1894). Higher frequencies are expressed in terms of kilohertz (kHz; 10^3 or 1,000 cycles per second) or megahertz (MHz; 10^6 or 1 million cycles per second.)

KINETIC ENERGY: The energy that an object possesses due to its motion, as with a sled when sliding down a hill. This is contrasted with potential energy.

LONGITUDINAL WAVE: A wave in which the movement of vibration is in the same direction as the wave itself. This is contrasted to a transverse wave.

MAXIMUM DISPLACEMENT: For an object in oscillation, maximum displacement is the furthest point from stable equilibrium.

OSCILLATION: A type of harmonic motion, typically periodic, in one or more dimensions.

PERIOD: The amount of time required for one cycle in oscillating motion.

PERIODIC MOTION: Motion that is repeated at regular intervals. These intervals are known as periods.

PERIODIC WAVE: A wave in which a uniform series of crests and troughs follow one after the other in regular succession.

POTENTIAL ENERGY: The energy that an object possesses due to its position, as, for instance, with a sled at the top of a hill. This is contrasted with kinetic energy.

RESONANCE: The condition in which force is applied to an object in oscillation at the point of maximum amplitude.

RESONANT FREQUENCY: A frequency that matches that of an oscillating object.

STABLE EQUILIBRIUM: A position in which, if an object were disturbed, it would tend to return to its original position. For an object in oscillation, stable equilibrium is in the middle of a cycle, between two points of maximum displacement.

TRANSVERSE WAVE: A wave in which the vibration or motion is perpendicular to the direction in which the wave is moving. This is contrasted to a longitudinal wave.

WAVE MOTION: A type of harmonic motion that carries energy from one place to another without actually moving any matter.

HOW RESONANCE CAN BREAK A BRIDGE

The power of resonance goes beyond shattering a glass or torturing eardrums with feedback; it can actually destroy large structures. There is an old folk saying that a cat can destroy a bridge if it walks across it in a certain way. This may or may not be true, but it is certainly conceivable that a group of soldiers marching across a bridge can cause it to crumble, even though it is capable of holding much more than their weight, if the rhythm of their synchronized footsteps resonates with the natural frequency of the bridge. For this reason, officers or sergeants typically order their troops to do something very unmilitary—to march out of step—when crossing a bridge.

The resonance between vibrations produced by wind and those of the structure itself brought down a powerful bridge in 1940, a highly dramatic illustration of physics in action that was captured on both still photographs and film. Located on Puget Sound near Seattle, Washington, the Tacoma Narrows Bridge was, at 2,800 ft (853 m) in length, the third-longest suspension bridge in the world. But on November 7, 1940, it gave way before winds of 42 mi (68 km) per hour.

It was not just the speed of these winds, but the fact that they produced oscillations of resonant frequency, that caused the bridge to twist and, ultimately, to crumble. In those few seconds of battle with the forces of nature, the bridge writhed and buckled until a large segment collapsed into the waters of Puget Sound. Fortunately, no one was killed, and a new, more stable bridge was later built in place of the one that had come to be known as "Galloping Gertie." The incident led to increased research and progress in understanding of aerodynamics, harmonic motion, and resonance.

WHERE TO LEARN MORE

Beiser, Arthur. *Physics,* 5th ed. Reading, MA: Addison-Wesley, 1991.

Berger, Melvin. *The Science of Music.* Illustrated by Yvonne Buchanan. New York: Crowell, 1989.

"*Bridges and Resonance*" (Web site). <http://instruction.ferris.edu/loub/media/BRIDGE/Bridge.html> (April 23, 2001).

"*Resonance*" (Web site). <http://hyperphysics.phyastr.gsu.edu/hbase/sound/reson.html> (April 26, 2001).

"*Resonance*" (Web site). <http://www.exploratorium.edu/xref/phenomena/resonance.html> (April 23, 2001).

"*Resonance.*" The Physics Classroom (Web site). <http://www.glenbrook.k12.il.us/gbssci/phys/Class/sound/u11l5a.html> (April 26, 2001).

"*Resonance Experiment*" (Web site). <http://131.123.17.138/> (April 26, 2001).

"*Resonance, Frequency, and Wavelength*" (Web site). <http://www.cpo.com/CPOCatalog/SW/sw_b1.html> (April 26, 2001).

Suplee, Curt. *Everyday Science Explained.* Washington, D.C.: National Geographic Society, 1996.

"*Tacoma Narrows Bridge Disaster*" (Web site). <http://www.enm.bris.ac.uk/research/nonlinear/tacoma/tacoma.html> (April 23, 2001).

INTERFERENCE

CONCEPT

When two or more waves interact and combine, they interfere with one another. But interference is not necessarily bad: waves may interfere constructively, resulting in a wave larger than the original waves. Or, they may interfere destructively, combining in such a way that they form a wave smaller than the original ones. Even so, destructive interference may have positive effects: without the application of destructive interference to the muffler on an automobile exhaust system, for instance, noise pollution from cars would be far worse than it is. Other examples of interference, both constructive and destructive, can be found wherever there are waves: in water, in sound, in light.

HOW IT WORKS

WAVES

Whenever energy ripples through space, there is a wave. In fact, wave motion can be defined as a type of harmonic motion (repeated movement of a particle about a position of equilibrium, or balance) that carries energy from one place to another without actually moving any matter. A wave on the ocean is an example of a mechanical wave, or one that involves matter; but though the matter moves in place, it is only the energy in the wave that experiences net movement.

Wave motion is related to oscillation, a type of harmonic motion in one or more dimensions. There is one critical difference, however: oscillation involves no net movement, only movement in place, whereas the harmonic motion of waves carries energy from one place to another. Yet, individual waves themselves are oscillating even as the overall wave pattern moves.

A transverse wave forms a regular up-and-down pattern in which the oscillation is perpendicular to the direction the wave is moving. Ocean waves are transverse, though they also have properties of longitudinal waves. In a longitudinal wave, of which a sound wave is the best example, oscillation occurs in the same direction as the wave itself.

PARAMETERS OF WAVE MOTION. Some waves, composed of pulses, do not follow regular patterns. However, the waves of principal concern in the present context are periodic waves, ones in which a uniform series of crests and troughs follow each other in regular succession. Periodic motion is movement repeated at regular intervals called periods. In the case of wave motion, a period (represented by the symbol T) is the amount of time required to complete one full cycle of the wave, from trough to crest and back to trough.

Period can be mathematically related to several other aspects of wave motion, including wave speed, frequency, and wavelength. Frequency (abbreviated f) is the number of waves passing through a given point during the interval of one second. It is measured in Hertz (Hz), named after nineteenth-century German physicist Heinrich Rudolf Hertz (1857-1894), and a Hertz is equal to one cycle of oscillation per second. Higher frequencies are expressed in terms of kilohertz (kHz; 10^3 or 1,000 cycles per second) or megahertz (MHz; 10^6 or 1 million cycles per second.) Wavelength (represented by the symbol abbreviated λ, the Greek letter lambda) is the distance between a crest and the adjacent crest, or a

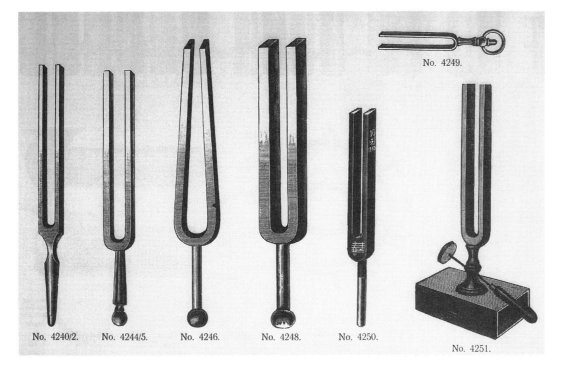

A PIANO TUNER, USING A TUNING FORK SUCH AS THE ONES SHOWN ABOVE, UTILIZES INTERFERENCE TO TUNE THE INSTRUMENT. *(Bettmann/Corbis. Reproduced by permission.)*

trough and an adjacent trough, of a wave. The higher the frequency, the shorter the wavelength.

Another parameter for describing wave motion—one that is mathematically independent from the quantities so far described—is amplitude, or the maximum displacement of particles from a position of stable equilibrium. For an ocean wave, amplitude is the distance from either the crest or the trough to the level that the ocean would maintain if it were perfectly still.

SUPERPOSITION AND INTERFERENCE

SUPERPOSITION. The principle of superposition holds that when several individual but similar physical events occur in close proximity, the resulting effect is the sum of the magnitude of the separate events. This is akin to the popular expression, "The whole is greater than the sum of the parts," and it has numerous applications in physics.

Where the strength of a gravitational field is being measured, for instance, superposition dictates that the strength of that field at any given point is the sum of the mass of the individual particles in that field. In the realm of electromag-netic force, the same statement applies, though the units being added are electrical charges or magnetic poles, rather than quantities of mass. Likewise, in an electrical circuit, the total current or voltage is the sum of the individual currents and voltages in that circuit.

Superposition applies only in equations for linear events—that is, phenomena that involve movement along a straight line. Waves are linear phenomena, and, thus, the principle describes the behavior of all waves when they come into contact with one another. If two or more waves enter the same region of space at the same time, then, at any instant, the total disturbance produced by the waves at any point is equal to the sum of the disturbances produced by the individual waves.

INTERFERENCE. The principle of superposition does not require that waves actually combine; rather, the net effect is as though they were combined. The actual combination or joining of two or more waves at a given point in space is called interference, and, as a result, the waves produce a single wave whose properties are determined by the properties of the individual waves.

If two waves of the same wavelength occupy the same space in such a way that their crests and

If this boat's wake were to cross the wake of another boat, the result would be both constructive and destructive interference. *(Photograph by Roger Ressmeyer/Corbis. Reproduced by permission.)*

troughs align, the wave they produce will have an amplitude greater than that possessed by either wave initially. This is known as constructive interference. The more closely the waves are in phase—that is, perfectly aligned—the more constructive the interference.

It is also possible that two or more waves can come together such that the trough of one meets the crest of the other, or vice versa. In this case, what happens is destructive interference, and the resulting amplitude is the difference between the values for the individual waves. If the waves are perfectly unaligned—in other words, if the trough of one exactly meets the crest of the other—their amplitudes cancel out, and the result is no wave at all.

RESONANCE

It is easy to confuse interference with resonance, and, therefore, a word should be said about the latter phenomenon. The term resonance describes a situation in which force is applied to an oscillator at the point of maximum amplitude. In this way, the motion of the outside force is perfectly matched to that of the oscillator, making possible a transfer of energy. As with interference, resonance implies alignment

between two physical entities; however, there are several important differences.

Resonance can involve waves, as, for instance, when sound waves resonate with the vibrations of an oscillator, causing a transfer of energy that sometimes produces dramatic results. (See essay on Resonance.) But in these cases, a wave is interacting with an oscillator, not a wave with a wave, as in situations of interference. Furthermore, whereas resonance entails a transfer of energy, interference involves a combination of energy.

TRANSFER VS. COMBINATION. The importance of this distinction is easy to see if one substitutes money for energy, and people for objects. If one passes on a sum of money to another person, a business, or an institution—as a loan, repayment of a loan, a purchase, or a gift—this is an example of a transfer. On the other hand, when married spouses each earn paychecks, their cash is combined.

Transfer thus indicates that the original holder of the cash (or energy) no longer has it. Yet, if the holder of the cash combines funds with those of another, both share rights to an amount of money greater than the amount each originally owned. This is analogous to constructive interference.

On the other hand, a husband and wife (or any other group of people who pool their cash) also share liabilities, and, thus, a married person may be subject to debt incurred by his or her spouse. If one spouse creates debt so great that the other spouse cannot earn enough to maintain the payments, this painful situation is analogous to destructive interference.

REAL-LIFE APPLICATIONS

MECHANICAL WAVES

One of the easiest ways to observe interference is by watching the behavior of mechanical waves. Drop a stone into a still pond, and watch how its waves ripple: this, as with most waveforms in water, is an example of a surface wave, or one that displays aspects of both transverse and longitudinal wave motion. Thus, as the concentric circles of a longitudinal wave ripple outward in one dimension, there are also transverse movements along a plane perpendicular to that of the longitudinal wave.

While the first wave is still rippling across the water, drop another stone close to the place where the first one was dropped. Now, there are two surface waves, crests and troughs colliding and interfering. In some places, they will interfere constructively, producing a wave—or rather, a portion of a wave—that is greater in amplitude than either of the original waves. At other places, there will be destructive interference, with some waves so perfectly out of phase that at one instant in time, a given spot on the water may look as though it had not been disturbed at all.

One of the interesting aspects of this interaction is the lack of uniformity in the instances of interference. As suggested in the preceding paragraph, it is usually not entire waves, but merely portions of waves, that interfere constructively or destructively. The result is that a seemingly simple event—dropping two stones into a still pond—produces a dazzling array of colliding circles, broken by outwardly undisturbed areas of destructive interference.

A similar phenomenon, though manifested by the interaction of geometric lines rather than concentric circles, occurs when two power boats pass each other on a lake. The first boat chops up the water, creating a wake that widens behind it: when seen from the air, the boat appears to be at the apex of a triangle whose sides are formed by rippling eddies of water.

Now, another boat passes through the wake of the first, only it is going in the opposite direction and producing its own ever-widening wake as it goes. As the waves from the two boats meet, some are in phase, but, more often than not, they are only partly in phase, or they possess differing wavelengths. Therefore, the waves at least partially cancel out one another in places, and in other places, reinforce one another. The result is an interesting patchwork of patterns when seen from the air.

SOUND WAVES

IN TUNE AND OUT OF TUNE. The relationships between musical notes can be intriguing, and though tastes in music vary, most people know when music is harmonious and when it is discordant. As discussed in the essay on frequency, this harmony or discord can be equated to the mathematical relationships between the frequencies of specific notes: the lower the numbers involved in the ratio, the more pleasing the sound.

The ratio between the frequency of middle C and that of its first harmonic—that is, the C note exactly one octave above it—is a nice, clean 1:2. If one were to play a song in the key of C-which, on a piano, involves only the "white notes" C-D-E-F-G-A-B—everything should be perfectly harmonious and (presumably) pleasant to the ear. But what if the piano itself is out of tune? Or what if one key is out of tune with the others?

The result, for anyone who is not tone-deaf, produces an overall impression of unpleasantness: it might be a bit hard to identify the source of this discomfort, but it is clear that something is amiss. At best, an out-of-tune piano might sound like something that belonged in a saloon from an old Western; at worst, the sound of notes that do not match their accustomed frequencies can be positively grating.

HOW A TUNING FORK WORKS. To rectify the situation, a professional piano tuner uses a tuning fork, an instrument that produces a single frequency—say, 264 Hz, which is the frequency of middle C. The piano tuner strikes the tuning fork, and at the same time strikes the appropriate key on the piano. If their frequencies are perfectly aligned, so is the sound of both; but, more likely, there will be interference, both constructive and destructive.

As time passes—measured in seconds or even fractions of seconds—the sounds of the tuning fork and that of the piano key will alternate between constructive and destructive interference. In the case of constructive interference, their combined sound will become louder than the individual sounds of either; and when the interference is destructive, the sound of both together will be softer than that produced by either the fork or the key.

The piano tuner listens for these fluctuations of loudness, which are called beats, and adjusts the tension in the appropriate piano string until the beats disappear completely. As long as there are beats, the piano string and the tuning fork will produce together a frequency that is the average of the two: if, for instance, the out-of-tune middle C string vibrates at 262 Hz, the resulting frequency will be 263 Hz.

DIFFERENCE TONES. Another interesting aspect of the interaction between notes is the "difference tone," created by discord, which the human ear perceives as a third tone. Though E and F are both part of the C scale, when struck together, the sound is highly discordant. In light of what was said above about ratios between frequencies, this dissonance is fitting, as the ratio here involves relatively high numbers—15:16.

When two notes are struck together, they produce a combination tone, perceived by the human ear as a third tone. If the two notes are harmonious, the "third tone" is known as a summation tone, and is equal to the combined frequencies of the two notes. But if the combination is dissonant, as in the case of E and F, the third tone is known as a difference tone, equal to the difference in frequencies. Since an E note vibrates at 330 Hz, and an F note at 352 Hz, the resulting difference tone is equal to 22 Hz.

DESTRUCTIVE INTERFERENCE IN SOUND WAVES. When music is played in a concert hall, it reverberates off the walls of the auditorium. Assuming the place is well designed acoustically, these bouncing sound waves will interfere constructively, and the auditorium comes alive with the sound of the music. In other situations, however, the sound waves may interfere destructively, and the result is a certain muffled deadness to the sound.

Clearly, in a music hall, destructive interference is a problem; but there are cases in which it can be a benefit—situations, that is, in which the purpose, indeed, is to deaden the sound. One example is an automobile muffler. A car's exhaust system makes a great deal of noise, and, thus, if a car does not have a proper muffler, it creates a great deal of noise pollution. A muffler counteracts this by producing a sound wave out of phase with that of the exhaust system; hence, it cancels out most of the noise.

Destructive interference can also be used to reduce sound in a room. Once again, a machine is calibrated to generate sound waves that are perfectly out of phase with the offending noise—say, the hum of another machine. The resulting effect conveys the impression that there is no noise in the room, though, in fact, the sound waves are still there; they have merely canceled each other out.

ELECTROMAGNETIC WAVES

In 1801, English physicist Thomas Young (1773-1829), known for Young's modulus of elasticity became the first scientist to identify interference in light waves. Challenging the corpuscular theory of light put forward by Sir Isaac Newton (1642-1727), Young set up an experiment in which a beam of light passed through two closely spaced pinholes onto a screen. If light was truly made of particles, he said, the beams would project two distinct points onto the screen. Instead, what he saw was a pattern of interference.

In fact, Newton was partly right, but Young's discovery helped advance the view of light as a wave, which is also partly right. (According to quantum theory, developed in the twentieth century, light behaves both as waves and as particles.) The interference in the visible spectrum that Young witnessed was manifested as bright and dark bands. These bands are known as fringes—variations in intensity not unlike the beats created in some instances of sound interference, described above.

OILY FILMS AND RAINBOWS. Many people have noticed the strangely beautiful pattern of colors generated when light interacts with an oily substance, as when light reflected on a soap bubble produces an astonishing array of shades. Sometimes, this can happen in situations not otherwise aesthetically pleasing: an oily film in a parking lot, left there by a car's leaky

KEY TERMS

AMPLITUDE: The maximum displacement of particles in oscillation from a position of stable equilibrium. For an ocean wave, amplitude is the distance from either the crest or the trough to the level that the ocean would maintain if it were perfectly still.

CONSTRUCTIVE INTERFERENCE: A type of interference that occurs when two or more waves combine in such a way that they produce a wave whose amplitude is greater than that of the original waves. If waves are perfectly in phase—in other words, if the crest and trough of one exactly meets the crest and trough of the other—then the resulting amplitude is the sum of the individual amplitudes of the separate waves.

CYCLE: In oscillation, a cycle occurs when the oscillating particle moves from a certain point in a certain direction, then switches direction and moves back to the original point. Typically, this is from the position of stable equilibrium to maximum displacement and back again to the stable equilibrium position. In a wave, a cycle is equivalent to the movement from trough to crest and back to trough.

DESTRUCTIVE INTERFERENCE: A type of interference that occurs when two or more waves combine to produce a wave whose amplitude is less than that of the original waves. If waves are perfectly out of phase—in other words, if the trough of one exactly meets the crest of the other, and vice versa—their amplitudes cancel out, and the result is no wave at all.

FREQUENCY: In wave motion, frequency is the number of waves passing through a given point during the interval of one second. The higher the frequency, the shorter the wavelength. Frequency is mathematically related to wave speed and period.

HARMONIC MOTION: The repeated movement of a particle about a position of equilibrium, or balance.

HERTZ: A unit for measuring frequency, named after nineteenth-century German physicist Heinrich Rudolf Hertz (1857-1894). High frequencies are expressed in terms of kilohertz (kHz; 10^3 or 1,000 cycles per second) or megahertz (MHz; 10^6 or 1 million cycles per second.)

INTERFERENCE: The combination of two or more waves at a given point in space to produce a wave whose properties are determined by the properties of the individual waves. This accords with the principle of superposition.

LONGITUDINAL WAVE: A wave in which the movement of vibration is in the same direction as the wave itself. This is contrasted to a transverse wave.

MAXIMUM DISPLACEMENT: For an object in oscillation, maximum displacement is the furthest point from stable equilibrium.

MECHANICAL WAVE: A type of wave—for example, a wave on the ocean—that involves matter. The matter itself may move in place, but as with all types of wave motion, there is no net movement of matter—only of energy.

OSCILLATION: A type of harmonic motion, typically periodic, in one or more dimensions.

KEY TERMS CONTINUED

PERIOD: For wave motion, a period is the amount of time required to complete one full cycle. Period is mathematically related to frequency, wavelength, and wave speed.

PERIODIC MOTION: Motion that is repeated at regular intervals. These intervals are known as periods.

PERIODIC WAVE: A wave in which a uniform series of crests and troughs follow one after the other in regular succession.

PHASE: When two waves of the same frequency and amplitude are perfectly aligned, they are said to be in phase.

PRINCIPLE OF SUPERPOSITION: A physical principle stating that when several individual, but similar, physical events occur in close proximity to one another, the resulting effect is the sum of the magnitude of the separate events. Interference is an example of superposition.

PULSE: An isolated, non-periodic disturbance that takes place in wave motion of a type other than that of a periodic wave.

RESONANCE: The condition in which force is applied to an object in oscillation at the point of maximum amplitude.

STABLE EQUILIBRIUM: A position in which, if an object were disturbed, it would tend to return to its original position. For an object in oscillation, stable equilibrium is in the middle of a cycle, between two points of maximum displacement.

SURFACE WAVE: A wave that exhibits the behavior of both a transverse wave and a longitudinal wave.

TRANSVERSE WAVE: A wave in which the vibration or motion is perpendicular to the direction in which the wave is moving. This is contrasted to a longitudinal wave.

WAVELENGTH: The distance between a crest and the adjacent crest, or the trough and an adjacent trough, of a wave. Wavelength, abbreviated λ (the Greek letter lambda) is mathematically related to wave speed, period, and frequency.

WAVE MOTION: A type of harmonic motion that carries energy from one place to another, without actually moving any matter.

---

crankcase, can produce a rainbow of colors if the sunlight hits it just right.

This happens because the thickness of the oil causes a delay in reflection of the light beam. Some colors pass through the film, becoming delayed and, thus, getting out of phase with the reflected light on the surface of the film. These shades destructively interfere to such an extent that the waves are cancelled, rendering them invisible. Other colors reflect off the surface so that they are perfectly in phase with the light traveling through the film, and appear as an attractive swirl of color on the surface of the oil.

The phenomenon of light-wave interference with oily or filmy surfaces has the effect of filtering light, and, thus, has a number of applications in areas relating to optics: sunglasses, lenses for binoculars or cameras, and even visors for astronauts. In each case, unfiltered light could be harmful or, at least, inconvenient for the user, and the destructive interference eliminates certain colors and unwanted reflections.

RADIO WAVES. Visible light is only a small part of the electromagnetic spectrum, whose broad range of wave phenomena are, likewise, subject to constructive or destructive interference. After visible light, the area of the spectrum most people experience during an average day is the realm of relatively low-frequency, long-wavelength radio waves and microwaves, the latter including television broadcast signals.

People who rely on an antenna for their TV reception are likely to experience interference at some point. However, an increasing number of Americans use either cable or satellite systems to pick up TV programs. These are much less susceptible to interference, due to the technology of coaxial cable, on the one hand, and digital compression, on the other. Thus, interference in television reception is a gradually diminishing problem.

Interference among radio signals continues to be a challenge, since most people still hear the radio via old-fashioned means rather than through new technology, such as the Internet. A number of interference problems are created by activity on the Sun, which has an enormously powerful electromagnetic field. Obviously, such interference is beyond the control of most radio listeners, but according to a Web page set up by WHKY Radio in Hickory, North Carolina, there are a number of things listeners can do to decrease interference in their own households.

Among the suggestions offered at the WHKY Web site is this: "Nine times out of ten, if your radio is near a computer, it will interfere with your radio. Computers send out all kinds of signals that your radio 'thinks' is a real radio signal. Try to locate your radio away from comput-

ers... especially the monitor." The Web site listed a number of other household appliances, as well as outside phenomena such as power lines or thunderstorms, that can contribute to radio interference.

WHERE TO LEARN MORE

Beiser, Arthur. Physics, 5th ed. Reading, MA: Addison-Wesley, 1991.

Bloomfield, Louis A. "How Things Work: Radio." How Things Work (Web site). <http://rabi.phys.virginia. edu/HTW//radio.html> (April 27, 2001).

Harrison, David. "Sound" (Web site). <http://www.newi. ac.uk/buckley/sound.html> (April 27, 2001).

Interference Handbook/Federal Communications Commission (Web site). <http://www.fcc.gov/cib/Publications/tvibook.html> (April 27, 2001).

Internet Resources for Sound and Light (Web site). <http://electro.sau.edu/SLResources.html> (April 25, 2001).

"Light—A-to-Z Science." DiscoverySchool.com (Web site). <http://school.discovery.com/homeworkhelp/ worldbook/atozscience/l/323260.html> (April 27, 2001).

Oxlade, Chris. Light and Sound. Des Plaines, IL: Heinemann Library, 2000.

"Sound Wave—Constructive and Destructive Interference" (Web site). <http://csgrad.cs.vt.edu/~chin/ interference.html> (April 27, 2001).

Topp, Patricia. This Strange Quantum World and You. Nevada City, CA: Blue Dolphin, 1999.

WHKY Radio and TV (Web site). <http://www.whky.com/antenna.html> (April 27, 2001).

DIFFRACTION

CONCEPT

Diffraction is the bending of waves around obstacles, or the spreading of waves by passing them through an aperture, or opening. Any type of energy that travels in a wave is capable of diffraction, and the diffraction of sound and light waves produces a number of effects. (Because sound waves are much larger than light waves, however, diffraction of sound is a part of daily life that most people take for granted.) Diffraction of light waves, on the other hand, is much more complicated, and has a number of applications in science and technology, including the use of diffraction gratings in the production of holograms.

HOW IT WORKS

COMPARING SOUND AND LIGHT DIFFRACTION

Imagine going to a concert hall to hear a band, and to your chagrin, you discover that your seat is directly behind a wide post. You cannot see the band, of course, because the light waves from the stage are blocked. But you have little trouble hearing the music, since sound waves simply diffract around the pillar. Light waves diffract slightly in such a situation, but not enough to make a difference with regard to your enjoyment of the concert: if you looked closely while sitting behind the post, you would be able to observe the diffraction of the light waves glowing slightly, as they widened around the post.

Suppose, now, that you had failed to obtain a ticket, but a friend who worked at the concert venue arranged to let you stand outside an open door and hear the band. The sound quality would be far from perfect, of course, but you would still be able to hear the music well enough. And if you stood right in front of the doorway, you would be able to see light from inside the concert hall. But, if you moved away from the door and stood with your back to the building, you would see little light, whereas the sound would still be easily audible.

WAVELENGTH AND DIFFRACTION. The reason for the difference—that is, why sound diffraction is more pronounced than light diffraction—is that sound waves are much, much larger than light waves. Sound travels by longitudinal waves, or waves in which the movement of vibration is in the same direction as the wave itself. Longitudinal waves radiate outward in concentric circles, rather like the rings of a bull's-eye.

The waves by which sound is transmitted are larger, or comparable in size to, the column or the door—which is an example of an aperture—and, hence, they pass easily through apertures and around obstacles. Light waves, on the other hand, have a wavelength, typically measured in nanometers (nm), which are equal to one-millionth of a millimeter. Wavelengths for visible light range from 400 (violet) to 700 nm (red): hence, it would be possible to fit about 5,000 of even the longest visible-light wavelengths on the head of a pin!

Whereas differing wavelengths in light are manifested as differing colors, a change in sound wavelength indicates a change in pitch. The higher the pitch, the greater the frequency, and, hence, the shorter the wavelength. As with light waves—though, of course, to a much lesser

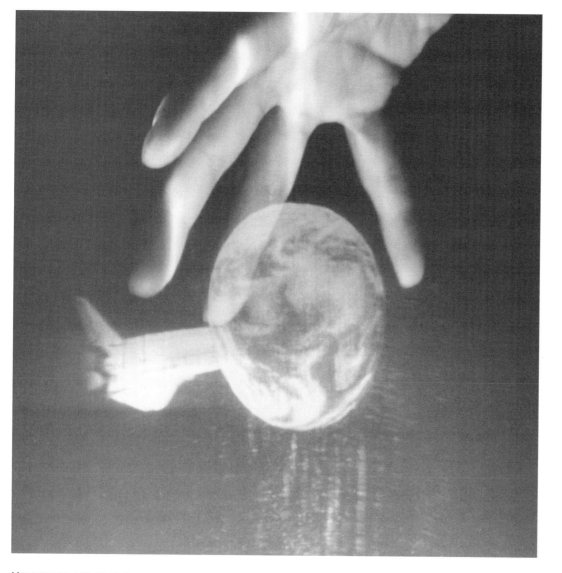

HOLOGRAMS ARE MADE POSSIBLE THROUGH THE PRINCIPLE OF DIFFRACTION. SHOWN HERE IS A HOLOGRAM OF A SPACE SHUTTLE ORBITING EARTH. *(Photograph by Roger Ressmeyer/Corbis. Reproduced by permission.)*

extent—short-wavelength sound waves are less capable of diffracting around large objects than are long-wavelength sound waves. Chances are, then, that the most easily audible sounds from inside the concert hall are the bass and drums; higher-pitched notes from a guitar or other instruments, such as a Hammond organ, are not as likely to reach a listener outside.

OBSERVING DIFFRACTION IN LIGHT

Due to the much wider range of areas in which light diffraction has been applied by scientists, diffraction of light and not sound will be the principal topic for the remainder of this essay. We

have already seen that wavelength plays a role in diffraction; so, too, does the size of the aperture relative to the wavelength. Hence, most studies of diffraction in light involve very small openings, as, for instance, in the diffraction grating discussed below.

But light does not only diffract when passing through an aperture, such as the concert-hall door in the earlier illustration; it also diffracts around obstacles, as, for instance, the post or pillar mentioned earlier. This can be observed by looking closely at the shadow of a flagpole on a bright morning. At first, it appears that the shadow is "solid," but if one looks closely enough, it becomes clear that, at the edges, there is a blur-

THE CHECKOUT SCANNERS IN GROCERY STORES USE HOLOGRAPHIC TECHNOLOGY THAT CAN READ A UNIVERSAL PRODUCT CODE (UPC) FROM ANY ANGLE. *(Photograph by Bob Rowan; Progressive Image/Corbis. Reproduced by permission.)*

ring from darkness to light. This "gray area" is an example of light diffraction.

Where the aperture or obstruction is large compared to the wave passing through or around it, there is only a little "fuzziness" at the edge, as in the case of the flagpole. When light passes through an aperture, most of the beam goes straight through without disturbance, with only the edges experiencing diffraction. If, however, the size of the aperture is close to that of the wavelength, the diffraction pattern will widen. Sound waves diffract at large angles through an open door, which, as noted, is comparable in size to a sound wave; similarly, when light is passed through extremely narrow openings, its diffraction is more noticeable.

Early Studies in Diffraction

Though his greatest contributions lay in his epochal studies of gravitation and motion, Sir Isaac Newton (1642-1727) also studied the production and propagation of light. Using a prism, he separated the colors of the visible light spectrum—something that had already been done by other scientists—but it was Newton who discerned that the colors of the spectrum could be recombined to form white light again.

Newton also became embroiled in a debate as the nature of light itself—a debate in which diffraction studies played an important role. Newton's view, known at the time as the corpuscular theory of light, was that light travels as a stream of particles. Yet, his contemporary, Dutch physicist and astronomer Christiaan Huygens (1629-1695), advanced the wave theory, or the idea that light travels by means of waves. Huygens maintained that a number of factors, including the phenomena of reflection and refraction, indicate that light is a wave. Newton, on the other hand, challenged wave theorists by stating that if light were actually a wave, it should be able to bend around corners—in other words, to diffract.

GRIMALDI IDENTIFIES DIF-FRACTION. Though it did not become widely known until some time later, in 1648—more than a decade before the particle-wave controversy erupted—Johannes Marcus von Kronland (1595-1667), a scientist in Bohemia (now part of the Czech Republic), discovered the diffraction of light waves. However, his findings were not recognized until some time later; nor did he give a name to the phenomenon he had observed. Then, in 1660, Italian physicist Francesco Grimaldi (1618-1663) conducted an experiment with diffraction that gained wide-spread attention.

Grimaldi allowed a beam of light to pass through two narrow apertures, one behind the other, and then onto a blank surface. When he did so, he observed that the band of light hitting the surface was slightly wider than it should be, based on the width of the ray that entered the first aperture. He concluded that the beam had been bent slightly outward, and gave this phenomenon the name by which it is known today: diffraction.

FRESNEL AND FRAUNHOFER DIFFRACTION. Particle theory continued to have its adherents in England, Newton's homeland, but by the time of French physicist Augustin Jean Fresnel (1788-1827), an increasing number of scientists on the European continent had come to accept the wave theory. Fresnel's work, which he published in 1818, served to advance that theory, and, in particular, the idea of light as a transverse wave.

In *Memoire sur la diffraction de la lumiere,* Fresnel showed that the transverse-wave model accounted for a number of phenomena, including diffraction, reflection, refraction, interference, and polarization, or a change in the oscillation patterns of a light wave. Four years after publishing this important work, Fresnel put his ideas into action, using the transverse model to create a pencil-beam of light that was ideal for lighthouses. This prism system, whereby all the light emitted from a source is refracted into a horizontal beam, replaced the older method of mirrors used since ancient times. Thus Fresnel's work revolutionized the effectiveness of lighthouses, and helped save lives of countless sailors at sea.

The term "Fresnel diffraction" refers to a situation in which the light source or the screen are close to the aperture; but there are situations in which source, aperture, and screen (or at least two of the three) are widely separated. This is known as Fraunhofer diffraction, after German physicist Joseph von Fraunhofer (1787-1826), who in 1814 discovered the lines of the solar spectrum (source) while using a prism (aperture). His work had an enormous impact in the area of spectroscopy, or studies of the interaction between electromagnetic radiation and matter.

REAL-LIFE APPLICATIONS

DIFFRACTION STUDIES COME OF AGE

Eventually the work of Scottish physicist James Clerk Maxwell (1831-1879), German physicist Heinrich Rudolf Hertz (1857-1894), and others confirmed that light did indeed travel in waves. Later, however, Albert Einstein (1879-1955) showed that light behaves both as a wave and, in certain circumstances, as a particle.

In 1912, a few years after Einstein published his findings, German physicist Max Theodor Felix von Laue (1879-1960) created a diffraction grating, discussed below. Using crystals in his grating, he proved that x rays are part of the electromagnetic spectrum. Laue's work, which earned him the Nobel Prize in physics in 1914, also made it possible to measure the length of x rays, and, ultimately, provided a means for studying the atomic structure of crystals and polymers.

SCIENTIFIC BREAKTHROUGHS MADE POSSIBLE BY DIFFRACTION STUDIES. Studies in diffraction advanced during the early twentieth century. In 1926, English physicist J. D. Bernal (1901-1971) developed the Bernal chart, enabling scientists to deduce the crystal structure of a solid by analyzing photographs of x-ray diffraction patterns. A decade later, Dutch-American physical chemist Peter Joseph William Debye (1884-1966) won the Nobel Prize in Chemistry for his studies in the diffraction of x rays and electrons in gases, which advanced understanding of molecular structure. In 1937, a year after Debye's Nobel, two other scientists—American physicist Clinton Joseph Davisson (1881-1958) and English physicist George Paget Thomson (1892-1975)—won the Prize in Physics for their discovery that crystals can bring about the diffraction of electrons.

Also, in 1937, English physicist William Thomas Astbury (1898-1961) used x-ray diffraction to discover the first information concerning nucleic acid, which led to advances in the study of DNA (deoxyribonucleic acid), the building-blocks of human genetics. In 1952, English biophysicist Maurice Hugh Frederick Wilkins (1916-) and molecular biologist Rosalind Elsie Franklin (1920-1958) used x-ray diffraction to photograph DNA. Their work directly influenced a breakthrough event that followed a year later: the discovery of the double-helix or double-spiral model of DNA by American molecular biologists James D. Watson (1928-) and Francis Crick (1916-). Today, studies in DNA are at the frontiers of research in biology and related fields.

DIFFRACTION GRATING

Much of the work described in the preceding paragraphs made use of a diffraction grating, first developed in the 1870s by American physicist Henry Augustus Rowland (1848-1901). A diffraction grating is an optical device that consists of not one but many thousands of apertures: Rowland's machine used a fine diamond point to rule glass gratings, with about 15,000 lines per in (2.2 cm). Diffraction gratings today can have as many as 100,000 apertures per inch. The apertures in a diffraction grating are not mere holes, but extremely narrow parallel slits that transform a beam of light into a spectrum.

Each of these openings diffracts the light beam, but because they are evenly spaced and the same in width, the diffracted waves experience constructive interference. (The latter phenomenon, which describes a situation in which two or more waves combine to produce a wave of greater magnitude than either, is discussed in the essay on Interference.) This constructive interference pattern makes it possible to view components of the spectrum separately, thus enabling a scientist to observe characteristics ranging from the structure of atoms and molecules to the chemical composition of stars.

X-RAY DIFFRACTION. Because they are much higher in frequency and energy levels, x rays are even shorter in wavelength than visible light waves. Hence, for x-ray diffraction, it is necessary to have gratings in which lines are separated by infinitesimal distances. These distances are typically measured in units called an angstrom, of which there are 10 million to a millimeter. Angstroms are used in measuring atoms, and, indeed, the spaces between lines in an x-ray diffraction grating are comparable to the size of atoms.

When x rays irradiate a crystal—in other words, when the crystal absorbs radiation in the form of x rays—atoms in the crystal diffract the rays. One of the characteristics of a crystal is that its atoms are equally spaced, and, because of this, it is possible to discover the location and distance between atoms by studying x-ray diffraction patterns. Bragg's law—named after the father-and-son team of English physicists William Henry Bragg (1862-1942) and William Lawrence Bragg (1890-1971)—describes x-ray diffraction patterns in crystals.

Though much about x-ray diffraction and crystallography seems rather abstract, its application in areas such as DNA research indicates that it has numerous applications for improving human life. The elder Bragg expressed this fact in 1915, the year he and his son received the Nobel Prize in physics, saying that "We are now able to look ten thousand times deeper into the structure of the matter that makes up our universe than when we had to depend on the microscope alone." Today, physicists applying x-ray diffraction use an instrument called a diffractometer, which helps them compare diffraction patterns with those of known crystals, as a means of determining the structure of new materials.

HOLOGRAMS

A hologram—a word derived from the Greek *holos,* "whole," and *gram,* "message"—is a three-dimensional (3-D) impression of an object, and the method of producing these images is known as holography. Holograms make use of laser beams that mix at an angle, producing an interference pattern of alternating bright and dark lines. The surface of the hologram itself is a sort of diffraction grating, with alternating strips of clear and opaque material. By mixing a laser beam and the unfocused diffraction pattern of an object, an image can be recorded. An illuminating laser beam is diffracted at specific angles, in accordance with Bragg's law, on the surfaces of the hologram, making it possible for an observer to see a three-dimensional image.

Holograms are not to be confused with ordinary three-dimensional images that use only visible light. The latter are produced by a method known as stereoscopy, which creates a single image from two, superimposing the images to create the impression of a picture with depth. Though stereoscopic images make it seem as though one can "step into" the picture, a hologram actually enables the viewer to glimpse the image from any angle. Thus, stereoscopic images can be compared to looking through the plate-glass window of a store display, whereas holo-

KEY TERMS

APERTURE: An opening.

DIFFRACTION: The bending of waves around obstacles, or the spreading of waves by passing them through an aperture.

ELECTROMAGNETIC SPECTRUM: The complete range of electromagnetic waves on a continuous distribution from a very low range of frequencies and energy levels, with a correspondingly long wavelength, to a very high range of frequencies and energy levels, with a correspondingly short wavelength. Included on the electromagnetic spectrum are long-wave and short-wave radio; microwaves; infrared, visible, and ultraviolet light; x rays, and gamma rays.

FREQUENCY: The number of waves passing through a given point during the interval of one second. The higher the frequency, the shorter the wavelength.

LONGITUDINAL WAVE: A wave in which the movement of vibration is in the same direction as the wave itself. A sound wave is an example of a longitudinal wave.

PRISM: A three-dimensional glass shape used for the diffusion of light rays.

PROPAGATION: The act or state of traveling from one place to another.

RADIATION: In a general sense, radiation can refer to anything that travels in a stream, whether that stream be composed of subatomic particles or electromagnetic waves.

REFLECTION: A phenomenon whereby a light ray is returned toward its source rather than being absorbed at the interface.

REFRACTION: The bending of a light ray that occurs when it passes through a dense medium, such as water or glass.

SPECTRUM: The continuous distribution of properties in an ordered arrangement across an unbroken range. Examples of spectra (the plural of "spectrum") include the colors of visible light, or the electromagnetic spectrum of which visible light is a part.

TRANSVERSE WAVE: A wave in which the vibration or motion is perpendicular to the direction in which the wave is moving.

WAVELENGTH: The distance between a crest and the adjacent crest, or the trough and an adjacent trough, of a wave. The shorter the wavelength, the higher the frequency.

---

grams convey the sensation that one has actually stepped into the store window itself.

DEVELOPMENTS IN HOLOGRAPHY. While attempting to improve the resolution of electron microscopes in 1947, Hungarian-English physicist and engineer Dennis Gabor (1900-1979) developed the concept of holography and coined the term "hologram." His work in this area could not progress by a great measure, however, until the creation of the laser in 1960. By the early 1960s, scientists were using

lasers to create 3-D images, and in 1971, Gabor received the Nobel Prize in physics for the discovery he had made a generation before.

Today, holograms are used on credit cards or other identification cards as a security measure, providing an image that can be read by an optical scanner. Supermarket checkout scanners use holographic optical elements (HOEs), which can read a universal product code (UPC) from any angle. Use of holograms in daily life and scientific research is likely to increase as scientists find new applications: for instance, holographic images will aid the design of everything from bridges to automobiles.

HOLOGRAPHIC MEMORY. One of the most fascinating areas of research in the field of holography is holographic memory. Computers use a binary code, a pattern of ones and zeroes that is translated into an electronic pulse, but holographic memory would greatly extend the capabilities of computer memory systems. Unlike most images, a hologram is not simply the sum of its constituent parts: the data in a holographic image is contained in every part of the image, meaning that part of the image can be destroyed without a loss of data.

To bring the story full-circle, holographic memory calls to mind an idea advanced by a scientist who, along with Huygens, was one of Newton's great professional rivals, German mathematician and philosopher Gottfried Wilhelm Leibniz (1646-1716). Though Newton is usually credited as the father of calculus, Leibniz developed his own version of calculus at around the same time.

As a philosopher, Leibniz had apparently had a number of strange ideas, which made him the butt of jokes among some sectors of European intellectual society: hence, the French writer and thinker Voltaire (François-Marie Arouet; 1694-1778) satirized him with the character Dr. Pangloss in *Candide* (1759). Few of Leibniz's ideas were more bizarre than that of the monad: an elementary particle of existence that reflected the whole of the universe.

In advancing the concept of a monad, Leibniz was not making a statement after the manner of a scientist: there was no proof that monads existed, nor was it possible to prove this in any scientific way. Yet, a hologram appears to be very much like a manifestation of Leibniz's imagined monads, and both the hologram and the monad relate to a more fundamental aspect of life: human memory. Neurological research in the late twentieth century suggested that the structure of memory in the human mind is holographic. Thus, for instance, a patient suffering an injury affecting 90% of the brain experiences only a 10% memory loss.

WHERE TO LEARN MORE

Barrett, Norman S. *Lasers and Holograms.* New York: F. Watts, 1985.

"Bragg's Law and Diffraction: How Waves Reveal the Atomic Structure of Crystals" (Web site). <http://www.journey.sunysb.edu/ProjectJava/Bragg/home.html> (May 6, 2001).

Burkig, Valerie. *Photonics: The New Science of Light.* Hillside, N.J.: Enslow Publishers, 1986.

"Diffraction of Sound" (Web site). <http://hyperphysics.phy-astr.gsu.edu/hbase/sound/diffrac.html> (May 6, 2001).

Gardner, Robert. *Experimenting with Light.* New York: F. Watts, 1991.

Graham, Ian. *Lasers and Holograms.* New York: Shooting Star Press, 1993.

Holoworld: Holography, Lasers, and Holograms (Web site). <http://www.holoworld.com> (May 6, 2001).

Proffen, T. H. and R. B. Neder. *Interactive Tutorial About Diffraction* (Web site). <http://www.uniwuerzburg.de/mineralogie/crystal/teaching/teaching.html> (May 6, 2001).

Snedden, Robert. *Light and Sound.* Des Plaines, IL: Heinemann Library, 1999.

"Wave-Like Behaviors of Light." The Physics Classroom (Web site). <http://www.glenbrook.k12.il.us/gbssci/phys/Class/light/u12l1a.html> (May 6, 2001).

DOPPLER EFFECT

CONCEPT

Almost everyone has experienced the Doppler effect, though perhaps without knowing what causes it. For example, if one is standing on a street corner and an ambulance approaches with its siren blaring, the sound of the siren steadily gains in pitch as it comes closer. Then, as it passes, the pitch suddenly lowers perceptibly. This is an example of the Doppler effect: the change in the observed frequency of a wave when the source of the wave is moving with respect to the observer. The Doppler effect, which occurs both in sound and electromagnetic waves—including light waves—has a number of applications. Astronomers use it, for instance, to gauge the movement of stars relative to Earth. Closer to home, principles relating to the Doppler effect find application in radar technology. Doppler radar provides information concerning weather patterns, but some people experience it in a less pleasant way: when a police officer uses it to measure their driving speed before writing a ticket.

HOW IT WORKS

WAVE MOTION AND ITS PROPERTIES

Sound and light are both examples of energy, and both are carried on waves. Wave motion is a type of harmonic motion that carries energy from one place to another without actually moving any matter. It is related to oscillation, a type of harmonic motion in one or more dimensions. Oscillation involves no net movement, only movement in place; yet individual points in the wave medium are oscillating even as the overall wave pattern moves.

The term periodic motion, or movement repeated at regular intervals called periods, describes the behavior of periodic waves—waves in which a uniform series of crests and troughs follow each other in regular succession. A period (represented by the symbol T) is the amount of time required to complete one full cycle of the wave, from trough to crest and back to trough.

Period is mathematically related to several other aspects of wave motion, including wave speed, frequency, and wavelength. Frequency (abbreviated f) is the number of waves passing through a given point during the interval of one second. It is measured in Hertz (Hz), named after nineteenth-century German physicist Heinrich Rudolf Hertz (1857-1894), and a Hertz is equal to one cycle of oscillation per second. Higher frequencies are expressed in terms of kilohertz (kHz; 10^3 or 1,000 cycles per second); megahertz (MHz; 10^6 or 1 million cycles per second); and gigahertz (GHz; 10^9 or 1 billion cycles per second.) Wavelength (represented by the symbol λ, the Greek letter lambda) is the distance between a crest and the adjacent crest, or a trough and an adjacent trough, of a wave. The higher the frequency, the shorter the wavelength.

Amplitude, though mathematically independent from the parameters discussed, is critical to the understanding of sound. Defined as the maximum displacement of a vibrating material, amplitude is the "size" of a wave. The greater the amplitude, the greater the energy the wave contains: amplitude indicates intensity, which, in the case of sound waves, is manifested as what people commonly call "volume." Similarly, the

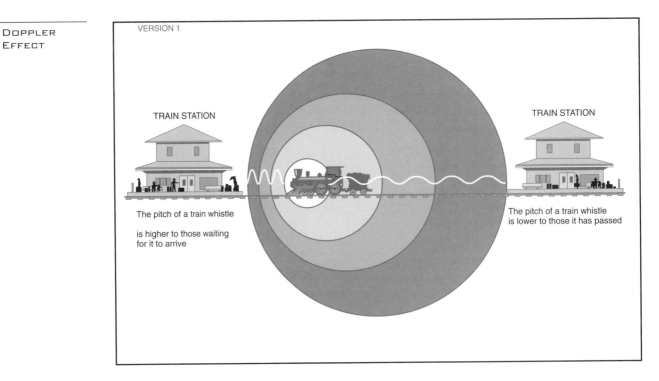

VERSION 1

TRAIN STATION

The pitch of a train whistle

is higher to those waiting
for it to arrive

TRAIN STATION

The pitch of a train whistle
is lower to those it has passed

THE DOPPLER EFFECT.

amplitude of a light wave determines the intensity of the light.

FRAME OF REFERENCE

A knowledge of the fundamentals involved in wave motion is critical to understanding the Doppler effect; so, too, is an appreciation of another phenomenon, which is as much related to human psychology and perception as it is to physics. Frame of reference is the perspective of an observer with regard to an object or event. Things may look different for one person in one frame of reference than they do to someone in another.

For example, if you are sitting across the table from a friend at lunch, and you see that he has a spot of ketchup to the right of his mouth, the tendency is to say, "You have some ketchup right here"—and then point to the left of your own mouth, since you are directly across the table from his right. Then he will rub the left side of his face with his napkin, missing the spot entirely, unless you say something like, "No— mirror image." The problem is that each of you has a different frame of reference, yet only your friend took this into account.

RELATIVE MOTION. Physicists often speak of relative motion, or the motion of one object in relation to another. For instance, the molecules in the human body are in a constant state of motion, but they are not moving relative to the body itself: they are moving relative to one another.

On a larger scale, Earth is rotating at a rate of about 1,000 MPH (1,600 km/h), and orbiting the Sun at 67,000 MPH (107,826 km/h)—almost three times as fast as humans have ever traveled in a powered vehicle. Yet no one senses the speed of Earth's movement in the way that one senses the movement of a car—or, indeed, the way the astronauts aboard *Apollo 11* in 1969 perceived that their spacecraft was moving at about 25,000 MPH (40,000 km/h). In the case of the car or the spacecraft, movement can be perceived in relation to other objects: road signs and buildings on the one hand, Earth and the Moon on the other. But humans have no frame of reference from which to perceive the movement of Earth itself.

If one were traveling in a train alongside another train at constant velocity, it would be impossible to perceive that either train was actually moving, unless one looked at a reference point, such as the trees or mountains in the background. Likewise, if two trains were sitting side by side, and one train started to move, the relative motion might cause a passenger in the unmoving train to believe that his or her train

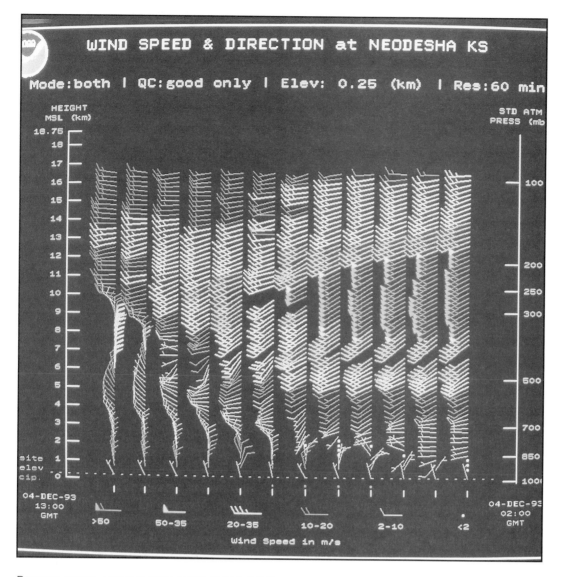

WIND SPEED & DIRECTION at NEODESHA KS

Mode:both | QC:good only | Elev: 0.25 (km) | Res:60 min

DOPPLER RADAR, LIKE THE WIND SPEED RADAR SHOWN HERE, HAVE BECOME A FUNDAMENTAL TOOL FOR METE-OROLOGISTS. *(Photograph by Roger Ressmeyer/Corbis. Reproduced by permission.)*

was the one moving. In fact, as Albert Einstein (1879-1955) demonstrated with his Theory of Relativity, all motion is relative: when we say that something is moving, we mean that it is moving in relation to something else.

DOPPLER'S DISCOVERY

Long before Einstein was born, Austrian physicist Christian Johann Doppler (1803-1853) made an important discovery regarding the relative motion of sound waves or light waves. While teaching in Prague, now the capital of the Czech Republic, but then a part of the Austro-Hungarian Empire, Doppler became fascinated with a common, but previously unexplained, phenomenon. When an observer is standing beside a railroad track and a train approaches, Doppler noticed, the train's whistle has a high pitch. As it passes by, however, the sound of the train whistle suddenly becomes much lower.

By Doppler's time, physicists had recognized the existence of sound waves, as well as the fact a sound's pitch is a function of frequency—in other words, the closer the waves are to one another, the higher the pitch. Taking this knowledge, he reasoned that if a source of sound is moving toward a listener, the waves in front of the source are compressed, thus creating a higher frequency. On the other hand, the waves behind the moving source are stretched out, resulting in a lower frequency.

After developing a mathematical formula to describe this effect, Doppler presented his findings in 1842. Three years later, he and Dutch meteorologist Christopher Heinrich Buys-Ballot (1817-1890) conducted a highly unusual experiment to demonstrate the theory. Buys-Ballot arranged for a band of trumpet players to perform on an open railroad flatcar, while riding past a platform on which a group of musicians with perfect pitch (that is, a finely tuned sense of hearing) sat listening.

The experiment went on for two days, the flatcar passing by again and again, while the horns blasted and the musicians on the platform recorded their observations. Though Doppler and Buys-Ballot must have seemed like crazy men to those who were not involved in the experiment, the result—as interpreted from the musicians' written impressions of the pitches they heard—confirmed Doppler's theory.

REAL-LIFE APPLICATIONS

SOUND COMPRESSION AND THE DOPPLER EFFECT

As stated in the introduction, one can observe the Doppler effect in a number of settings. If a person is standing by the side of a road and a car approaches at a significant rate of speed, the frequency of the sound waves grows until the car passes the observer, then the frequency suddenly drops. But Doppler, of course, never heard the sound of an automobile, or the siren of a motorized ambulance or fire truck.

In his day, the horse-drawn carriage still constituted the principal means of transportation for short distances, and such vehicles did not attain the speeds necessary for the Doppler effect to become noticeable. Only one mode of transportation in the mid-nineteenth century made it possible to observe and record the effect: a steam-powered locomotive. Therefore, let us consider the Doppler effect as Doppler himself did—in terms of a train passing through a s tation.

THE SOUND OF A TRAIN WHISTLE. When a train is sitting in a station prior to leaving, it blows its whistle, but listeners standing nearby notice nothing unusual. There is no difference—except perhaps in degree of intensity—between the sound heard by someone on the platform, and the sound of the train as heard by someone standing behind the caboose. This is because a stationary train is at the center of the sound waves it produces, which radiate in concentric circles (like a bulls-eye) around it.

As the train begins to move, however, it is no longer at the center of the sound waves emanating from it. Instead, the circle of waves is moving forward, along with the train itself, and, thus, the locomotive compresses waves toward the front. If someone is standing further ahead along the track, that person hears the compressed sound waves. Due to their compression, these have a much higher frequency than the waves produced by a stationary train.

At the same time, someone standing behind the train—a listener on the platform at the station, watching the train recede into the distance—hears the sound waves that emanate from behind the train. It is the same train making the same sound, but because the train has compressed the sound waves in front of it, the waves behind it are spread out, producing a sound of much lower frequency. Thus, the sound of the train, as perceived by two different listeners, varies with frame of reference.

THE SONIC BOOM: A RELATED EFFECT. Some people today have had the experience of hearing a jet fly high overhead, producing a shock wave known as a sonic boom. A sonic boom, needless to say, is certainly not something of which Doppler would have had any knowledge, nor is it an illustration of the Doppler effect, per se. But it is an example of sound compression, and, therefore, it deserves attention here.

The speed of sound, unlike the speed of light, is dependant on the medium through which it travels. Hence, there is no such thing as a fixed "speed of sound"; rather, there is only a speed at which sound waves are transmitted through a given type of material. Its speed through a gas, such as air, is proportional to the square root of the pressure divided by the density. This, in turn, means that the higher the altitude, the slower the speed of sound: for the altitudes at which jets fly, it is about 660 MPH (1,622 km/h).

As a jet moves through the air, it too produces sound waves which compress toward the front, and widen toward the rear. Since sound waves themselves are really just fluctuations in pressure, this means that the faster a jet goes, the greater the pressure of the sound waves bunched up in front of it. Jet pilots speak of "breaking the sound barrier," which is more than just a figure of speech. As the craft approaches the speed of sound, the pilot becomes aware of a wall of high pressure to the front of the plane, and as a result of this high-pressure wall, the jet experiences enormous turbulence.

The speed of sound is referred to as Mach 1, and at a speed of between Mach 1.2 and Mach 1.4, even stranger things begin to happen. Now the jet is moving faster than the sound waves emanating from it, and, therefore, an observer on the ground sees the jet move by well before hearing the sound. Of course, this would happen to some extent anyway, since light travels so much faster than sound; but the difference between the arrival time of the light waves and the sound waves is even more noticeable in this situation.

Meanwhile, up in the air, every protruding surface of the aircraft experiences intense pressure: in particular, sound waves tend to become highly compressed along the aircraft's nose and tail. Eventually these compressed sound waves build up, resulting in a shock wave. Down on the ground, the shock wave manifests as a "sonic boom"—or rather, two sonic booms—one from the nose of the craft, and one from the tail. People in the aircraft do not hear the boom, but the shock waves produced by the compressed sound can cause sudden changes in pressure, density, and temperature that can pose dangers to the operation of the airplane. To overcome this problem, designers of supersonic aircraft have developed planes with wings that are swept back, so they fit within the cone of pressure.

DOPPLER RADAR AND OTHER SENSING TECHNOLOGY

The Doppler effect has a number of applications relating to the sensing of movement. For instance, physicians and medical technicians apply it to measure the rate and direction of blood flow in a patient's body, along with ultrasound. As blood moves through an artery, its top speed is 0.89 MPH (0.4 m/s)—not very fast, yet fast enough, given the small area in which move-

ment is taking place, for the Doppler effect to be observed. A beam of ultrasound is pointed toward an artery, and the reflected waves exhibit a shift in frequency, because the blood cells are acting as moving sources of sound waves—just like the trains Doppler observed.

Not all applications of the Doppler effect fall under the heading of "technology": some can be found in nature. Bats use the Doppler effect to hunt for prey. As a bat flies, it navigates by emitting whistles and listening for the echoes. When it is chasing down food, its brain detects a change in pitch between the emitted whistle, and the echo it receives. This tells the bat the speed of its quarry, and the bat adjusts its own speed accordingly.

DOPPLER RADAR. Police officers may not enjoy the comparison—given the public's general impression of bats as evil, bloodthirsty creatures—but in using radar as a basis to check for speeding violations, the police are applying a principle similar to that used by bats. Doppler radar, which uses the Doppler effect to calculate the speed of moving objects, is a form of technology used not only by law-enforcement officers, but also by meteorologists.

The change in frequency experienced as a result of the Doppler effect is exactly twice the ratio between the velocity of the target (for instance, a speeding car) and the speed with which the radar pulse is directed toward the target. From this formula, it is possible to determine the velocity of the target when the frequency change and speed of radar propagation are known. The police officer's Doppler radar performs these calculations; then all the officer has to do is pull over the speeder and write a ticket.

Meteorologists use Doppler radar to track the movement of storm systems. By detecting the direction and velocity of raindrops or hail, for instance, Doppler radar can be used to determine the motion of winds and, thus, to predict weather patterns that will follow in the next minutes or hours. But Doppler radar can do more than simply detect a storm in progress: Doppler technology also aids meteorologists by interpreting wind direction, as an indicator of coming storms.

THE DOPPLER EFFECT IN LIGHT WAVES

So far the Doppler effect has been discussed purely in terms of sound waves; but Doppler

KEY TERMS

AMPLITUDE: The maximum displacement of a vibrating material. In wave motion, amplitude is the "size" of a wave, an indicator of the energy and intensity of the wave.

CYCLE: One complete oscillation. In wave motion, this is equivalent to the movement of a wave from trough to crest and back to trough. For a sound wave, in particular, a cycle is one complete vibration.

DOPPLER EFFECT: The change in the observed frequency of a wave when the source of the wave is moving with respect to the observer. It is named after Austrian physicist Johann Christian Doppler (1803-1853), who discovered it.

FRAME OF REFERENCE: The perspective an observer has with regard to an object or action. Frame of reference affects perception of various physical properties, and plays a significant role in the Doppler effect.

FREQUENCY: In wave motion, frequency is the number of waves passing through a given point during the interval of one second. The higher the frequency, the shorter the wavelength. Measured in Hertz, frequency is mathematically related to wave speed, wavelength, and period.

HARMONIC MOTION: The repeated movement of a particle about a position of equilibrium, or balance.

HERTZ: A unit for measuring frequency, named after nineteenth-century German physicist Heinrich Rudolf Hertz (1857-1894). High frequencies are expressed in terms of kilohertz (kHz; 10^3 or 1,000 cycles per second); megahertz (MHz; 10^6 or 1 million cycles per second); and gigahertz (GHz; 10^9 or 1 billion cycles per second.)

INTENSITY: Intensity is the rate at which a wave moves energy per unit of cross-sectional area. Where sound waves are concerned, intensity is commonly known as "volume."

OSCILLATION: A type of harmonic motion, typically periodic, in one or more dimensions.

PERIOD: For wave motion, a period is the amount of time required to complete one full cycle. Period is mathematically related to frequency, wavelength, and wave speed.

PERIODIC MOTION: Motion that is repeated at regular intervals. These intervals are known as periods.

PERIODIC WAVE: A wave in which a uniform series of crests and troughs follow one after the other in regular succession.

RELATIVE MOTION: The motion of one object in relation to another.

WAVELENGTH: The distance between a crest and the adjacent crest, or the trough and an adjacent trough, of a wave. Wavelength, symbolized λ (the Greek letter lambda) is mathematically related to wave speed, period, and frequency.

WAVE MOTION: A type of harmonic motion that carries energy from one place to another without actually moving any matter.

himself maintained that it could be applied to light waves as well, and experimentation conducted in 1901 proved him correct. This was far from an obvious point, since light is quite different from sound.

Not only does light travel much, much faster—186,000 mi (299,339 km) a second—but unlike sound, light does not need to travel through a medium. Whereas sound cannot be transmitted in outer space, light is transmitted by radiation, a form of energy transfer that can be directed as easily through a vacuum as through matter.

The Doppler effect in light can be demonstrated by using a device called a spectroscope, which measures the spectral lines from an object of known chemical composition. These spectral lines are produced either by the absorption or emission of specific frequencies of light by electrons in the source material. If the light waves appear at the blue, or high-frequency end of the visible light spectrum, this means that the object is moving toward the observer. If, on the other hand, the light waves appear at the red, or low-frequency end of the spectrum, the object is moving away.

HUBBLE AND THE RED SHIFT. In 1923, American astronomer Edwin Hubble (1889-1953) observed that the light waves from distant galaxies were shifted so much to the red end of the light spectrum that they must be moving away from the Milky Way, the galaxy in which Earth is located, at a high rate. At the same time, nearer galaxies experienced much less of a red shift, as this phenomenon came to be known, meaning that they were moving away at relatively slower speeds.

Six years later, Hubble and another astronomer, Milton Humason, developed a mathematical formula whereby astronomers could determine the distance to another galaxy by measuring that galaxy's red shifts. The formula came to be known as Hubble's constant, and it established the relationship between red shift and the velocity at which a galaxy or object was receding from Earth. From Hubble's work, it became clear that the universe was expanding, and research by a number of physicists and astronomers led to the development of the "big bang" theory—the idea that the universe emerged almost instantaneously, in some sort of explosion, from a compressed state of matter.

WHERE TO LEARN MORE

Beiser, Arthur. *Physics*, 5th ed. Reading, MA: Addison-Wesley, 1991.

Bryant-Mole, Karen. *Sound and Light.* Crystal Lake, IL: Rigby Interactive Library, 1997.

Challoner, Jack. *Sound and Light.* New York: Kingfisher, 2001.

Dispenzio, Michael A. *Awesome Experiments in Light and Sound.* Illustrated by Catherine Leary. New York: Sterling Publishing Company, 1999.

"The Doppler Effect." The Physics Classroom (Web site). <http://www.glenbrook.k12.il.us/gbssci/phys/Class/waves/u10l3d.html> (April 29, 2001).

Maton, Anthea. *Exploring Physical Science.* Upper Saddle River, N.J.: Prentice Hall, 1997.

Russell, David A. *"The Doppler Effect and Sonic Booms" Kettering University* (Web site). <http://www.kettering.edu/~drussell/Demos/doppler/doppler.html> (April 29, 2001).

Snedden, Robert. *Light and Sound.* Des Plaines, IL: Heinemann Library, 1999.

"Sound Wave—Doppler Effect" (Web site). <http://csgrad.cs.vt.edu/~chin/doppler.html> (April 29, 2001).

"Wave Motion—Doppler Effect" (Web site). <http://members.aol.com/cepeirce/b21.html> (April 29, 2001).

SOUND

ACOUSTICS

ULTRASONICS

ACOUSTICS

CONCEPT

The area of physics known as acoustics is devoted to the study of the production, transmission, and reception of sound. Thus, wherever sound is produced and transmitted, it will have an effect somewhere, even if there is no one present to hear it. The medium of sound transmission is an all-important, key factor. Among the areas addressed within the realm of acoustics are the production of sounds by the human voice and various instruments, as well as the reception of sound waves by the human ear.

HOW IT WORKS

WAVE MOTION AND SOUND WAVES

Sound waves are an example of a larger phenomenon known as wave motion, and wave motion is, in turn, a subset of harmonic motion—that is, repeated movement of a particle about a position of equilibrium, or balance. In the case of sound, the "particle" is not an item of matter, but of energy, and wave motion is a type of harmonic movement that carries energy from one place to another without actually moving any matter.

Particles in waves experience oscillation, harmonic motion in one or more dimensions. Oscillation itself involves little movement, though some particles do move short distances as they interact with other particles. Primarily, however, it involves only movement in place. The waves themselves, on the other hand, move across space, ending up in a position different from the one in which they started.

A transverse wave forms a regular up-and-down pattern in which the oscillation is perpendicular to the direction the wave is moving. This is a fairly easy type of wave to visualize: imagine a curve moving up and down along a straight line. Sound waves, on the other hand, are longitudinal waves, in which oscillation occurs in the same direction as the wave itself.

These oscillations are really just fluctuations in pressure. As a sound wave moves through a medium such as air, these changes in pressure cause the medium to experience alternations of density and rarefaction (a decrease in density). This, in turn, produces vibrations in the human ear or in any other object that receives the sound waves.

PROPERTIES OF SOUND WAVES

CYCLE AND PERIOD. The term cycle has a definition that varies slightly, depending on whether the type of motion being discussed is oscillation, the movement of transverse waves, or the motion of a longitudinal sound wave. In the latter case, a cycle is defined as a single complete vibration.

A period (represented by the symbol T) is the amount of time required to complete one full cycle. The period of a sound wave can be mathematically related to several other aspects of wave motion, including wave speed, frequency, and wavelength.

THE SPEED OF SOUND IN VARIOUS MEDIA. People often refer to the "speed of sound" as though this were a fixed value like the speed of light, but, in fact, the speed of sound is a function of the medium through which it travels. What people ordinarily mean by the "speed of sound" is the speed of sound through air at a specific temperature. For sound

BECAUSE THE SOUND GENERATED BY A JET ENGINE CAN DAMAGE A PERSON'S HEARING, AIRPORT GROUND CREWS ALWAYS WEAR PROTECTIVE HEADGEAR. *(Photograph by Patrick Bennett/Corbis. Reproduced by permission.)*

traveling at sea level, the speed at 32°F (0°C) is 740 MPH (331 m/s), and at 68°F (20°C), it is 767 MPH (343 m/s).

In the essay on aerodynamics, the speed of sound for aircraft was given at 660 MPH (451 m/s). This is much less than the figures given above for the speed of sound through air at sea level, because obviously, aircraft are not flying at sea level, but well above it, and the air through which they pass is well below freezing temperature.

The speed of sound through a gas is proportional to the square root of the pressure divided by the density. According to Gay-Lussac's law, pressure is directly related to temperature, meaning that the lower the pressure, the lower the temperature—and vice versa. At high altitudes, the temperature is low, and, therefore, so is the pressure; and, due to the relatively small gravitational pull that Earth exerts on the air at that height, the density is also low. Hence, the speed of sound is also low.

It follows that the higher the pressure of the material, and the greater the density, the faster sound travels through it: thus sound travels faster through a liquid than through a gas. This might seem a bit surprising: at first glance, it would seem that sound travels fastest through air, but

only because we are just more accustomed to hearing sounds that travel through that medium. The speed of sound in water varies from about 3,244 MPH (1,450 m/s) to about 3,355 MPH (1500 m/s). Sound travels even faster through a solid—typically about 11,185 MPH (5,000 m/s)—than it does through a liquid.

FREQUENCY. Frequency (abbreviated *f*) is the number of waves passing through a given point during the interval of one second. It is measured in Hertz (Hz), named after nineteenth-century German physicist Heinrich Rudolf Hertz (1857-1894) and a Hertz is equal to one cycle of oscillation per second. Higher frequencies are expressed in terms of kilohertz (kHz; 10^3 or 1,000 cycles per second) or megahertz (MHz; 10^6 or 1 million cycles per second.)

The human ear is capable of hearing sounds from 20 to approximately 20,000 Hz—a relatively small range for a mammal, considering that bats, whales, and dolphins can hear sounds at a frequency up to 150 kHz. Human speech is in the range of about 1 kHz, and the 88 keys on a piano vary in frequency from 27 Hz to 4,186 Hz. Each note has its own frequency, with middle C (the "white key" in the very middle of a piano keyboard) at 264 Hz. The quality of harmony or dissonance when two notes are played together is a

PIANO STRINGS GENERATE SOUND AS THEY ARE SET INTO VIBRATION BY THE HAMMERS. THE HAMMERS, IN TURN, ARE ATTACHED TO THE BLACK-AND-WHITE KEYS ON THE OUTSIDE OF THE PIANO. *(Photograph by Bob Krist/Corbis. Reproduced by permission.)*

function of the relationship between the frequencies of the two.

Frequencies below the range of human audibility are called infrasound, and those above it are referred to as ultrasound. There are a number of practical applications for ultrasonic technology in medicine, navigation, and other fields.

WAVELENGTH. Wavelength (represented by the symbol λ, the Greek letter lambda) is the distance between a crest and the adjacent crest, or a trough and an adjacent trough, of a wave. The higher the frequency, the shorter the wavelength, and vice versa. Thus, a frequency of 20 Hz, at the bottom end of human audibility, has a very large wavelength: 56 ft (17 m). The top end frequency of 20,000 Hz is only 0.67 inches (17 mm).

There is a special type of high-frequency sound wave beyond ultrasound: hypersound, which has frequencies above 10^7 MHz, or 10 trillion Hz. It is almost impossible for hypersound waves to travel through all but the densest media, because their wavelengths are so short. In order to be transmitted properly, hypersound requires an extremely tight molecular structure; otherwise, the wave would get lost between molecules.

Wavelengths of visible light, part of the electromagnetic spectrum, have a frequency much higher even than hypersound waves: about 10^9 MHz, 100 times greater than for hypersound. This, in turn, means that these wavelengths are incredibly small, and this is why light waves can easily be blocked out by using one's hand or a curtain.

The same does not hold for sound waves, because the wavelengths of sounds in the range of human audibility are comparable to the size of ordinary objects. To block out a sound wave, one needs something of much greater dimensions—width, height, and depth—than a mere cloth curtain. A thick concrete wall, for instance, may be enough to block out the waves. Better still would be the use of materials that absorb sound, such as cork, or even the use of machines that produce sound waves which destructively interfere with the offending sound.

AMPLITUDE AND INTENSITY. Amplitude is critical to the understanding of sound, though it is mathematically independent from the parameters so far discussed. Defined as the maximum displacement of a vibrating material, amplitude is the "size" of a wave. The greater the amplitude, the greater the energy the wave

THE HUMAN VOICE, WHETHER IN SPEECH OR IN SONG, IS A REMARKABLE SOUND-PRODUCING INSTRUMENT: AT ANY GIVEN MOMENT AS A PERSON IS TALKING OR SINGING, PARTS OF THE VOCAL CORDS ARE OPENED, AND PARTS ARE CLOSED. SHOWN HERE IS OPERA SUPERSTAR JOAN SUTHERLAND. *(Hulton-Deutsch Collection/Corbis. Reproduced by permission.)*

contains: amplitude indicates intensity, commonly known as "volume," which is the rate at which a wave moves energy per unit of a cross-sectional area.

Intensity can be measured in watts per square meter, or W/m^2. A sound wave of minimum intensity for human audibility would have a value of 10^{-12}, or 0.000000000001, W/m^2. As a basis of comparison, a person speaking in an ordinary tone of voice generates about 10^{-4}, or 0.0001, watts. On the other hand, a sound with an intensity of 1 W/m^2 would be powerful enough to damage a person's ears.

REAL-LIFE APPLICATIONS

DECIBEL LEVELS

For measuring the intensity of a sound as experienced by the human ear, we use a unit other than the watt per square meter, because ears do not respond to sounds in a linear, or straight-line, progression. If the intensity of a sound is doubled, a person perceives a greater intensity, but nothing approaching twice that of the original sound. Instead, a different system—known in mathematics as a logarithmic scale—is applied.

In measuring the effect of sound intensity on the human ear, a unit called the decibel (abbreviated dB) is used. A sound of minimal audibility (10^{-12} W/m^2) is assigned the value of 0 dB, and 10 dB is 10 times as great—10^{-11} W/m^2. But 20 dB is not 20 times as intense as 0 dB; it is 100 times as intense, or 10^{-10} W/m^2. Every increase of 10 dB thus indicates a tenfold increase in intensity. Therefore, 120 dB, the maximum decibel level that a human ear can endure without experiencing damage, is not 120 times as great as the minimal level for audibility, but 10^{12} (1 trillion) times as great—equal to 1 W/m^2, referred to above as the highest safe intensity level.

Of course, sounds can be much louder than 120 dB: a rock band, for instance, can generate sounds of 125 dB, which is 5 times the maximum safe decibel level. A gunshot, firecracker, or a jet—if one is exposed to these sounds at a sufficiently close proximity—can be as high as 140 dB, or 20 times the maximum safe level. Nor is 120 dB safe for prolonged periods: hearing experts indicate that regular and repeated exposure to even 85 dB (5 less than a lawn mower) can cause permanent damage to one's hearing.

PRODUCTION OF SOUND WAVES

MUSICAL INSTRUMENTS. Sound waves are vibrations; thus, in order to produce sound, vibrations must be produced. For a stringed instrument, such as a guitar, harp, or piano, the strings must be set into vibration, either by the musician's fingers or the mechanism that connects piano keys to the strings inside the case of the piano.

In other woodwind instruments and horns, the musician causes vibrations by blowing into the mouthpiece. The exact process by which the vibrations emerge as sound differs between woodwind instruments, such as a clarinet or saxophone on the one hand, and brass instruments, such as a trumpet or trombone on the other. Then there is a drum or other percussion instrument, which produces vibrations, if not musical notes.

ELECTRONIC AMPLIFICATION.

Sound is a form of energy: thus, when an automobile or other machine produces sound incidental to its operation, this actually represents energy that is lost. Energy itself is conserved, but not all of the energy put into the machine can ever be realized as useful energy; thus, the automobile loses some energy in the form of sound and heat.

The fact that sound is energy, however, also means that it can be converted to other forms of energy, and this is precisely what a microphone does: it receives sound waves and converts them to electrical energy. These electrical signals are transmitted to an amplifier, and next to a loudspeaker, which turns electrical energy back into sound energy—only now, the intensity of the sound is much greater.

Inside a loudspeaker is a diaphragm, a thin, flexible disk that vibrates with the intensity of the sound it produces. When it pushes outward, the diaphragm forces nearby air molecules closer together, creating a high-pressure region around the loudspeaker. (Remember, as stated earlier, that sound is a matter of fluctuations in pressure.) The diaphragm is then pushed backward in response, freeing up an area of space for the air molecules. These, then, rush toward the diaphragm, creating a low-pressure region behind the high-pressure one. The loudspeaker thus sends out alternating waves of high and low pressure, vibrations on the same frequency of the original sound.

THE HUMAN VOICE.

As impressive as the electronic means of sound production are (and of course the description just given is highly simplified), this technology pales in comparison to the greatest of all sound-producing mechanisms: the human voice. Speech itself is a highly complex physical process, much too involved to be discussed in any depth here. For our present purpose, it is important only to recognize that speech is essentially a matter of producing vibrations on the vocal cords, and then transmitting those vibrations.

Before a person speaks, the brain sends signals to the vocal cords, causing them to tighten. As speech begins, air is forced across the vocal cords, and this produces vibrations. The action of the vocal cords in producing these vibrations is, like everything about the miracle of speech, exceedingly involved: at any given moment as a person is talking, parts of the vocal cords are opened, and parts are closed.

The sound of a person's voice is affected by a number of factors: the size and shape of the sinuses and other cavities in the head, the shape of the mouth, and the placement of the teeth and tongue. These factors influence the production of specific frequencies of sound, and result in differing vocal qualities. Again, the mechanisms of speech are highly complicated, involving action of the diaphragm (a partition of muscle and tissue between the chest and abdominal cavities), larynx, pharynx, glottis, hard and soft palates, and so on. But, it all begins with the production of vibrations.

PROPAGATION: DOES IT MAKE A SOUND?

As stated in the introduction, acoustics is concerned with the production, transmission (sometimes called propagation), and reception of sound. Transmission has already been examined in terms of the speed at which sound travels through various media. One aspect of sound transmission needs to be reiterated, however: for sound to be propagated, there must be a medium.

There is an age-old "philosophical" question that goes something like this: If a tree falls in the woods and there is no one to hear it, does it make a sound? In fact, the question is not a matter of philosophy at all, but of physics, and the answer is, of course, "yes." As the tree falls, it releases energy in a number of forms, and part of this energy is manifested as sound waves.

Consider, on the other hand, this rephrased version of the question: "If a tree falls in a vacuum—an area completely devoid of matter, including air—does it make a sound?" The answer is now a qualified "no": certainly, there is a release of energy, as before, but the sound waves cannot be transmitted. Without air or any other matter to carry the waves, there is literally no sound.

Hence, there is a great deal of truth to the tagline associated with the 1979 science-fiction film *Alien*: "In space, no one can hear you scream." Inside an astronaut's suit, there is pressure and an oxygen supply; without either, the astronaut would perish quickly. The pressure and air inside the suit also allow the astronaut to hear sounds within the suit, including communica-

KEY TERMS

ACOUSTICS: An area of physics devoted to the study of the production, transmission, and reception of sound.

AMPLITUDE: The maximum displacement of a vibrating material. In wave motion, amplitude is the "size" of a wave, and for sound waves, amplitude indicates the intensity or volume of sound.

CYCLE: For a sound wave, a cycle is a single complete vibration.

DECIBEL: A unit for measuring intensity of sound. Decibels, abbreviated dB, are calibrated along a logarithmic scale whereby every increase of 10 dB indicates an increase in intensity by a factor of 10. Thus if the level of intensity is increased from 30 to 60 dB, the resulting intensity is not twice as great as that of the earlier sound—it is 1,000 times as great.

ENERGY: The ability to perform work, which is the exertion of force over a given distance. Work is the product of force and distance, where force and distance are exerted in the same direction.

FREQUENCY: In wave motion, frequency is the number of waves passing through a given point during the interval of one second. The higher the frequency, the shorter the wavelength. Measured in Hertz, frequency is mathematically related to wave speed, wavelength, and period.

HARMONIC MOTION: The repeated movement of a particle about a position of equilibrium, or balance.

HERTZ: A unit for measuring frequency, named after nineteenth-century German physicist Heinrich Rudolf Hertz (1857-1894). High frequencies are expressed in terms of kilohertz (kHz; 10^3 or 1,000 cycles per second) or megahertz (MHz; 10^6 or 1 million cycles per second.)

INTENSITY: Intensity is the rate at which a wave moves energy per unit of cross-sectional area. Where sound waves are concerned, intensity is commonly known as "volume."

---

tions via microphone from other astronauts. But, if there were an explosion in the vacuum of deep space outside the spacecraft, no one inside would be able to hear it.

RECEPTION OF SOUND

RECORDING. Earlier the structure of electronic amplification was described in very simple terms. Some of the same processes—specifically, the conversion of sound to electrical energy—are used in the recording of sound. In sound recording, when a sound wave is emitted, it causes vibrations in a diaphragm attached to an electrical condenser. This causes variations in the electrical current passed on by the condenser.

These electrical pulses are processed and ultimately passed on to an electromagnetic "recording head." The magnetic field of the recording head extends over the section of tape being recorded: what began as loud sounds now produce strong magnetic fields, and soft sounds produce weak fields. Yet, just as electronic means of sound production and transmission are still not as impressive as the mechanisms of the human voice, so electronic sound reception and recording technology is a less magnificent device than the human ear.

HOW THE EAR HEARS. As almost everyone has noticed, a change in altitude (and, hence, of atmospheric pressure) leads to a

KEY TERMS CONTINUED

LONGITUDINAL WAVE: A wave in which the movement of vibration is in the same direction as the wave itself. A sound wave is an example of a longitudinal wave.

MATTER: Physical substance that has mass; occupies space; is composed of atoms; and is ultimately convertible to energy.

MEDIUM: Material through which sound travels. (It cannot travel through a vacuum.) The most common medium (plural, media) of sound transmission experienced in daily life is air, but in fact sound can travel through any type of matter.

OSCILLATION: The vibration experienced by individual waves even as the wave itself is moving through space. Oscillation is a type of harmonic motion, typically periodic, in one or more dimensions.

PERIOD: For wave motion, a period is the amount of time required to complete one full cycle. Period is mathematically related to frequency, wavelength, and wave speed.

PERIODIC MOTION: Motion that is repeated at regular intervals. These intervals are known as periods.

RAREFACTION: A decrease in density.

ULTRASOUND : Sound waves with a frequency above 20,000 Hertz, which makes them inaudible to the human ear.

VACUUM: An area entirely devoid of matter, including air.

WAVELENGTH: The distance between a crest and the adjacent crest, or the trough and an adjacent trough, of a wave. Wavelength, symbolized by λ (the Greek letter lambda) is mathematically related to wave speed, period, and frequency.

WAVE MOTION: A type of harmonic motion that carries energy from one place to another without actually moving any matter.

strange "popping" sensation in the ears. Usually, this condition can be overcome by swallowing, or even better, by yawning. This opens the Eustachian tube, a passageway that maintains atmospheric pressure in the ear. Useful as it is, the Eustachian tube is just one of the human ear's many parts.

The "funny" shape of the ear helps it to capture and amplify sound waves, which pass through the ear canal and cause the eardrum to vibrate. Though humans can hear sounds over a much wider range, the optimal range of audibility is from 3,000 to 4,000 Hz. This is because the structure of the ear canal is such that sounds in this frequency produce magnified pressure fluc-tuations. Thanks to this, as well as other specific properties, the ear acts as an amplifier of sounds.

Beyond the eardrum is the middle ear, an intricate sound-reception device containing some of the smallest bones in the human body—bones commonly known, because of their shapes, as the hammer, anvil, and stirrup. Vibrations pass from the hammer to the anvil to the stirrup, through the membrane that covers the oval window, and into the inner ear.

Filled with liquid, the inner ear contains the semicircular canals responsible for providing a sense of balance or orientation: without these, a person literally "would not know which way is up." Also, in the inner ear is the cochlea, an organ shaped like a snail. Waves of pressure from the fluids of the inner ear are passed through the cochlea to the auditory nerve, which then transmits these signals to the brain.

The basilar membrane of the cochlea is a particularly wondrous instrument, responsible in large part for the ability to discriminate between sounds of different frequencies and intensities. The surface of the membrane is covered with thousands of fibers, which are highly sensitive to disturbances, and it transmits information concerning these disturbances to the auditory nerve. The brain, in turn, forms a relation between the position of the nerve ending and the frequency of the sound. It also equates the degree of disturbance in the basilar membrane with the intensity of the sound: the greater the disturbance, the louder the sound.

WHERE TO LEARN MORE

Adams, Richard C. and Peter H. Goodwin. *Engineering Projects for Young Scientists.* New York: Franklin Watts, 2000.

Beiser, Arthur. *Physics*, 5th ed. Reading, MA: Addison-Wesley, 1991.

Friedhoffer, Robert. *Sound.* Illustrated by Richard Kaufman and Linda Eisenberg; photographs by Timothy White. New York: F. Watts, 1992.

Gardner, Robert. *Science Projects About Sound.* Berkeley Heights, NJ: Enslow Publishers, 2000.

Internet Resources for Sound and Light (Web site). <http://electro.sau.edu/SLResources.html> (April 25, 2001).

"Music and Sound Waves" (Web site). <http://www.silcom.com/~aludwig/musicand.htm> (April 28, 2001).

Oxlade, Chris. *Light and Sound.* Des Plaines, IL: Heinemann Library, 2000.

Physics Tutorial System: Sound Waves Modules (Web site). <http://csgrad.cs.vt.edu/~chin/chin_sound.html> (April 25, 2001).

"Sound Waves and Music." The Physics Classroom (Web site). <http://www.glenbrook.k12.il.us/gbssci/phys/Class/sound/soundtoc.h tml> (April 28, 2001).

"What Are Sound Waves?" (Web site). <http://rustam.uwp.edu/GWWM/sound_waves.html> (April 28, 2001).

ULTRASONICS

CONCEPT

The word ultrasonic combines the Latin roots ultra, meaning "beyond," and sonic, or sound. The field of ultrasonics thus involves the use of sound waves outside the audible range for humans. These sounds have applications for imaging, detection, and navigation—from helping prospective parents get a glimpse of their unborn child to guiding submarines through the oceans. Ultrasonics can be used to join materials, as for instance in welding or the homogenization of milk, or to separate them, as for example in extremely delicate cleaning operations. Among the broad sectors of society that regularly apply ultrasonic technology are the medical community, industry, the military, and private citizens.

HOW IT WORKS

In the realm of physics, ultrasonics falls under the category of studies in sound. Sound itself fits within the larger heading of wave motion, which is in turn closely related to vibration, or harmonic (back-and-forth) motion. Both wave motion and vibration involve the regular repetition of a certain form of movement; and in both, potential energy (think of the energy in a sled at the top of a hill) is continually converted to kinetic energy (like the energy of a sled as it is sliding down the hill) and back again.

Wave motion carries energy from one place to another without actually moving any matter. Waves themselves may consist of matter, as for instance in the case of a wave on a plucked string or the waves on the ocean. This type of wave is called a mechanical wave, but again, the matter itself does not undergo any net displacement over horizontal space: contrary to what our eyes tell us, molecules of water in an ocean wave move up and down, but they do not actually travel with the wave itself. Only the energy is moved.

SOUND WAVES

Then there are waves of pulses, such as light, sound, radio, or electromagnetic waves. Sound travels by means of periodic waves, a period being the amount of time it takes a complete wave, from trough to crest and back again, to pass through a given point. These periodic waves are typified by a sinusoidal pattern. To picture a sinusoidal wave, one need only imagine an x-axis crossed at regular intervals by a curve that rises above the line to point y before moving downward, below the axis, to point y. This may be expressed also as a graph of sin x versus x. In any case, the wave varies by equal distances upward and downward as it moves along the x-axis in a regular, unvarying pattern.

Periodic waves have three notable interrelated characteristics. One of these is speed, typically calculated in seconds. Another is wavelength, or the distance between a crest and the adjacent crest, or a trough and the adjacent trough, along a plane parallel to that of the wave itself. Finally, there is frequency, the number of waves passing through a given point during the interval of one second.

Frequency is measured in terms of cycles per second, or Hertz (Hz), named in honor of the nineteenth-century German physicist Heinrich Hertz. If a wave has a frequency of 100 Hz, this means that 100 waves are passing through a given point during the interval of one second. Higher frequencies are expressed in terms of kilohertz

SUBMARINES, SUCH AS THE U.S. NAVY SUBMARINE PICTURED HERE, RELY HEAVILY ON SONAR. *(Corbis. Reproduced by permission.)*

(kHz; 10^3 or 1,000 cycles per second) or megahertz (MHz; 10^6 or 1 million cycles per second.)

Clearly, frequency is a function of the wave's speed or velocity, and the same relationship—though it is not so obvious intuitively—exists between wavelength and speed. Over the interval of one second, a given number of waves pass a certain point (frequency), and each wave occupies a certain distance (wavelength). Multiplied by one another, these two properties equal the velocity of the wave.

An additional characteristic of waves (though one that is not related mathematically to the three named above) is amplitude, or maximum displacement, which can be described as the distance from the x-axis to either the crest or the trough. Amplitude is related to the intensity or the amount of energy in the wave.

These four qualities are easiest to imagine on a transverse wave, described earlier with reference to the x-axis—a wave, in other words, in which vibration or harmonic motion occurs perpendicular to the direction in which the wave is moving.

Such a wave is much easier to picture, for the purposes of illustrating concepts such as frequency, than a longitudinal wave; but in fact, sound waves are longitudinal. A longitudinal wave is one in which the individual segments vibrate in the same direction as the wave itself. The shock waves of an explosion, or the concentric waves of a radio transmission as it goes out from the station to all points within receiving distance, are examples of longitudinal waves. In this type of wave pattern, the crests and troughs are not side by side in a line; they radiate outward. Wavelength is the distance between each concentric circle or semicircle (that is, wave), and amplitude the "width" of each wave, which one may imagine by likening it to the relative width of colors on a rainbow.

Having identified its shape, it is reasonable to ask what, exactly, a sound wave is. Simply put, sound waves are changes in pressure, or an alternation between condensation and rarefaction. Imagine a set of longitudinal waves—represented as concentric circles—radiating from a sound source. The waves themselves are relatively higher in pressure, or denser, than the "spaces" between them, though this is just an illustration for the purposes of clarity: in fact the "spaces" are waves of lower pressure that alternate with higher-pressure waves.

Vibration is integral to the generation of sound. When the diaphragm of a loudspeaker

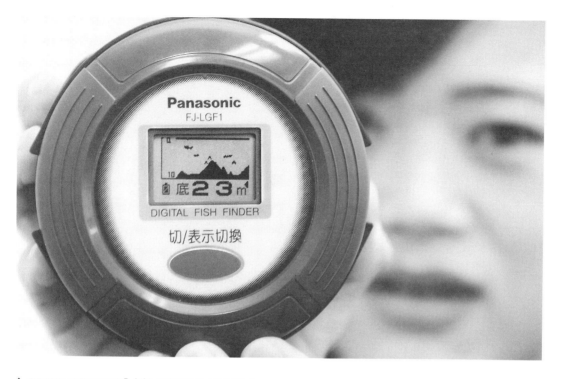

A FISHING REVOLUTION? IT'S POSSIBLE, THANKS TO THIS ELECTRONIC FISH-FINDER DEVELOPED BY MATSUSHITA. THE DEVICE INCLUDES A DETACHABLE SONAR SENSOR. *(AFP/Corbis. Reproduced by permission.)*

pushes outward, it forces nearby air molecules closer together, creating a high-pressure region all around the loudspeaker. The diaphragm is pushed backward in response, thus freeing up a volume of space for the air molecules. These then rush toward the diaphragm, creating a low-pressure region behind the high-pressure one. As a result, the loudspeaker sends out alternating waves of high pressure (condensation) and low pressure (rarefaction). Furthermore, as sound waves pass through a medium—air, for the purposes of this discussion—they create fluctuations between density and rarefaction. These result in pressure changes that cause the listener's eardrum to vibrate with the same frequency as the sound wave, a vibration that the ear's inner mechanisms translate and pass on to the brain.

THE SPEED OF SOUND: CONSIDER THE MEDIUM

The speed of sound varies with the hardness of the medium through which it passes: contrary to what you might imagine, it travels faster through liquids than through gasses such as air, and faster through solids than through liquids. By definition, molecules are closer together in harder material, and thus more quickly responsive to signals from neighboring particles. In granite, for

instance, sound travels at 19,680 ft per second (6,000 mps), whereas in air, the speed of sound is only 1,086 ft per second (331 mps). It follows that sound travels faster in water—5,023 ft per second (1,531 mps), to be exact—than in air. It should be clear, then, that there is a correlation between density and the ease with which a sound travels. Thus, sound cannot travel in a vacuum, giving credence to the famous tagline from the 1979 science fiction thriller *Alien:* "In space, no one can hear you scream."

When sound travels through a medium such as air, however, two factors govern its audibility: intensity or volume (related to amplitude and measured in decibels, or dB) and frequency. There is no direct correlation between intensity and frequency, though for a person to hear a very low-frequency sound, it must be above a certain decibel level. (At all frequencies, however, the threshold of discomfort is around 120 decibels.)

In any case, when discussing ultrasonics, frequency and not intensity is of principal concern. The range of audibility for the human ear is from 20 Hz to 20,000 Hz, with frequencies below that range dubbed infrasound and those above it referred to as ultrasound. (There is a third category, hypersound, which refers to frequencies above 10^{13}, or 10 trillion Hz. It is almost impos-

TINY LISTENING DEVICES, LIKE THE ONE SHOWN HERE, HAVE A HOST OF MODERN APPLICATIONS, FROM ELECTRON-IC EAVESDROPPING TO ULTRASONIC STEREO SYSTEMS. *(Photograph by Jeffrey L. Rotman/Corbis. Reproduced by permission.)*

sible for hypersound waves to travel through most media, because its wavelengths are so short.)

What Makes the Glass Shatter

The lowest note of the eighty-eight keys on a piano is 27 Hz and the highest 4,186 Hz. This places the middle and upper register of the piano well within the optimal range for audibility, which is between 3,000 and 4,000 Hz. Clearly, the higher the note, the higher the frequency—but it is not high frequency, per se, that causes a glass to shatter when a singer or a violinist hits a certain note. All objects, or at least all rigid ones, possess their own natural frequency of vibration or oscillation. This frequency depends on a number of factors, including material composition and shape, and its characteristics are much more complex than those of sound frequency described above. In any case, a musician cannot cause a glass to shatter simply by hitting a very high note; rather, the note must be on the exact frequency at which the glass itself oscillates. Under such conditions, all the energy from the voice or musical instrument is transferred to the glass, a sudden burst that overloads the object and causes it to shatter.

To create ultrasonic waves, technicians use a transducer, a device that converts energy into ultrasonic sound waves. The most basic type of transducer is mechanical, involving oscillators or vibrating blades powered either by gas or the pressure of gas or liquids—that is, pneumatic or hydrodynamic pressure, respectively. The vibrations from these mechanical devices are on a relatively low ultrasonic frequency, and most commonly they are applied in industry for purposes such as drying or cleaning.

An electromechanical transducer, which has a much wider range of applications, converts electrical energy, in the form of current, to mechanical energy—that is, sound waves. This it does either by a magnetostrictive or a piezoelectric device. The term "magnetostrictive" comes from *magneto,* or magnetic, and *strictio,* or "drawing together." This type of transducer involves the magnetization of iron or nickel, which causes a change in dimension by forcing the atoms together. This change in dimension in turn produces a high-frequency vibration. Again, the frequency is relatively low in ultrasonic terms, and likewise the application is primarily industrial, for purposes such as cleaning and machining.

Most widely used is a transducer equipped with a specially cut piezoelectric quartz crystal. Piezoelectricity involves the application of mechanical pressure to a nonconducting crystal, which results in polarization of electrical charges, with all positive charges at one end of the crystal and all negative charges at the other end. By successively compressing and stretching the crystal at an appropriate frequency, an alternating electrical current is generated that can be converted into mechanical energy—specifically, ultrasonic waves.

Scientists use different shapes and materials (including quartz and varieties of ceramic) in fashioning piezoelectric crystals: for instance, a concave shape is best for an ultrasonic wave that will be focused on a very tight point. Piezoelectric transducers have a variety of applications in ultrasonic technology, and are capable of acting as receivers for ultrasonic vibrations.

REAL-LIFE APPLICATIONS

PETS AND PESTS: ULTRASONIC BEHAVIOR MODIFICATION

Some of the simplest ultrasonic applications build on the fact that the upper range of audibility for human beings is relatively low among animals. Cats, by comparison, have an infrasound threshold only slightly higher than that of humans (100 Hz), but their ultrasound range of audibility is much greater—32,000 Hz instead of a mere 20,000. This explains why a cat sometimes seems to respond mysteriously to noises its owner is incapable of hearing.

For dogs, the difference is even more remarkable: their lower threshold is 40 Hz, and their high end 46,000 Hz, giving them a range more than twice that of humans. It has been said Paul McCartney, who was fond of his sheepdog Martha, arranged for the Beatles' sound engineer at Abbey Road Studios to add a short 20,000-Hz tone at the very end of *Sgt. Peppers' Lonely Hearts Club Band* in 1967. Thus—if the story is true—the Beatles' human fans would never hear the note, but it would be a special signal to Martha and all the dogs of England.

On a more practical level, a dog whistle is an extremely simple ultrasonic or near-ultrasonic device, one that obviously involves no transducers. The owner blows the whistle, which utters a tone nearly inaudible to humans but—like McCartney's 20,000-Hz tone—well within a dog's range. In fact, the Acme Silent Dog Whistle, which the company has produced since 1935, emits a tone that humans can hear (the listed range is 5,800 to 12,400 Hz), but which dogs can hear much better.

There are numerous products on the market that use ultrasonic waves for animal behavior modification of one kind or another; however, most such items are intended to repel rather than attract the animal. Hence, there are ultrasonic devices to discourage animals from relieving themselves in the wrong places, as well as some which keep unwanted dogs and cats away.

Then there is one of the most well-known uses of ultrasound for pets, which, rather than keeping other animals out, is designed to keep one's own animals in the yard. Many people know this item as an "Invisible Fence," though in fact that term is a registered trademark of the Invisible Fence Company. The "Invisible Fence" and similar products literally create a barrier of sound, using both radio signals and ultrasonics. The pet is outfitted with a collar that contains a radio receiver, and a radio transmitter is placed in some centrally located place on the owner's property—a basement, perhaps, or a garage. The "fence" itself is "visible," though usually buried, and consists of an antenna wire at the perimeter of the property. The transmitter sends a signal to the wire, which in turn signals the pet's collar. A tiny computer in the collar emits an ultrasonic sound if the animal tries to stray beyond the boundaries.

Not all animals have a higher range of hearing than humans: elephants, for instance, cannot hear tones above 12,000 Hz. On the other hand, some are drastically more sensitive acoustically than dogs: bats, whales, and dolphins all have an upper range of 150,000 Hz, though both have a low-end threshold of 1,000. Mice, at 100,000 Hz, are also at the high end, while a number of other pests—rodents and insects—fall into the region between 40,000 and 100,000 Hz. This fact has given rise to another type of ultrasonic device, for repelling all kinds of unwanted household creatures by bombarding them with ear-splitting tones.

An example of this device is the Transonic 1X-L, which offers three frequency ranges: "loud mode" (1,000-50,000 Hz); "medium mode" (10,000-50,000 Hz); and "quiet mode" (20,000-50,000 Hz). The lowest of these can be used for repelling pest birds and small animals, the medium range for insects, and the "quiet mode" for rodents.

ULTRASONIC DETECTION IN MEDICINE

Medicine represents one of the widest areas of application for ultrasound. Though the machinery used to provide parents-to-be with an image of their unborn child is the most well-known form of medical ultrasound, it is far from the only one. Developed in 1957 by British physician Ian Donald (1910-1987), also a pioneer in the use of ultrasonics to detect flaws in machinery, ultrasound was first used to diagnosis a patient's heart condition. Within a year, British hospitals began using it with pregnant women.

High-frequency waves penetrate soft tissue with ease, but they bounce off of harder tissue such as organs and bones, and thus send back a message to the transducer. Because each type of tissue absorbs or deflects sound differently, according to its density, the ultrasound machine can interpret these signals, creating an image of what it "sees" inside the patient's body. The technician scans the area to be studied with a series of ultrasonic waves in succession, and this results in the creation of a moving picture. It is this that creates the sight so memorable in the lives of many a modern parent: their first glimpse of their child in its mother's womb.

Though ultrasound enables physicians and nurses to determine the child's sex, this is far from being the only reason it is used. It also gives them data concerning the fetus's size; position (for instance, if the head is in a place that suggests the baby will have to be delivered by means of cesarean section); and other abnormalities.

The beauty of ultrasound is that it can provide this information without the danger posed by x rays or incisions. Doctors and ultrasound technicians use ultrasonic technology to detect body parts as small as 0.004 in (0.1 mm), making it possible to conduct procedures safely, such as locating foreign objects in the eye or measuring the depth of a severe burn. Furthermore, ultrasonic microscopes can image cellular structures to within 0.2 microns (0.002 mm).

Ultrasonic heart examination can locate tumors, valve diseases, and accumulations of fluid. Using the Doppler effect—the fact that a sound's perceived frequency changes as its source moves past the observer—physicians observe shifts in the frequency of ultrasonic measurements to determine the direction of blood flow in the body. Not only can ultrasound be used to differentiate tumors from healthy tissue, it can sometimes be used to destroy those tumors. In some cases, ultrasound actually destroys cancer cells, making use of a principle called cavitation— a promising area of ultrasound research.

Perhaps the best example of cavitation occurs when you are boiling a pot of water: bubbles—temporary cavities in the water itself—rise up from the bottom to the surface, then collapse, making a popping sound as they do. Among the research areas combining cavitation and ultrasonics are studies of light emissions produced in the collapse of a cavity created by an ultrasonic wave. These emissions are so intense that for an infinitesimal moment, they produce heat of staggering proportions—hotter than the surface of the Sun, some scientists maintain. (Again, it should be stressed that this occurs during a period too small to measure with any but the most sophisticated instruments.)

As for the use of cavitation in attacking cancer cells, ultrasonic waves can be used to create microscopic bubbles which, when they collapse, produce intense shock waves that destroy the cells. Doctors are now using a similar technique against gallstones and kidney stones. Other medical uses of ultrasound technology include ultrasonic heat for treating muscle strain, or—in a process similar to some industrial applications—the use of 25,000-Hz signals to clean teeth.

SONAR AND OTHER DETECTION DEVICES

Airplane pilots typically use radar, but the crew of a ocean-going vessel relies on sonar (SOund Navigation and Ranging) to guide their vessel through the ocean depths. This technology takes advantage of the fact that sound waves travel well under water—much better, in fact, than light waves. Whereas a high-powered light would be of limited value underwater, particularly in the murky realms of the deep sea, sonar provides excellent data on the water's depth, as well as the location of shipwrecks, large obstacles—and, for commercial or even recreational fishermen—the presence of fish.

At the bottom of the craft's hull is a transducer, which emits an ultrasonic pulse. These sound waves travel through the water to the bottom, where they bounce back. Upon receiving the echo, the transducer sends this information to an onboard computer, which converts data on the amount of time the signal took, providing a reading of distance that gives an accurate measurement of the vessel's clearance. For instance, it takes one second for sound waves from a depth of 2,500 ft (750 m) to return to the ship. The onboard computer converts this data into a rough picture of what lies below: the ocean floor, and schools of fish or other significant objects between it and the ship.

Even more useful is a scanning sonar, which adds dimension to the scope of the ship's ultrasonic detection: not only does the sonar beam move forward along with the vessel, but it moves from side-to-side, providing a picture of a wider area along the ship's path. Sonar in general, and particularly scanning sonar, is of particular importance to a submarine's crew. Despite the fact that the periscope is perhaps the most notable feature of these underwater craft, from the viewpoint of a casual observer, in fact, the purely visual data provided by the periscope is of limited value—and that value decreases as the sub descends. It is thanks to sonar (which produces the pinging sound one so often hears in movie scenes depicting the submarine control room), combined with nuclear technology, that makes it possible for today's U.S. Navy submarines to stay submerged for months.

Sonar is perhaps the most dramatic use of ultrasonic technology for detection; less well-known—but equally intriguing, especially for its connection with clandestine activity—is the use of ultrasonics for electronic eavesdropping. Private detectives, suspicious spouses, and no doubt international spies from the CIA or Britain's MI5, use ultrasonic waves to listen to conversations in places where they cannot insert a microphone. For example, an operative might want to listen in on an encounter taking place on the seventh floor of a building with heavy security, meaning it would be impossible to plant a microphone either inside the room or on the window ledge.

Instead, the operative uses ultrasonic waves, which a transducer beams toward the window of the room being monitored. If people are speaking inside the room, this will produce vibrations on the window the transducer can detect, although the sounds would not be decipherable as conversation by a person with unaided perception. Speech vibrations from inside produce characteristic effects on the ultrasonic waves beamed back to the transducer and the operative's monitoring technology. The transducer then converts these reflected vibrations into electrical signals, which analysts can then reconstruct as intelligible sounds.

Much less dramatic, but highly significant, is the use of ultrasonic technology for detection in industry. Here the purpose is to test materials for faults, holes, cracks, or signs of corrosion. Again, the transducer beams an ultrasonic signal, and the way in which the material reflects this signal can alert the operator to issues such as metal fatigue or a faulty weld. Another method is to subject the material or materials to stress, then look for characteristic acoustic emissions from the stressed materials. (The latter is a developing field of acoustics known as acoustic emission.)

Though industrial detection applications can be used on materials such as porcelain (to test for microscopic cracks) or concrete (to evaluate how well it was poured), ultrasonics is particularly effective on metal, in which sound moves more quickly and freely than any other type of wave. Not only does ultrasonics provide an opportunity for thorough, informative, but nondestructive testing, it also allows technicians to penetrate areas where they otherwise could not go—or, in the case of ultrasonic inspection of the interior of a nuclear reactor while in operation—would not and should not go.

BINDING AND LOOSENING: A HOST OF INDUSTRIAL APPLICATIONS

Materials testing is but one among myriad uses for ultrasonics in industry, applications that can be described broadly as "binding and loosening"—either bringing materials together, or pulling them apart.

For instance, ultrasound is often used to bind, or coagulate, loose particles of dust, mist, or smoke. This makes it possible to clean a factory smokestack before it exhales pollutants into the atmosphere, or to clear clumps of fog and mist off a runway. Another form of "binding" is the use of ultrasonic vibrations to heat and weld together materials. Ultrasonics provides an even, localized flow of molten material, and is effective both on plastics and metals.

Ultrasonic soldering implements the principle of cavitation, producing microscopic bubbles in molten solder, a process that removes metal oxides. Hence, this is a case of both "binding" (soldering) and "loosening"—removing impurities from the area to be soldered. The dairy industry, too, uses ultrasonics for both purposes: ultrasonic waves break up fat globules in milk, so that the fat can be mixed together with the milk in the well-known process of homogenization. Similarly, ultrasonic pasteurization facilitates the separation of the milk from harmful bacteria and other microorganisms.

The uses of ultrasonics to "loosen" include ultrasonic humidification, wherein ultrasonic vibrations reduce water to a fine spray. Similarly, ultrasonic cleaning uses ultrasound to break down the attraction between two different types of materials. Though it is not yet practical for home use, the technology exists today to use ultrasonics for laundering clothes without using water: the ultrasonic vibrations break the bond between dirt particles and the fibers of a garment, shaking loose the dirt and subjecting the fabric to far less trauma than the agitation of a washing machine does.

As noted earlier, dentists use ultrasound for cleaning teeth, another example of loosening the bond between materials. In most of these forms of ultrasonic cleaning, a critical part of the process is the production of microscopic shock waves in the process of cavitation. The frequency of sound waves in these operations ranges from 15,000 Hz (15 kHz) to 2 million Hz (2 MHz). Ultrasonic cleaning has been used on metals, plastics, and ceramics, as well as for cleaning precision instruments used in the optical, surgical, and dental fields. Nor is it just for small objects: the electronics, automotive, and aircraft industries make heavy use of ultrasonic cleaning for a variety of machines.

Ultrasonic "loosening" makes it possible to drill though extremely hard or brittle materials, including tungsten carbide or precious stones. Just as a dental hygienist cleaning a person's teeth bombards the enamel with gentle abrasives, this form of high-intensity drilling works hand-in-hand with the use of abrasive materials such as silicon carbide or aluminum oxide.

A WORLD OF APPLICATIONS

Scientists often use ultrasound in research, for instance to break up high molecular weight polymers, thus creating new plastic materials. Indeed, ultrasound also makes it possible to determine the molecular weight of liquid polymers, and to conduct other forms of investigation on the physical properties of materials.

Ultrasonics can also speed up certain chemical reactions. Hence, it has gained application in agriculture, thanks to research which revealed that seeds subjected to ultrasound may germinate more rapidly and produce higher yields. In addition to its uses in the dairy industry, noted above, ultrasonics is of value to farmers in the related beef industry, who use it to measure cows' fat layers before taking them to market.

In contrast to the use of ultrasonics for electronic eavesdropping, as noted earlier, today ultrasonic technology is available to persons who think someone might be spying on them: now they can use ultrasonics to detect the presence of electronic eavesdropping, and thus circumvent it. Closer to home is another promising application of ultrasonics for remote sensing of sounds: ultrasonic stereo speakers.

These make use of research dating back to the 1960s, which showed that ultrasound waves of relatively low frequency can carry audible sound to pinpointed locations. In 1996, Woody Norris had perfected the technology necessary to reduce distortion, and soon he and his son Joe began selling the ultrasonic speakers through the elder Norris's company, American Technology Corporation of San Diego, California.

Eric Niiler in *Business Week* described a demonstration: "Joe Norris twists a few knobs on a receiver, takes aim with a 10-inch-square gold-covered flat speaker, and blasts an invisible beam.... Thirty feet away, the tinny but easily recognizable sound of Vivaldi's *Four Seasons* rushes over you. Step to the right or left, however, and it fades away. The exotic-looking speaker emits `sound beams' that envelop the listener but are silent to those nearby. `We use the air as our virtual speakers,' says Norris...." Niiler went to note several other applications suggested by Norris: "Airline passengers could listen to their own

KEY TERMS

FREQUENCY: The number of waves passing through a given point during the interval of one second. The higher the frequency, the shorter the wavelength.

HERTZ: A unit for measuring frequency, equal to one cycle per second. If a sound wave has a frequency of 20,000 Hz, this means that 20,000 waves are passing through a given point during the interval of one second. Higher frequencies are expressed in terms of kilohertz (kHz; 103 or 1,000 cycles per second) or megahertz (MHz; 106 or 1 million cycles per second.) Hence 20,000 Hz—the threshold of ultrasonic sound—would be rendered as 20 kHz.

INFRASOUND: Sound of a frequency between 20 Hz, which places it outside the range of audibility for human beings. Its opposite is ultrasound.

LONGITUDINAL WAVE: A wave in which the individual segments vibrate in the same direction as the wave itself. This is in contrast to a transverse wave, or one in which the vibration or harmonic motion occurs perpendicular to the direction in which the wave is moving. Waves on the ocean are an example of transverse waves; by contrast, the shock waves of an explosion, the concentric waves of a radio transmission, and sound waves are all examples of longitudinal waves.

TRANSDUCER: A device that converts energy into ultrasonic sound waves.

ULTRASOUND: Sound waves with a frequency above 20,000 Hz, which makes them inaudible to the human ear. Its opposite is infrasound.

WAVELENGTH: The distance, measured on a plane parallel to that of the wave itself, between a crest and the adjacent crest, or the trough and an adjacent trough. On a longitudinal wave, this is simply the distance between waves, which constitute a series of concentric circles radiating from the source.

WAVE MOTION: Activity that carries energy from one place to another without actually moving any matter.

---

music channels*sans* headphones without disturbing neighbors. Troops could confuse the enemy with `virtual' artillery fire, or talk to each other without having their radio communications picked up by eavesdroppers."

WHERE TO LEARN MORE

Beiser, Arthur. *Physics*, 5th ed. Reading, MA: Addison-Wesley, 1991.

Crocker, Malcolm J. *Encyclopedia of Acoustics.* New York: John Wiley & Sons, 1997.

Knight, David C. *Silent Sound: The World of Ultrasonics.* New York: Morrow, 1980.

Langone, John. *National Geographic's How Things Work.* Washington, D.C.: National Geographic Society, 1999.

Medical Ultrasound WWW Directory (Web site). <http://www.ultrasoundinsider.com> (February 16, 2001).

Niiler, Eric; edited by Alex Salkever. "Now Here This—If You're in the Sweet Spot." *Business Week*, October 16, 2000.

Meire, Hylton B. and Pat Farrant. *Basic Ultrasound.* Chichester, N.Y.: John Wiley & Sons, 1995.

Suplee, Curt. *Everyday Science Explained.* Washington, D.C.: National Geographic Society, 1996.

LIGHT AND ELECTROMAGNETISM

MAGNETISM

ELECTROMAGNETIC SPECTRUM

LIGHT

LUMINESCENCE

MAGNETISM

CONCEPT

Most people are familiar with magnets primarily as toys, or as simple objects for keeping papers attached to a metal surface such as a refrigerator door. In fact the areas of application for magnetism are much broader, and range from security to health care to communication, transportation, and numerous other aspects of daily life. Closely related to electricity, magnetism results from specific forms of alignment on the part of electron charges in certain varieties of metal and alloy.

HOW IT WORKS

Magnetism, along with electricity, belongs to a larger phenomenon, electromagnetism, or the force generated by the passage of an electric current through matter. When two electric charges are at rest, it appears to the observer that the force between them is merely electric. If the charges are in motion, however—and in this instance motion or rest is understood in relation to the observer—then it appears as though a different sort of force, known as magnetism, exists between them.

In fact, the difference between magnetism and electricity is purely artificial. Both are manifestations of a single fundamental force, with "magnetism" simply being an abstraction that people use for the changes in electromagnetic force created by the motion of electric charges. It is a distinction on the order of that between water and wetness; nonetheless, it is often useful and convenient to discuss the two phenomena as though they were separate.

At the atomic level, magnetism is the result of motion by electrons, negatively charged sub-

atomic particles, relative to one another. Rather like planets in a solar system, electrons both revolve around the atom's nucleus and rotate on their own axes. (In fact the exact nature of their movement is much more complex, but this analogy is accurate enough for the present purposes.) Both types of movement create a magnetic force field between electrons, and as a result the electron takes on the properties of a tiny bar magnet with a north pole and south pole. Surrounding this infinitesimal magnet are lines of magnetic force, which begin at the north pole and curve outward, describing an ellipse as they return to the south pole.

In most atomic elements, the structure of the atom is such that the electrons align in a random manner, rather like a bunch of basketballs bumping into one another as they float in a swimming pool. Because of this random alignment, the small magnetic fields cancel out one another. Two such self-canceling particles are referred to as paired electrons, and again, the analogy to bar magnets is an appropriate one: if one were to shake a bag containing an even number of bar magnets, they would all wind up in pairs, joined at opposing (north-south) poles.

There are, however, a very few elements in which the fields line up to create what is known as a net magnetic dipole, or a unity of direction—rather like a bunch of basketballs simultaneously thrown from in the same direction at the same time. These elements, among them iron, cobalt, and nickel, as well as various alloys or mixtures, are commonly known as magnetic metals or natural magnets.

It should be noted that in magnetic metals, magnetism comes purely from the alignment of

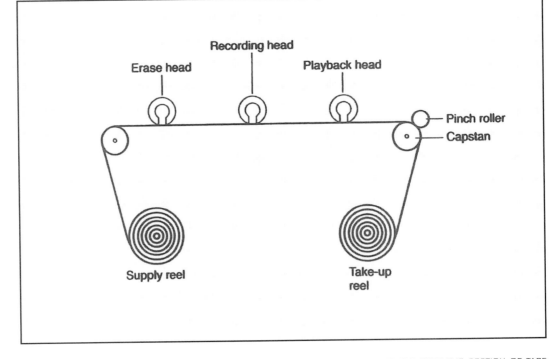

THE RECORDING HEAD IS A SMALL ELECTROMAGNET WHOSE MAGNETIC FIELD EXTENDS OVER THE SECTION OF TAPE BEING RECORDED. LOUD SOUNDS PRODUCE STRONG MAGNETIC FIELDS, AND SOFT ONES WEAK FIELDS. *(Illustration by Hans & Cassidy. The Gale Group.)*

forces exerted by electrons as they spin on their axes, whereas the forces created by their orbital motion around the nucleus tend to cancel one another out. But in magnetic rare earth elements such as cerium, magnetism comes both from rotational and orbital forms of motion. Of principal concern in this discussion, however, is the behavior of natural magnets on the one hand, and of nonmagnetic materials on the other.

There are five different types of magnetism—diamagnetism, paramagnetism, ferromagnetism, ferrimagnetism, and antiferromagnetism. Actually, these terms describe five different types of response to the process of magnetization, which occurs when an object is placed in a magnetic field.

A magnetic field is an area in which a magnetic force acts on a moving charged particle such that the particle would experience no force if it moved in the direction of the magnetic field—in other words, it would be "drawn," as a ten-penny nail is drawn to a common bar or horseshoe (U-shaped) magnet. An electric current is an example of a moving charge, and indeed one of the best ways to create a magnetic field is with a current. Often this is done by means of a solenoid, a current-carrying wire coil

through which the material to be magnetized is passed, much as one would pass an object through the interior of a spring.

All materials respond to a magnetic field; they just respond in different ways. Some nonmagnetic substances, when placed within a magnetic field, slightly reduce the strength of that field, a phenomenon known as diamagnetism. On the other hand, there are nonmagnetic substances possessing an uneven number of electrons per atom, and in these instances a slight increase in magnetism, known as paramagnetism, occurs. Paramagnetism always has to overcome diamagnetism, however, and hence the gain in magnetic force is very small. In addition, the thermal motion of atoms and molecules prevents the objects' magnetic fields from coming into alignment with the external field. Lower temperatures, on the other hand, enhance the process of paramagnetism.

In contrast to diamagnetism and paramagnetism, ferro-, ferri-, and antiferromagnetism all describe the behavior of natural magnets when exposed to a magnetic field. The name ferromagnetism suggests a connection with iron, but in fact the term can apply to any of those materials in which the magnitude of the object's magnetic

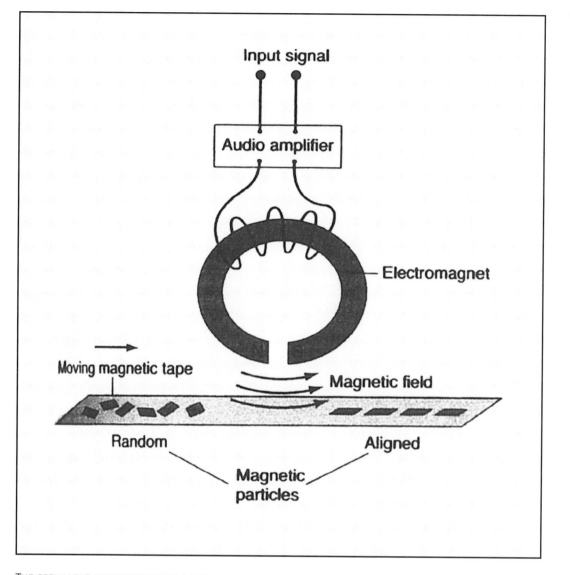

Input signal

Audio amplifier

Electromagnet

Moving magnetic tape

Magnetic field

Random

Aligned

Magnetic
particles

THE PERMANENT MAGNETIZATION OF A NATURAL MAGNET IS DIFFICULT TO REVERSE, BUT REVERSAL OF A TAPE'S MAG-
NETIZATION—IN OTHER WORDS, ERASING THE TAPE—IS EASY. AN ERASE HEAD, AN ELECTROMAGNET OPERATING AT
A FREQUENCY TOO HIGH FOR THE HUMAN EAR TO HEAR, SIMPLY SCRAMBLES THE MAGNETIC PARTICLES ON A PIECE
OF TAPE. *(Illustration by Hans & Cassidy. The Gale Group.)*

field increases greatly when it is placed within an external field. When a natural magnet becomes magnetized (that is, when a metal or alloy comes into contact with an external magnetic field), a change occurs at the level of the domain, a group of atoms equal in size to about 5×10^{-5} meters across—just large enough to be visible under a microscope.

In an unmagnetized sample, there may be an alignment of unpaired electron spins within a domain, but the direction of the various domains' magnetic forces in relation to one another is random. Once a natural magnet is placed within an external magnetic force field,

however, one of two things happens to the domains. Either they all come into alignment with the field or, in certain types of material, those domains in alignment with the field grow while the others shrink to nonexistence.

The first of these processes is called domain alignment or ferromagnetism, the second domain growth or ferrimagnetism. Both processes turn a natural magnet into what is known as a permanent magnet—or, in common parlance, simply a "magnet." The latter is then capable of temporarily magnetizing a ferromagnetic item, as for instance when one rubs a paper clip against a permanent magnet and then uses the magnetized clip to lift other paper clips. Of the two varieties, however, a ferromagnetic metal is stronger because it requires a more powerful magnetic force field in order to become magnetized. Most powerful of all is a saturated ferromagnetic metal, one in which all the unpaired electron spins are aligned.

Once magnetized, it is very hard for a ferromagnetic metal to experience demagnetization, or antiferromagnetism. Again, there is a connection between temperature and magnetism, with heat acting as a force to reduce the strength of a magnetic field. Thus at temperatures above 1,418°F (770°C), the atoms within a domain take on enough kinetic energy to overpower the forces holding the electron spins in alignment. In addition, mechanical disturbances—for instance, battering a permanent magnet with a hammer—can result in some reduction of magnetic force.

Many of the best permanent magnets are made of steel, which, because it is an alloy of iron with carbon and other elements, has an irregular structure that lends itself well to the ferromagnetic process of domain alignment. Iron, by contrast, will typically lose its magnetization when an external magnetic force field is removed; but this actually makes it a better material for some varieties of electromagnet.

The latter, in its simplest form, consists of an iron rod inside a solenoid. When a current is passing through the solenoid, it creates a magnetic force field, activating the iron rod and turning it into an electromagnet. But as soon as the current is turned off, the rod loses its magnetic force. Not only can an electromagnet thus be controlled, but it is often stronger than a permanent magnet: hence, for instance, giant electromagnets are used for lifting cars in junkyards.

REAL-LIFE
APPLICATIONS

FINDING THE WAY: MAGNETS IN COMPASSES

A north-south bar magnet exerts exactly the same sort of magnetic field as a solenoid. Lines of magnetic run through it in one direction, from "south" to "north," and upon leaving the north pole of the magnet, these lines describe an ellipse as they curve back around to the south pole. In view of this model, it is also easy to comprehend why a pair of opposing poles attracts one another, and a pair of like poles—for whom the lines of force are moving away from each other—repels. This is a fact particularly applicable to the operation of MAGLEV trains, as discussed later.

A magnetic compass works because Earth itself is like a giant bar magnet, complete with vast arcs of magnetic force, called the geomagnetic field, surrounding the planet. The first scientist to recognize the magnetic properties of Earth was the English physicist William Gilbert (1544-1603). Scientists today believe that the source of Earth's magnetism lies in a core of molten iron some 4,320 mi (6,940 km) across, constituting half the planet's diameter. Within this core run powerful electric currents that ultimately create the geomagnetic field.

Just as a powerful magnet causes all the domains in a magnetic metal to align with it, a bar magnet placed in a magnetic field will rotate until it lines up with the field's direction. The same thing happens when one suspends a magnet from a string: it lines up with Earth's magnetic field, and points in a north-south direction. The Chinese of the first century B.C., though unaware of the electromagnetic forces that caused this to happen, discovered that a strip of magnetic metal always tended to point toward geographic north.

This led ultimately to the development of the magnetic compass, which typically consists of a magnetized iron needle suspended over a card marked with the four cardinal directions. The needle is attached to a pivoting mechanism at its center, which allows it to move freely so that the "north" end will always point the user northward.

The magnetic compass proved so important that it is typically ranked alongside paper, printing, and gunpowder as one of premodern China's four great gifts to the West. Prior to the compass, mariners had to depend purely on the position of the Sun and other, less reliable, means of determining direction; hence the invention quite literally helped open up the world. But there is a somewhat irksome anomaly lurking in the seeming simplicity of the magnetic compass.

In fact magnetic north is not the same thing as true north; or, to put it another way, if one continued to follow a compass northward, it would lead not to the Earth's North Pole, but to a point identified in 1984 as 77°N, 102°18' W—that is, in the Queen Elizabeth Islands of far northern Canada. The reason for this is that Earth's magnetic field describes a current loop whose center is 11° off the planet's equator, and thus the north and south magnetic poles—which are on a plane perpendicular to that of the Earth's magnetic field—are 11° off of the planet's axis.

The magnetic field of Earth is changing position slowly, and every few years the United States Geological Survey updates magnetic declination, or the shift in the magnetic field. In addition, Earth's magnetic field is slowly weakening as well. The behavior, both in terms of weakening and movement, appears to be similar to changes taking place in the magnetic field of the Sun.

MAGNETS FOR DETECTION: BURGLAR ALARMS, MAGNETOMETERS, AND MRI

A compass is a simple magnetic instrument, and a burglar alarm is not much more complex. A magnetometer, on the other hand, is a much more sophisticated piece of machinery for detecting the strength of magnetic fields. Nonetheless, the magnetometer bears a relation to its simpler cousins: like a compass, certain kinds of magnetometers respond to a planet's magnetic field; and like a burglar alarm, other varieties of magnetometer are employed for security.

At heart, a burglar alarm consists of a contact switch, which responds to changes in the environment and sends a signal to a noisemaking device. The contact switch may be mechanical—a simple fastener, for instance—or magnetic. In the latter case, a permanent magnet may be installed in the frame of a window or door, and a piece of magnetized material in the window or door itself. Once the alarm is activated, it will respond to any change in the magnetic field—i.e., when someone slides open the door or window, thus breaking the connection between magnet and metal.

Though burglar alarms may vary in complexity, and indeed there may be much more advanced systems using microwaves or infrared rays, the application of magnetism in home security is a simple matter of responding to changes in a magnetic field. In this regard, the principle governing magnetometers used at security checkpoints is even simpler. Whether at an airport or at the entrance to some other high-security venue, whether handheld or stationary, a magnetometer merely detects the presence of magnetic metals. Since the vast majority of firearms, knife-blades, and other weapons are made of iron or steel, this provides a fairly efficient means of detection.

At a much larger scale, magnetometers used by astronomers detect the strength and sometimes the direction of magnetic fields surrounding Earth and other bodies in space. This variety of magnetometer dates back to 1832, when mathematician and scientist Carl Friedrich Gauss (1777-1855) developed a simple instrument consisting of a permanent bar magnet suspended horizontally by means of a gold wire. By measuring the period of the magnet's oscillation in Earth's magnetic field (or magnetosphere), Gauss was able to measure the strength of that field. Gauss's name, incidentally, would later be applied to the term for a unit of magnetic force. The gauss, however, has in recent years been largely replaced by the tesla, named after Nikola Tesla (1856-1943), which is equal to one newton/ampere meter (1 N/A•m) or 10^4 (10,000) gauss.

As for magnetometers used in astronomical research, perhaps the most prominent—and certainly one of the most distant—ones is on *Galileo*, a craft launched by the U.S. National Aeronautics and Space Administration (NASA) toward Jupiter on October 15, 1989. Among other instruments on board *Galileo*, which has been in orbit around the solar system's largest planet since 1995, is a magnetometer for measuring Jupiter's magnetosphere and that of its surrounding asteroids and moons.

Closer to home, but no less impressive, is another application of magnetism for the purposes of detection: magnetic resonance imagining, or MRI. First developed in the early 1970s, MRI permits doctors to make intensive diagnoses without invading the patient's body either with a surgical knife or x rays.

The heart of the MRI machine is a large tube into which the patient is placed in a supine position. A technician then activates a powerful magnetic field, which causes atoms within the patient's body to spin at precise frequencies. The machine then beams radio signals at a frequency matching that of the atoms in the cells (e.g., cancer cells) being sought. Upon shutting off the radio signals and magnetic field, those atoms emit bursts of energy that they have absorbed from the radio waves. At that point a computer scans the body for frequencies matching specific types of atoms, and translates these into three-dimensional images for diagnosis.

MAGNETS FOR PROJECTING SOUND: MICROPHONES, LOUD-SPEAKERS, CAR HORNS, AND ELECTRIC BELLS

The magnets used in *Galileo* or an MRI machine are, needless to say, very powerful ones, and as noted earlier, the best way to create a super-strong, controllable magnet is with an electrical current. When that current is properly coiled around a magnetic metal, this creates an electro-magnet, which can be used in a variety of applications.

As discussed above, the most powerful elec-tromagnets typically use nonpermanent magnets so as to facilitate an easy transition from an extremely strong magnetic field to a weak or nonexistent one. On the other hand, permanent magnets are also used in loudspeakers and simi-lar electromagnetic devices, which seldom require enormous levels of power.

In discussing the operation of a loudspeaker, it is first necessary to gain a basic understanding of how a microphone works. The latter contains a capacitor, a system for storing charges in the form of an electrical field. The capacitor's nega-tively charged plate constitutes the microphone's diaphragm, which, when it is hit by sound waves, vibrates at the same frequency as those waves. Current flows back and forth between the diaphragm and the positive plate of the capaci-tor, depending on whether the electrostatic or electrical pull is increasing or decreasing. This in turn produces an alternating current, at the same frequency as the sound waves, which travels through a mixer and then an amplifier to the speaker.

A loudspeaker typically contains a circular permanent magnet, which surrounds an electri-cal coil and is in turn attached to a cone-shaped diaphragm. Current enters the speaker ultimate-ly from the microphone, alternating at the same frequency as the source of the sound (a singer's voice, for instance). As it enters the coil, this cur-rent induces an alternating magnetic field, which causes the coil to vibrate. This in turn vibrates

the cone-shaped diaphragm, and the latter repro-duces sounds generated at the source.

A car horn also uses magnetism to create sound by means of vibration. When a person presses down on the horn embedded in his or her steering wheel, this in turn depresses an iron bar that passes through an electromagnet surround-ed by wires from the car's battery. The bar moves up and down within the electromagnetic field, causing the diaphragm to vibrate and producing a sound that is magnified greatly when released through a bell-shaped horn.

Electromagnetically induced vibration is also the secret behind another noise-making device, a vibrating electric doorbell used in many apartments. The button that a visitor presses is connected directly to a power source, which sends current flowing through a spring sur-rounding an electromagnet. The latter generates a magnetic field, drawing toward it an iron arma-ture attached to a hammer. The hammer then strikes the bell. The result is a mechanical reac-tion that pushes the armature away from the electromagnet, but the spring forces the arma-ture back against the electromagnet again. This cycle of contact and release continues for as long as the button is depressed, causing a continual ringing of the bell.

RECORDING AND READING DATA USING MAGNETS: FROM RECORDS AND TAPES TO DISK DRIVES

Just as magnetism plays a critical role in project-ing the volume of sound, it is also crucial to the recording and retrieval of sound and other data. Of course terms such as "retrieval" and "data" have an information-age sound to them, but the idea of using magnetism to record sound is an old one—much older than computers or com-pact discs (CDs). The latter, of course, replaced cassettes in the late 1980s as the preferred mode for listening to recorded music, just as cassettes had recently made powerful gains against phono-graph records.

Despite the fact that cassettes entered the market much later than records, however, recording engineers from the mid-twentieth cen-tury onward typically used magnetic tape for master recordings of songs. This master would then be used to create a metal master record disk by means of a cutting head that responded to vibrations from the master tape; then, the record company could produce endless plastic copies of the metal record.

In recording a tape—whether a stereo master or a mere home recording of a conversation—the principles at work are more or less the same. As noted in the earlier illustration involving a microphone and loudspeaker, sound comes through a microphone in the form of alternating current. The strength of this current in turn affects the "recording head," a small electromagnet whose magnetic field extends over the section of tape being recorded. Loud sounds produce strong magnetic fields, and soft ones weak fields.

All of this information becomes embedded on the cassette tape through a process of magnetic alignment not so different from the process described earlier for creating a permanent magnet. But whereas the permanent magnetization of a natural magnet is difficult to reverse, reversal of a tape's magnetization—in other words, erasing the tape—is easy. An erase head, an electromagnet operating at a frequency too high for the human ear to hear, simply scrambles the magnetic particles on a piece of tape.

A CD, as one might expect, is much more closely related to a computer disk-drive than it is to earlier forms of recording technology. The disk drive receives electronic on-off signals from the computer, and translates these into magnetic codes that it records on the surface of a floppy disk. The disk drive itself includes two electric motors: a disk motor, which rotates the disk at a high speed, and a head motor, which moves the computer's read-write head across the disk. (It should be noted that most electric motors, including the universal motors used in a variety of household appliances, also use electromagnets.)

A third motor, called a stepper motor, ensures that the drive turns at a precise rate of speed. The stepper motor contains its own magnet, in this case a permanent one of cylindrical shape that sends signals to rows of metal teeth surrounding it, and these teeth act as gears to regulate the drive's speed. Likewise a CD player, which actually uses laser beams rather than magnetic fields to retrieve data from a disc, also has a drive system that regulates the speed at which the disk spins.

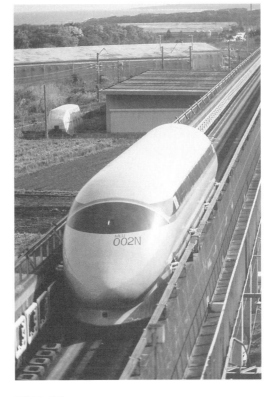

MAGLEV TRAINS, LIKE THIS EXPERIMENTAL ONE IN MIYAZAKI, JAPAN, MAY REPRESENT THE FUTURE OF MASS TRANSIT. *(Photograph by Michael S. Yamashita/Corbis. Reproduced by permission.)*

MAGLEV TRAINS: THE FUTURE OF TRANSPORT?

One promising application of electromagnetic technology relates to a form of transportation that might, at first glance, appear to be old news: trains. But MAGLEV, or magnetic levitation, trains are as far removed from the old steam engines of the Union Pacific as the space shuttle is from the Wright brothers' experimental airplane.

As discussed earlier, magnetic poles of like direction (i.e., north-north or south-south) repel one another such that, theoretically at least, it is possible to keep one magnet suspended in the air over another magnet. Actually it is impossible to produce these results with simple bar magnets, because their magnetic force is too small; but an electromagnet can create a magnetic field powerful enough that, if used properly, it exerts enough repulsive force to lift extremely heavy objects. Specifically, if one could activate train tracks with a strong electromagnetic field, it might be possible to "levitate" an entire train. This in turn

KEY TERMS

ELECTROMAGNET: A type of magnet in which an object is charged by an electrical current. Typically the object used is made of iron, which quickly loses magnetic force when current is reduced. Thus an electromagnet can be turned on or off, and its magnetic force altered, making it potentially much more powerful than a natural magnet.

ELECTROMAGNETISM: The unified electrical and magnetic force field generated by the passage of an electric current through matter.

ELECTRONS: Negatively charged subatomic particles whose motion relative to one another creates magnetic force.

MAGNETIC FIELD: Wherever a magnetic force acts on a moving charged particle, a magnetic field is said to exist. Magnetic fields are typically measured by a unit called a tesla.

NATURAL MAGNET: A chemical element in which the magnetic fields created by electrons' relative motion align uniformly to create a net magnetic dipole, or unity of direction. Such elements, among them iron, cobalt, and nickel, are also known as magnetic metals.

PERMANENT MAGNET: A magnetic material in which groups of atoms, known as domains, are brought into alignment, and in which magnetization cannot be changed merely by attempting to realign the domains. Permanent magnetization is reversible only at very high temperatures—for example, 1,418°F (770°C) in the case of iron.

would make possible a form of transport that could move large numbers of people in relative comfort, thus decreasing the environmental impact of automobiles, and do so at much higher speeds than a car could safely attain.

Actually the idea of MAGLEV trains goes back to a time when trains held complete supremacy over automobiles as a mode of transportation: specifically, 1907, when rocket pioneer Robert Goddard (1882-1945) wrote a story describing a vehicle that traveled by means of magnetic levitation. Just five years later, French engineer Emile Bachelet produced a working model for a MAGLEV train. But the amount of magnetic force required to lift such a vehicle made it impractical, and the idea fell to the wayside.

Then, in the 1960s, the advent of superconductivity—the use of extremely low temperatures, which facilitate the transfer of electrical current through a conducting material with virtually no resistance—made possible electromagnets of staggering force. Researchers began building MAGLEV prototypes using superconducting coils with strong currents to create a powerful magnetic field. The field in turn created a repulsive force capable of lifting a train several inches above a railroad track. Electrical current sent through guideway coils on the track allowed for enormous propulsive force, pushing trains forward at speeds up to and beyond 250 MPH (402 km/h).

Initially, researchers in the United States were optimistic about MAGLEV trains, but safety concerns led to the shelving of the idea for several decades. Meanwhile, other industrialized nations moved forward with MAGLEVs: in Japan, engineers built a 27-mi (43.5-km) experimental MAGLEV line, while German designers experimented with attractive (as opposed to repulsive) force in their Transrapid 07. MAGLEV trains gained a new defender in the United States with now-retired Senator Daniel Patrick Moynihan (D-NY), who as chairman of a Senate subcommittee overseeing the interstate highway system introduced legislation to fund MAGLEV research. The 1998 transportation bill allocated $950 million toward the Magnetic Levitation Prototype Development Program. As part of this program, in January 2001 the U.S. Department of Transportation selected projects in Maryland and Pennsylvania as the two finalists in the competition to build the first MAGLEV train service in the United States. The goal is to have the service in place by approximately 2010.

WHERE TO LEARN MORE

Barr, George. *Science Projects for Young People.* New York: Dover, 1964.

Beiser, Arthur. *Physics,* 5th ed. Reading, MA: Addison-Wesley, 1991.

Hann, Judith. *How Science Works.* Pleasantville, NY: Reader's Digest, 1991.

Macaulay, David. *The New Way Things Work.* Boston: Houghton Mifflin, 1998.

Molecular Expressions: Electricity and Magnetism: Interactive Java Tutorials (Web site). <http://micro.magnet.fsu.edu/electromag/java/> (January 26, 2001).

Topical Group on Magnetism (Web site). <http://www.aps.org/units/gmag/> (January 26, 2001).

VanCleave, Janice. *Magnets.* New York: John Wiley & Sons, 1993.

Wood, Robert W. *Physics for Kids: 49 Easy Experiments with Electricity and Magnetism.* New York: Tab, 1990.

ELECTROMAGNETIC SPECTRUM

CONCEPT

One of the most amazing aspects of physics is the electromagnetic spectrum—radio waves, microwaves, infrared light, visible light, ultraviolet light, x rays, and gamma rays—as well as the relationship between the spectrum and electromagnetic force. The applications of the electromagnetic spectrum in daily life begin the moment a person wakes up in the morning and "sees the light." Yet visible light, the only familiar part of the spectrum prior to the eighteenth and nineteenth centuries, is also its narrowest region. Since the beginning of the twentieth century, uses for other bands in the electromagnetic spectrum have proliferated. At the low-frequency end are radio, short-wave radio, and television signals, as well as the microwaves used in cooking. Higher-frequency waves, all of which can be generally described as light, provide the means for looking deep into the universe—and deep into the human body.

HOW IT WORKS

ELECTROMAGNETISM

The ancient Romans observed that a brushed comb would attract particles, a phenomenon now known as static electricity and studied within the realm of electrostatics in physics. Yet, the Roman understanding of electricity did not extend any further, and as progress was made in the science of physics—after a period of more than a thousand years, during which scientific learning in Europe progressed very slowly—it developed in areas that had nothing to do with the strange force observed by the Romans.

The fathers of physics as a serious science, Galileo Galilei (1564-1642) and Sir Isaac Newton (1642-1727), were concerned with gravitation, which Newton identified as a fundamental force in the universe. For nearly two centuries, physicists continued to believe that there was only one type of force. Yet, as scientists became increasingly aware of molecules and atoms, anomalies began to arise—in particular, the fact that gravitation alone could not account for the strong forces holding atoms and molecules together to form matter.

FOUNDATIONS OF ELECTRO-MAGNETIC THEORY. At the same time, a number of thinkers conducted experiments concerning the nature of electricity and magnetism, and the relationship between them. Among these were several giants in physics and other disciplines—including one of America's greatest founding fathers. In addition to his famous (and highly dangerous) experiment with lightning, Benjamin Franklin (1706-1790) also contributed the names "positive" and "negative" to the differing electrical charges discovered earlier by French physicist Charles Du Fay (1698-1739).

In 1785, French physicist and inventor Charles Coulomb (1736-1806) established the basic laws of electrostatics and magnetism. He maintained that there is an attractive force that, like gravitation, can be explained in terms of the inverse of the square of the distance between objects. That attraction itself, however, resulted not from gravity, but from electrical charge, according to Coulomb.

A few years later, German mathematician Johann Karl Friedrich Gauss (1777-1855) developed a mathematical theory for finding the magnetic potential of any point on Earth, and his contemporary, Danish physicist Hans Christian Oersted (1777-1851), became the first scientist to establish the existence of a clear relationship between electricity and magnetism. This led to the foundation of electromagnetism, the branch of physics devoted to the study of electrical and magnetic phenomena.

French mathematician and physicist André Marie Ampère (1775-1836) concluded that magnetism is the result of electricity in motion, and, in 1831, British physicist and chemist Michael Faraday (1791-1867) published his theory of electromagnetic induction. This theory shows how an electrical current in one coil can set up a current in another through the development of a magnetic field. This enabled Faraday to develop the first generator, and for the first time in history, humans were able to convert mechanical energy systematically into electrical energy.

MAXWELL AND ELECTROMAG-NETIC FORCE. A number of other figures contributed along the way; but, as yet, no one had developed a "unified theory" explaining the relationship between electricity and magnetism. Then, in 1865, Scottish physicist James Clerk Maxwell (1831-1879) published a groundbreaking paper, "On Faraday's Lines of Force," in which he outlined a theory of electromagnetic force—the total force on an electrically charged particle, which is a combination of forces due to electrical and/or magnetic fields around the particle.

Maxwell had thus discovered a type of force in addition to gravity, and this reflected a "new" type of fundamental interaction, or a basic mode by which particles interact in nature. Newton had identified the first, gravitational interaction, and in the twentieth century, two other forms of fundamental interaction—strong nuclear and weak nuclear—were identified as well.

In his work, Maxwell drew on the studies conducted by his predecessors, but added a new statement: that electrical charge is conserved. This statement, which did not contradict any of the experimental work done by the other physicists, was based on Maxwell's predictions regarding what should happen in situations of electromagnetism; subsequent studies have supported his predictions.

ELECTROMAGNETIC RADIATION

So far, what we have seen is the foundation for modern understanding of electricity and mag-netism. This understanding grew enormously in the late nineteenth and early twentieth centuries, thanks both to the theoretical work of physicists, and the practical labors of inventors such as Thomas Alva Edison (1847-1931) and Serbian-American electrical engineer Nikola Tesla (1856-1943). But our concern in the present context is with electromagnetic radiation, of which the waves on the electromagnetic spectrum are a particularly significant example.

Energy can travel by conduction or convection, two principal means of heat transfer. But the energy Earth receives from the Sun—the energy conveyed through the electromagnetic spectrum—is transferred by another method, radiation. Whereas conduction of convection can only take place where there is matter, which provides a medium for the energy transfer, radiation requires no medium. Thus, electromagnetic energy passes from the Sun to Earth through the vacuum of empty space.

ELECTROMAGNETIC WAVES. The connection between electromagnetic radiation and electromagnetic force is far from obvious. Even today, few people not scientifically trained understand that there is a clear relationship between electricity and magnetism—let alone a connection between these and visible light. The breakthrough in establishing that connection can be attributed both to Maxwell and to German physicist Heinrich Rudolf Hertz (1857-1894).

Maxwell had suggested that electromagnetic force carried with it a certain wave phenomenon, and predicted that these waves traveled at a certain speed. In his *Treatise on Electricity and Magnetism* (1873), he predicted that the speed of these waves was the same as that of light—186,000 mi (299,339 km) per second—and theorized that the electromagnetic interaction included not only electricity and magnetism, but light as well. A few years later, while studying the behavior of electrical currents, Hertz confirmed Maxwell's proposition regarding the wave phenomenon by showing that an electrical current generated some sort of electromagnetic radiation.

In addition, Hertz found that the flow of electrical charges could be affected by light under certain conditions. Ultraviolet light had already been identified, and Hertz shone an ultraviolet beam on the negatively charged side of a gap in a

THE HUBBLE SPACE TELESCOPE INCLUDES AN ULTRAVIOLET LIGHT INSTRUMENT CALLED THE GODDARD HIGH RES-
OLUTION SPECTROGRAPH THAT IT IS CAPABLE OF OBSERVING EXTREMELY DISTANT OBJECTS. *(Photograph by Roger Ress-
meyer/Corbis. Reproduced by permission.)*

current loop. This made it easier for an electrical spark to jump the gap. Hertz could not explain this phenomenon, which came to be known as the photoelectric effect. Indeed, no one else could explain it until quantum theory was developed in the early twentieth century. In the meantime, however, Hertz's discovery of electromagnetic waves radiating from a current loop led to the invention of radio by Italian physicist and engineer Guglielmo Marconi (1874-1937) and others.

LIGHT: WAVES OR PARTICLES?

At this point, it is necessary to jump backward in history, to explain the progression of scientists'

understanding of light. Advancement in this area took place over a long period of time: at the end of the first millennium A.D., the Arab physicist Alhasen (Ibn al-Haytham; c. 965-1039) showed that light comes from the Sun and other self-illuminated bodies—not, as had been believed up to that time—from the eye itself. Thus, studies in optics, or the study of light and vision, were—compared to understanding of electromagnetism itself—relatively advanced by 1666, when Newton discovered the spectrum of colors in light. As Newton showed, colors are arranged in a sequence, and white light is a combination of all colors.

Newton put forth the corpuscular theory of light—that is, the idea that light is made up of particles—but his contemporary Christiaan Huygens (1629-1695), a Dutch physicist and astronomer, maintained that light appears in the form of a wave. For the next century, adherents of Newton's corpuscular theory and of Huygens's wave theory continued to disagree. Physicists on the European continent began increasingly to accept wave theory, but corpuscular theory remained strong in Newton's homeland.

Thus, it was ironic that the physicist whose work struck the most forceful blow against corpuscular theory was himself an Englishman: Thomas Young (1773-1829), who in 1801 demonstrated interference in light. Young directed a beam of light through two closely spaced pinholes onto a screen, reasoning that if light truly were made of particles, the beams would project two distinct points onto the screen. Instead, what he saw was a pattern of interference—a wave phenomenon.

By the time of Hertz, wave theory had become dominant; but the photoelectric effect also exhibited aspects of particle behavior. Thus, for the first time in more than a century, particle theory gained support again. Yet, it was clear that light had certain wave characteristics, and this raised the question—which is it, a wave or a set of particles streaming through space?

The work of German physicist Max Planck (1858-1947), father of quantum theory, and of Albert Einstein (1879-1955), helped resolve this apparent contradiction. Using Planck's quantum principles, Einstein, in 1905, showed that light appears in "bundles" of energy, which travel as waves but behave as particles in certain situations. Eighteen years later, American physicist Arthur Holly Compton (1892-1962) showed that, depending on the way it is tested, light appears as either a particle or a wave. These particles he called photons.

Wave Motion and Electromagnetic Waves

The particle behavior of electromagnetic energy is beyond the scope of the present discussion, though aspects of it are discussed elsewhere. For the present purposes, it is necessary only to view the electromagnetic spectrum as a series of waves, and in the paragraphs that follow, the

rudiments of wave motion will be presented in short form.

A type of harmonic motion that carries energy from one place to another without actually moving any matter, wave motion is related to oscillation, harmonic—and typically periodic—motion in one or more dimensions. Oscillation involves no net movement, but only movement in place; yet individual waves themselves are oscillating, even as the overall wave pattern moves.

The term periodic motion, or movement repeated at regular intervals called periods, describes the behavior of periodic waves: waves in which a uniform series of crests and troughs follow each other in regular succession. Periodic waves are divided into longitudinal and transverse waves, the latter (of which light waves are an example) being waves in which the vibration or motion is perpendicular to the direction in which the wave is moving. Unlike longitudinal waves, such as those that carry sound energy, transverse waves are fairly easy to visualize, and assume the shape that most people imagine when they think of waves: a regular up-and-down pattern, called "sinusoidal" in mathematical terms.

PARAMETERS OF WAVE MOTION. A period (represented by the symbol T) is the amount of time required to complete one full cycle of the wave, from trough to crest and back to trough. Period is mathematically related to several other aspects of wave motion, including wave speed, frequency, and wavelength.

Frequency (abbreviated f) is the number of waves passing through a given point during the interval of one second. It is measured in Hertz (Hz), named after Hertz himself: a single Hertz (the term is both singular and plural) is equal to one cycle of oscillation per second. Higher frequencies are expressed in terms of kilohertz (kHz; 10^3 or 1,000 cycles per second); megahertz (MHz; 10^6 or 1 million cycles per second); and gigahertz (GHz; 10^9 or 1 billion cycles per second.)

Wavelength (represented by the symbol λ, the Greek letter lambda) is the distance between a crest and the adjacent crest, or a trough and an adjacent trough, of a wave. The higher the frequency, the shorter the wavelength; and, thus, it is possible to describe waves in terms of either. According to quantum theory, however, electromagnetic waves can also be described in terms of photon energy level, or the amount of energy in each photon. Thus, the electromagnetic spectrum, as we shall see, varies from relatively long-wavelength, low-frequency, low-energy radio waves on the one end to extremely short-wavelength, high-frequency, high-energy gamma rays on the other.

The other significant parameter for describing a wave—one mathematically independent from those so far discussed—is amplitude. Defined as the maximum displacement of a vibrating material, amplitude is the "size" of a wave. The greater the amplitude, the greater the energy the wave contains: amplitude indicates intensity. The amplitude of a light wave, for instance, determines the intensity of the light.

A RIGHT-HAND RULE. Physics textbooks use a number of "right-hand rules": devices for remembering certain complex physical interactions by comparing the lines of movement or force to parts of the right hand. In the present context, a right-hand rule makes it easier to visualize the mutually perpendicular directions of electromagnetic waves, electric field, and magnetic field.

A field is a region of space in which it is possible to define the physical properties of each point in the region at any given moment in time. Thus, an electrical field and magnetic field are simply regions in which electrical and magnetic components, respectively, of electromagnetic force are exerted.

Hold out your right hand, palm perpendicular to the floor and thumb upright. Your fingers indicate the direction that an electromagnetic wave is moving. Your thumb points in the direction of the electrical field, as does the heel of your hand: the electrical field forms a plane perpendicular to the direction of wave propagation. Similarly, both your palm and the back of your hand indicate the direction of the magnetic field, which is perpendicular both to the electrical field and the direction of wave propagation.

THE ELECTROMAGNETIC SPECTRUM

As stated earlier, an electromagnetic wave is transverse, meaning that even as it moves forward, it oscillates in a direction perpendicular to the line of propagation. An electromagnetic wave can thus be defined as a transverse wave with mutually perpendicular electrical and magnetic fields that emanate from it.

The electromagnetic spectrum is the complete range of electromagnetic waves on a continuous distribution from a very low range of frequencies and energy levels, with a correspondingly long wavelength, to a very high range of frequencies and energy levels, with a correspondingly short wavelength. Included on the electromagnetic spectrum are radio waves and microwaves; infrared, visible, and ultraviolet light; x rays, and gamma rays. Though each occupies a definite place on the spectrum, the divisions between them are not firm: as befits the nature of a spectrum, one simply "blurs" into another.

FREQUENCY RANGE OF THE ELECTROMAGNETIC SPECTRUM. The range of frequencies for waves in the electromagnetic spectrum is from approximately 10^2 Hz to more than 10^{25} Hz. These numbers are an example of scientific notation, which makes it possible to write large numbers without having to include a string of zeroes. Without scientific notation, the large numbers used for discussing properties of the electromagnetic spectrum can become bewildering.

The first number given, for extremely low-frequency radio waves, is simple enough—100—but the second would be written as 1 followed by 25 zeroes. (A good rule of thumb for scientific notation is this: for any power n of 10, simply attach that number of zeroes to 1. Thus 10^6 is 1 followed by 6 zeroes, and so on.) In any case, 10^{25} is a much simpler figure than 10,000,000,000,000,000,000,000,000—or 10 trillion trillion. As noted earlier, gigahertz, or units of 1 billion Hertz, are often used in describing extremely high frequencies, in which case the number is written as 10^{16} GHz. For simplicity's sake, however, in the present context, the simple unit of Hertz (rather than kilo-, mega-, or gigahertz) is used wherever it is convenient to do so.

WAVELENGTHS ON THE ELECTROMAGNETIC SPECTRUM. The range of wavelengths found in the electromagnetic spectrum is from about 10^8 centimeters to less than 10^{-15} centimeters. The first number, equal to 1 million meters (about 621 mi), obviously expresses a great length. This figure is for radio waves of extremely low frequency; ordinary radio waves of the kind used for actual radio broadcasts are closer to 10^5 centimeters (about 328 ft).

For such large wavelengths, the use of centimeters might seem a bit cumbersome; but, as with the use of Hertz for frequencies, centimeters provide a simple unit that can be used to measure all wavelengths. Some charts of the electromagnetic spectrum nonetheless give figures in meters, but for parts of the spectrum beyond microwaves, this, too, can become challenging. The ultra-short wavelengths of gamma rays, after all, are equal to one-trillionth of a centimeter. By comparison, the angstrom—a unit so small it is used to measure the diameter of an atom—is 10 million times as large.

ENERGY LEVELS ON THE ELECTROMAGNETIC SPECTRUM. Finally, in terms of photon energy, the unit of measurement is the electron volt (eV), which is used for quantifying the energy in atomic particles. The range of photon energy in the electromagnetic spectrum is from about 10^{-13} to more than 10^{10} electron volts. Expressed in terms of joules, an electron volt is equal to $1.6 \cdot 10^{-19}$ J.

To equate these figures to ordinary language would require a lengthy digression; suffice it to say that even the highest ranges of the electromagnetic spectrum possess a small amount of energy in terms of joules. Remember, however, that the energy level identified is for a photon—a light particle. Again, without going into a great deal of detail, one can just imagine how many of these particles, which are much smaller than atoms, would fit into even the smallest of spaces. Given the fact that electromagnetic waves are traveling at a speed equal to that of light, the amount of photon energy transmitted in a single second is impressive, even for the lower ranges of the spectrum. Where gamma rays are concerned, the energy levels are positively staggering.

REAL-LIFE APPLICATIONS

THE RADIO SUB-SPECTRUM

Among the most familiar parts of the electromagnetic spectrum, in modern life at least, is radio. In most schematic representations of the spectrum, radio waves are shown either at the left end or the bottom, as an indication of the fact that these are the electromagnetic waves with the lowest frequencies, the longest wavelengths, and the smallest levels of photon energy. Included in this broad sub-spectrum, with frequencies up to about 10^7 Hertz, are long-wave radio, short-wave radio, and microwaves. The areas of communication affected are many: broadcast radio, television, mobile phones, radar—and even highly specific forms of technology such as baby monitors.

Though the work of Maxwell and Hertz was foundational to the harnessing of radio waves for human use, the practical use of radio had its beginnings with Marconi. During the 1890s, he made the first radio transmissions, and, by the end of the century, he had succeeded in transmitting telegraph messages across the Atlantic Ocean—a feat which earned him the Nobel Prize for physics in 1909.

Marconi's spark transmitters could send only coded messages, and due to the broad, long-wavelength signals used, only a few stations could broadcast at the same time. The development of the electron tube in the early years of the twentieth century, however, made it possible to transmit narrower signals on stable frequencies. This, in turn, enabled the development of technology for sending speech and music over the airwaves.

BROADCAST RADIO

THE DEVELOPMENT OF AM AND FM. A radio signal is simply a carrier: the process of adding information—that is, complex sounds such as those of speech or music—is called modulation. The first type of modulation developed was AM, or amplitude modulation, which Canadian-American physicist Reginald Aubrey Fessenden (1866-1932) demonstrated with the first United States radio broadcast in 1906. Amplitude modulation varies the instantaneous amplitude of the radio wave, a function of the radio station's power, as a means of transmitting information.

By the end of World War I, radio had emerged as a popular mode of communication: for the first time in history, entire nations could hear the same sounds at the same time. During the 1930s, radio became increasingly important, both for entertainment and information. Families in the era of the Great Depression would gather around large "cathedral radios"—so named for their size and shape—to hear comedy programs, soap operas, news programs, and speeches by important public figures such as President Franklin D. Roosevelt.

Throughout this era—indeed, for more than a half-century from the end of the first World War to the height of the Vietnam Conflict in the mid-1960s—AM held a dominant position in radio. This remained the case despite a number of limitations inherent in amplitude modulation: AM broadcasts flickered with popping noises from lightning, for instance, and cars with AM radios tended to lose their signal when going under a bridge. Yet, another mode of radio transmission was developed in the 1930s, thanks to American inventor and electrical engineer Edwin H. Armstrong (1890-1954). This was FM, or frequency modulation, which varied the radio signal's frequency rather than its amplitude.

Not only did FM offer a different type of modulation; it was on an entirely different frequency range. Whereas AM is an example of a long-wave radio transmission, FM is on the microwave sector of the electromagnetic spectrum, along with television and radar. Due to its high frequency and form of modulation, FM offered a "clean" sound as compared with AM. The addition of FM stereo broadcasts in the 1950s offered still further improvements; yet despite the advantages of FM, audiences were slow to change, and FM did not become popular until the mid- to late 1960s.

SIGNAL PROPAGATION. AM signals have much longer wavelengths, and smaller frequencies, than do FM signals, and this, in turn, affects the means by which AM signals are propagated. There are, of course, much longer radio wavelengths; hence, AM signals are described as intermediate in wavelength. These intermediate-wavelength signals reflect off highly charged layers in the ionosphere between 25 and 200 mi (40-332 km) above Earth's surface. Short-wavelength signals, such as those of FM, on the other hand, follow a straight-line path. As a result, AM broadcasts extend much farther than FM, particularly at night.

At a low level in the ionosphere is the D layer, created by the Sun when it is high in the sky. The D layer absorbs medium-wavelength signals during the day, and for this reason, AM signals do not travel far during daytime hours. After the Sun goes down, however, the D layer soon fades, and this makes it possible for AM signals to reflect off a much higher layer of the ion-

osphere known as the F layer. (This is also sometimes known as the Heaviside layer, or the Kennelly-Heaviside layer, after English physicist Oliver Heaviside and British-American electrical engineer Arthur Edwin Kennelly, who independently discovered the ionosphere in 1902.) AM signals "bounce" off the F layer as though it were a mirror, making it possible for a listener at night to pick up a signal from halfway across the country.

The Sun has other effects on long-wave and intermediate-wave radio transmissions. Sunspots, or dark areas that appear on the Sun in cycles of about 11 years, can result in a heavier buildup of the ionosphere than normal, thus impeding radio-signal propagation. In addition, occasional bombardment of Earth by charged particles from the Sun can also disrupt transmissions.

Due to the high frequencies of FM signals, these do not reflect off the ionosphere; instead, they are received as direct waves. For this reason, an FM station has a fairly short broadcast range, and this varies little with regard to day or night. The limited range of FM stations as compared to AM means that there is much less interference on the FM dial than for AM.

DISTRIBUTION OF RADIO FREQUENCIES

In the United States and most other countries, one cannot simply broadcast at will; the airwaves are regulated, and, in America, the governing authority is the Federal Communications Commission (FCC). The FCC, established in 1934, was an outgrowth of the Federal Radio Commission, founded by Congress seven years earlier. The FCC actually "sells air," charging companies a fee to gain rights to a certain frequency. Those companies may in turn sell that air to others for a profit.

At the time of the FCC's establishment, AM was widely used, and the federal government assigned AM stations the frequency range of 535 kHz to 1.7 MHz. Thus, if an AM station today is called, for instance, "AM 640," this means that it operates at 640 kHz on the dial. The FCC assigned the range of 5.9 to 26.1 MHz to short-wave radio, and later the area of 26.96 to 27.41 MHz to citizens' band (CB) radio. Above these are microwave regions assigned to television stations, as well as FM, which occupies the range from 88 to 108 MHz.

The organization of the electromagnetic spectrum's radio frequencies—which, of course, is an entirely arbitrary, humanmade process—is fascinating. It includes assigned frequencies for everything from garage-door openers to deep-space radio communications. The FCC recognizes seven divisions of radio carriers, using a system that is not so much based on rational rules as it is on the way that the communications industries happened to develop over time.

THE SEVEN FCC DIVISIONS. Most of what has so far been described falls under the heading of "Public Fixed Radio Services": AM and FM radio, other types of radio such as shortwave, television, various other forms of microwave broadcasting, satellite systems, and communication systems for federal departments and agencies. "Public Mobile Services" include pagers, air-to-ground service (for example, aircraft-to-tower communications), offshore service for sailing vessels, and rural radio-telephone service. "Commercial Mobile Radio Services" is the realm of cellular phones, and "Personal Communications Service" that of the newer wireless technology that began to challenge cellular for market dominance in the late 1990s.

"Private Land Mobile Radio Service" (PMR) and "Private Operational-Fixed Microwave Services" (OFS) are rather difficult to distinguish, the principal difference being that the former is used exclusively by profit-making businesses, and the latter mostly by nonprofit institutions. An example of PMR technology is the dispatching radios used by taxis, but this is only one of the more well-known forms of internal electronic communications for industry. For instance, when a film production company is shooting a picture and the director needs to speak to someone at the producer's trailer a mile away, she may use PMR radio technology. OFS was initially designated purely for nonprofit use, and is used often by schools; but banks and other profit-making institutions often use OFS because of its low cost.

Finally, there is the realm of "Personal Radio Services," created by the FCC in 1992. This branch, still in its infancy, will probably one day include video-on-demand, interactive polling, online shopping and banking, and other activities classified under the heading of Interactive Video and Data Services, or IVDS. Unlike other types of video technology, these will all be wireless, and, therefore, represent a telecommunications revolution all their own.

MICROWAVES

MICROWAVE COMMUNICATION. Though microwaves are treated separately from radio waves, in fact, they are just radio signals of a very short wavelength. As noted earlier, FM signals are actually carried on microwaves, and, as with FM in particular, microwave signals in general are very clear and very strong, but do not extend over a great geographical area. Nor does microwave include only high-frequency radio and television; in fact, any type of information that can be transmitted via telephone wires or coaxial cables can also be sent via a microwave circuit.

Microwaves have a very narrow, focused beam: thus, the signal is amplified considerably when an antenna receives it. This phenomenon, known as "high antenna gain," means that microwave transmitters need not be highly powerful to produce a strong signal. To further the reach of microwave broadcasts, transmitters are often placed atop mountain peaks, hilltops, or tall buildings. In the past, a microwave-transmitting network such as NBC (National Broadcasting Company) or CBS (Columbia Broadcasting System) required a network of ground-based relay stations to move its signal across the continent. The advent of satellite broadcasting in the 1960s, however, changed much about the way signals are beamed: today, networks typically replace, or at least augment, ground-based relays with satellite relays.

The first worldwide satellite TV broadcast, in the summer of 1967, featured the Beatles singing their latest song "All You Need Is Love." Due to the international character of the broadcast, with an estimated 200 million viewers, John Lennon and Paul McCartney wrote a song with simple, universal lyrics, and the result was just another example of electronic communication uniting large populations. Indeed, the phenomenon of rock music, and of superstardom as people know it today, would be impossible without many of the forms of technology discussed here. Long before the TV broadcast, the Beatles had come to fame through the playing of their music on the radio waves—and, thus, they owed much to Maxwell, Hertz, and Marconi.

MICROWAVE OVENS. The same microwaves that transmit FM and television signals—to name only the most obviously applications of microwave for communication—can also be harnessed to cook food. The microwave oven, introduced commercially in 1955, was an outgrowth of military technology developed a decade before.

During World War II, the Raytheon Manufacturing Company had experimented with a magnetron, a device for generating extremely short-wavelength radio signals as a means of improving the efficiency of military radar. While working with a magnetron, a technician named Percy Spencer was surprised to discover that a candy bar in his pocket had melted, even though he had not felt any heat. This led him to considering the possibilities of applying the magnetron to peacetime uses, and a decade later, Raytheon's "radar range" hit the market.

Those early microwave ovens had none of varied power settings to which modern users of the microwave—found today in two-thirds of all American homes—are accustomed. In the first microwaves, the only settings were "on" and "off," because there were only two possible adjustments: either the magnetron would produce, or not produce, microwaves. Today, it is possible to use a microwave for almost anything that involves the heating of food that contains water—from defrosting a steak to popping popcorn.

As noted much earlier, in the general discussion of electromagnetic radiation, there are three basic types of heat transfer: conduction, convection, and radiation. Without going into too much detail here, conduction generally involves heat transfer between molecules in a solid; convection takes place in a fluid (a gas such as air or a liquid such as water); and radiation, of course, requires no medium.

A conventional oven cooks through convection, though conduction also carries heat from the outer layers of a solid (for example, a turkey) to the interior. A microwave, on the other hand, uses radiation to heat the outer layers of the food; then conduction, as with a conventional oven, does the rest. The difference is that the microwave heats only the food—or, more specifically, the water, which then transfers heat throughout the item being heated—and not the dish or plate. Thus, many materials, as long as they do not contain water, can be placed in a microwave oven without being melted or burned. Metal, though it contains no water, is unsafe because the microwaves bounce off the metal surfaces, creating a microwave buildup that can produce sparks and damage the oven.

In a microwave oven, microwaves emitted by a small antenna are directed into the cooking compartment, and as they enter, they pass a set of turning metal fan blades. This is the stirrer, which disperses the microwaves uniformly over the surface of the food to be heated. As a microwave strikes a water molecule, resonance causes the molecule to align with the direction of the wave. An oscillating magnetron causes the microwaves to oscillate as well, and this, in turn, compels the water molecules to do the same. Thus, the water molecules are shifting in position several million times a second, and this vibration generates energy that heats the water.

RADIO WAVES FOR MEASUREMENT AND RANGING

RADAR. Radio waves can be used to send communication signals, or even to cook food; they can also be used to find and measure things. One of the most obvious applications in this regard is radar, an acronym for *RAdio Detection And Ranging*.

Radio makes it possible for pilots to "see" through clouds, rain, fog, and all manner of natural phenomena—not least of which is darkness. It can also identify objects, both natural and manmade, thus enabling a peacetime pilot to avoid hitting another craft or the side of a mountain. On the other hand, radar may help a pilot in wartime to detect the presence of an enemy. Nor is radar used only in the skies, or for military purposes, such as guiding missiles: on the ground, it is used to detect the speeds of objects such as automobiles on an interstate highway, as well as to track storms.

In the simplest model of radar operation, the unit sends out microwaves toward the target, and the waves bounce back off the target to the unit. Though the speed of light is reduced somewhat, due to the fact that waves are traveling through air rather than through a vacuum, it is, nonetheless, possible to account for this difference. Hence, the distance to the target can be calculated using the simple formula $d = vt$, where d is distance, v is velocity, and t is time.

Typically, a radar system includes the following: a frequency generator and a unit for controlling the timing of signals; a transmitter and, as with broadcast radio, a modulator; a duplexer, which switches back and forth between transmission and reception mode; an antenna; a receiver, which detects and amplifies the signals bounced back to the antenna; signal and data processing units; and data display units. In a monostatic unit—one in which the transmitter and receiver are in the same location—the unit has to be continually switched between sending and receiving modes. Clearly, a bistatic unit—one in which the transmitter and receiver antennas are at different locations—is generally preferable; but on an airplane, for instance, there is no choice but to use a monostatic unit.

In order to determine the range to a target—whether that target be a mountain, an enemy aircraft, or a storm—the target itself must first be detected. This can be challenging, because only a small portion of the transmitted pulse comes back to the receiving antenna. At the same time, the antenna receives reflections from a number of other objects, and it can be difficult to determine which signal comes from the target. For an aircraft in a wartime situation, these problems are compounded by the use of enemy countermeasures such as radar "jamming." Still another difficulty facing a military flyer is the fact that the use of radar itself—that is, the transmission of microwaves—makes the aircraft detectable to opposing forces.

TELEMETRY. Telemetry is the process of making measurements from a remote location and transmitting those measurements to receiving equipment. The earliest telemetry systems, developed in the United States during the 1880s, monitored the distribution and use of electricity in a given region, and relayed this information back to power companies using telephone lines. By the end of World War I, electric companies used the power lines themselves as information relays, and though such electrical telemetry systems remain in use in some sectors, most modern telemetry systems apply radio signals.

An example of a modern telemetry application is the use of an input device called a transducer to measure information concerning an astronaut's vital signs (heartbeat, blood pressure, body temperature, and so on) during a manned space flight. The transducer takes this information and converts it into an electrical impulse, which is then beamed to the space monitoring station on Earth. Because this signal carries information, it must be modulated, but there is little danger of interference with broadcast transmissions on Earth. Typically, signals from spacecraft are sent in a range above 10^{10} Hz, far above the frequencies of most microwave transmissions for commercial purposes.

LIGHT: INVISIBLE, VISIBLE, AND INVISIBLE AGAIN

Between about 10^{13} and 10^{17} Hz on the electromagnetic spectrum is the range of light: infrared, visible, and ultraviolet. Light actually constitutes a small portion of the spectrum, and the area of visible light is very small indeed, extending from about $4.3 \cdot 10^{14}$ to $7.5 \cdot 10^{14}$ Hz. The latter, incidentally, is another example of scientific notation: not only is it easier not to use a string of zeroes, but where a coefficient or factor (for example, 4.3 or 7.5) is other than a multiple of 10, it is preferable to use what are called significant figures—usually a single digit followed by a decimal point and up to 3 decimal places.

Infrared light lies just below visible light in frequency, and this is easy to remember because of the name: red is the lowest in frequency of all the colors. Similarly, ultraviolet lies beyond the highest-frequency color, violet. Visible light itself, by far the most familiar part of the spectrum—especially prior to the age of radio communications—is discussed in detail elsewhere.

INFRARED LIGHT. Though we cannot see infrared light, we feel it as heat. German-English astronomer William Herschel (1738-1822), first scientist to detect infrared radiation from the Sun, demonstrated its existence in 1800 by using a thermometer. Holding a prism, a three-dimensional glass shape used for diffusing beams of light, he directed a beam of sunlight toward the thermometer, which registered the heat of the infrared rays.

Eighty years later, English scientist Sir William Abney (1843-1920) developed infrared photography, a method of capturing infrared radiation, rather than visible light, on film. By the mid-twentieth century, infrared photography had come into use for a variety of purposes. Military forces, for instance, may use infrared to

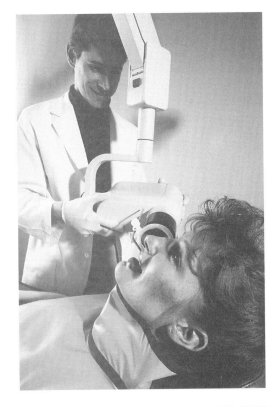

"SOFT" X-RAY MACHINES, SUCH AS THIS ONE BEING USED BY A DENTIST TO PHOTOGRAPH THE PATIENT'S TEETH, OPERATE AT RELATIVELY LOW FREQUENCIES AND THUS DON'T HARM THE PATIENT. *(Photograph by Richard T. Nowitz/Corbis. Reproduced by permission.)*

detect the presence of enemy troops. Medicine makes use of infrared photography for detecting tumors, and astronomers use infrared to detect stars too dim to be seen using ordinary visible light.

The uses of infrared imaging in astronomy, as a matter of fact, are many. The development in the 1980s of infrared arrays, two-dimensional grids which produce reliable images of infrared phenomena, revolutionized infrared astronomy. Because infrared penetrates dust much more easily than does visible light, infrared astronomy makes it easier to see regions of the universe where stars—formed from collapsing clouds of gas and dust—are in the process of developing. Because hydrogen molecules emit infrared radiation, infrared astronomy helps provide clues regarding the distribution of this highly significant chemical element throughout the universe.

ULTRAVIOLET LIGHT. Very little of the Sun's ultraviolet light penetrates Earth's atmosphere—a fortunate thing, since ultraviolet (UV) radiation can be very harmful to human skin. A suntan, as a matter of fact, is actually the skin's defense against these harmful UV rays. Due to the fact that Earth is largely opaque, or resistant, to ultraviolet light, the most significant technological applications of UV radiation are found in outer space.

In 1978 the United States, in cooperation with several European space agencies, launched the International Ultraviolet Explorer (IUE), which measured the UV radiation from tens of thousands of stars, nebulae, and galaxies. Despite the progress made with IUE, awareness of its limitations—including a mirror of only 17 in (45 cm) on the telescope itself—led to the development of a replacement in 1992.

This was the Extreme Ultraviolet Explorer (EUVE), which could observe UV phenomena over a much higher range of wavelengths than those observed by IUE. In addition, the Hubble Space Telescope, launched by the United States in 1990, includes a UV instrument called the Goddard High Resolution Spectrograph. With a mirror measuring 8.5 ft (2.6 m), it is capable of observing objects much more faint than those detected earlier by IUE.

Ultraviolet astronomy is used to study the winds created by hot stars, as well as stars still in the process of forming, and even stars that are dying. It is also useful for analyzing the densely packed, highly active sectors near the centers of galaxies, where both energy and temperatures are extremely high.

X RAYS

Though they are much higher in frequency than visible light—with wavelengths about 1,000 times shorter than for ordinary light rays—x rays are a familiar part of modern life due to their uses in medicine. German scientist Wilhelm Röntgen (1845-1923) developed the first x-ray device in 1895, and, thus, the science of using x-ray machines is called roentgenology.

The new invention became a curiosity, with carnivals offering patrons an opportunity to look at the insides of their hands. And just as many people today fear the opportunities for invasion of privacy offered by computer technology, many at the time worried that x rays would allow robbers and peeping toms to look into people's houses. Soon, however, it became clear that the most important application of x rays lay in medicine.

AMPLITUDE: The maximum displacement of a vibrating material. In wave motion, amplitude is the "size" of a wave, an indicator of the energy and intensity of the wave.

CYCLE: One complete oscillation. In wave motion, this is equivalent to the movement of a wave from trough to crest and back to trough.

ELECTROMAGNETIC FORCE: The total force on an electrically charged particle, which is a combination of forces due to electrical and/or magnetic fields around the particle. Electromagnetic force reflects electromagnetic interaction, one of the four fundamental interactions in nature.

ELECTROMAGNETIC SPECTRUM: The complete range of electromagnetic waves on a continuous distribution from a very low range of frequencies and energy levels, with a correspondingly long wavelength, to a very high range of frequencies and energy levels, with a correspondingly short wavelength. Included on the electromagnetic spectrum are long-wave and short-wave radio; microwaves; infrared, visible, and ultraviolet light; x rays, and gamma rays.

ELECTROMAGNETIC WAVE: A transverse wave with electrical and magnetic fields that emanate from it. The directions of these fields are perpendicular to one another, and both are perpendicular to the line of propagation for the wave itself.

ELECTROMAGNETISM: The branch of physics devoted to the study of electrical and magnetic phenomena.

FIELD: A region of space in which it is possible to define the physical properties of each point in the region at any given moment in time.

FREQUENCY: In wave motion, frequency is the number of waves passing through a given point during the interval of one second. The higher the frequency, the shorter the wavelength. Measured in Hertz, frequency is mathematically related to wave speed, wavelength, and period.

FUNDAMENTAL INTERACTION: The basic mode by which particles interact. There are four known fundamental interactions in nature: gravitational, electromagnetic, strong nuclear, and weak nuclear.

HARMONIC MOTION: The repeated movement of a particle about a position of equilibrium, or balance.

HERTZ: A unit for measuring frequency, named after nineteenth-century German physicist Heinrich Rudolf Hertz (1857-1894). High frequencies are expressed in terms of kilohertz (kHz; 10^3 or 1,000 cycles per second); megahertz (MHz; 10^6 or 1 million cycles per second); and gigahertz (GHz; 10^9 or 1 billion cycles per second.)

INTENSITY: Intensity is the rate at which a wave moves energy per unit of cross-sectional area.

OSCILLATION: A type of harmonic motion, typically periodic, in one or more dimensions.

PERIOD: For wave motion, a period is the amount of time required to complete one full cycle. Period is mathematically related to frequency, wavelength, and wave speed.

PERIODIC MOTION: Motion that is repeated at regular intervals. These intervals are known as periods.

PERIODIC WAVE: A wave in which a uniform series of crests and troughs follow one after the other in regular succession.

PHOTON: A particle of electromagnetic radiation carrying a specific amount of energy, measured in electron volts (eV). For parts of the electromagnetic spectrum with a low frequency and long wavelength, photon energy is relatively low; but for parts with a high frequency and short wavelength, the value of photon energy is very high.

PROPAGATION: The act or state of traveling from one place to another.

RADIATION: The transfer of energy by means of electromagnetic waves, which require no physical medium (for example, water or air) for the transfer. Earth receives the Sun's energy, via the electromagnetic spectrum, by means of radiation.

SCIENTIFIC NOTATION: A method used by scientists for writing extremely large numbers. This usually involves a coefficient, or factor, of a single digit followed by a decimal point and up to three decimal places, multiplied by 10 to a given exponent. Thus, instead of writing 75,120,000, the preferred scientific notation is $7.512 \cdot 10^7$. To visualize the value of very large multiples of 10, it is helpful to remember that the value of 10 raised to any power n is the same as 1 followed by that number of zeroes. Hence 10^{25}, for instance, is simply 1 followed by 25 zeroes.

TRANSVERSE WAVE: A wave in which the vibration or motion is perpendicular to the direction in which the wave is moving.

WAVELENGTH: The distance between a crest and the adjacent crest, or the trough and an adjacent trough, of a wave. Wavelength, symbolized λ (the Greek letter lambda) is mathematically related to wave speed, period, and frequency.

WAVE MOTION: A type of harmonic motion that carries energy from one place to another without actually moving any matter.

HOW A MEDICAL X-RAY MACHINE WORKS. Due to their very short wavelengths, x rays can pass through substances of low density—for example, fat and other forms of soft tissue—without their movement being interrupted. But in materials of higher density, such as bone, atoms are packed closely together, and this provides x rays with less space through which to travel. As a result, x-ray images show dark areas where the rays traveled completely through the target, and light images of dense materials that blocked the movement of the rays.

Medical x-ray machines are typically referred to either as "hard" or "soft." Soft x rays are the ones with which most people are more familiar. Operating at a relatively low frequency, these are used to photograph bones and internal organs, and provided the patient does not receive prolonged exposure to the rays, they cause little damage. Hard x rays, on the other hand, are designed precisely to cause damage—not to the patient, but to cancer cells. Because they use high voltage and high-frequency rays, hard x rays can be quite dangerous to the patient as well.

OTHER APPLICATIONS. X-ray crystallography, developed in the early twentieth century, is devoted to the study of the interference patterns produced by x rays passing through materials that are crystalline in Structure. Each of these discoveries, in turn, transformed daily life: insulin, by offering hope to diabetics, penicillin, by providing a treatment for a number of previously fatal illnesses, and DNA, by enabling scientists to make complex assessments of genetic information.

In addition to the medical applications, the scanning capabilities of x-ray machines make them useful for security. A healthy person receives an x ray at a doctor's office only once in a while; but everyone who carries items past a certain point in a major airport must submit to x-ray security scanning. If one is carrying a purse or briefcase, for instance, this is placed on a moving belt and subjected to scanning by a low-power device that can reveal the contents.

GAMMA RAYS

At the furthest known reaches of the electromagnetic spectrum are gamma rays, ultra high-frequency, high-energy, and short- wavelength forms of radiation. Human understanding of gamma rays, including the awesome powers they contain, is still in its infancy.

In 1979, a wave of enormous energy passed over the Solar System. Though its effects on Earth were negligible, instruments aboard several satellites provided data concerning an enormous quantity of radiation caused by gamma rays. As to the source of the rays themselves, believed to be a product of nuclear fusion on some other body in the universe, scientists knew nothing.

The Compton Gamma Ray Observatory Satellite, launched by NASA (National Aeronautics and Space Administration) in 1991, detected a number of gamma-ray bursts over the next two years. The energy in these bursts was staggering: just one of these, scientists calculated, contained more than a thousand times as much energy as the Sun will generate in its entire lifetime of 10 billion years.

Some astronomers speculate that the source of these gamma-ray bursts may ultimately be a distant supernova, or exploding star. If this is the case, scientists may have found the supernova; but do not expect to see it in the night sky. It is not known just how long ago it exploded, but its light appeared on Earth some 340,000 years ago, and during that time it was visible in daylight for more than two years. So great was its power that the effects of this stellar phenomenon are still being experienced.

WHERE TO LEARN MORE

Branley, Franklyn Mansfield. *The Electromagnetic Spectrum: Key to the Universe.* Illustrated by Leonard D. Dank. New York: Crowell, 1979.

Electromagnetic Spectrum (Web site). <http://www.jsc. mil/images/speccht.jpg> (April 25, 2001).

The Electromagnetic Spectrum (Web site). <http://library.thinkquest.org/11119/index.htm> (April 30, 2001).

The Electromagnetic Spectrum (Web site). <http://www. ospi.wednet.edu:8001/curric/space/lfs/emspectr. html> (April 30, 2001).

Electromagnetic Spectrum/NASA: National Aeronautics and Space Administration (Web site). <http:// imagine.gsfc.nasa.gov/docs/science/know_11/ emspectrum.html> (April 30, 2001).

"How the Radio Spectrum Works." How Stuff Works (Web site). <http://www.howstuffworks.com/ radio~spectrum.htm> (April 25, 2001).

Internet Resources for Sound and Light (Web site). <http://electro.sau.edu/SLResources.html> (April 25, 2001).

Nassau, Kurt. *Experimenting with Color.* New York: F. Watts, 1997.

"Radio Electronics Pages" ePanorama.net (Web site). <http://www.epanorama.net/radio.html> (April 25, 2001).

Skurzynski, Gloria. *Waves: The Electromagnetic Universe.* Washington, D.C.: National Geographic Society, 1996.

LIGHT

CONCEPT

Light exists along a relatively narrow bandwidth of the electromagnetic spectrum, and the region of visible light is more narrow still. Yet, within that realm are an almost infinite array of hues that quite literally give color to the entire world of human experience. Light, of course, is more than color: it is energy, which travels at incredible speeds throughout the universe. From prehistoric times, humans harnessed light's power through fire, and later, through the invention of illumination devices such as candles and gas lamps. In the late nineteenth century, the first electric-powered forms of light were invented, which created a revolution in human existence. Today, the power of lasers, highly focused beams of high-intensity light, make possible a number of technologies used in everything from surgery to entertainment.

HOW IT WORKS

Early Progress in Understanding of Light

The first useful observations concerning light came from ancient Greece. The Greeks recognized that light travels through air in rays, a term from geometry describing that part of a straight line that extends in one direction only. Upon entering some denser medium, such as glass or water, as Greek scientists noticed, the ray experiences refraction, or bending. Another type of incidence, or contact, between a light ray and any surface, is reflection, whereby a light ray returns, rather than being absorbed at the interface.

The Greeks worked out the basic laws governing reflection and refraction, observing, for instance, that in reflection, the angle of incidence is approximately equal to the angle of reflection. Unfortunately, they also subscribed to the erroneous concept of intromission—the belief that light rays originate in the eye and travel toward objects, making them visible. Some 1,500 years after the high point of Greek civilization, Arab physicist Alhasen (Ibn al-Haytham; c. 965-1039), sometimes called the greatest scientist of the Middle Ages, showed that light comes from a source such as the Sun, and reflects from an object to the eyes.

The next great era of progress in studies of light began with the Renaissance (c. 1300-c. 1600.) However, the most profound scientific achievements in this area belonged not to scientists, but to painters, who were fascinated by color, shading, shadows, and other properties of light. During the early seventeenth century, Galileo Galilei (1564-1642) and German astronomer Johannes Kepler (1571-1630) built the first refracting telescopes, while Dutch physicist and mathematician Willebrord Snell (1580-1626) further refined the laws of refraction.

The Spectrum

Sir Isaac Newton (1642-1727) was as intrigued with light as he was with gravity and the other concepts associated with his work. Though it was not as epochal as his contributions to mechanics, Newton's work in optics, an area of physics that studies the production and propagation of light, was certainly significant.

In Newton's time, physicists understood that a prism could be used for the diffusion of light rays—in particular, to produce an array of colors from a beam of white light. The prevailing belief was that white was a single color like the others, but Newton maintained that it was a combination of all other colors. To prove this, he directed a beam of white light through a prism, then allowed the diffused colors to enter another prism, at which point they recombined as white light.

Newton gave to the array of colors in visible light the term spectrum, (plural, "spectra") meaning the continuous distribution of properties in an ordered arrangement across an unbroken range. The term can be used for any set of characteristics for which there is a gradation, as opposed to an excluded middle. An ordinary light switch provides an example of a situation in which there is an excluded middle: there is nothing between "on" and "off." A dimmer switch, on the other hand, is a spectrum, because a very large number of gradations exist between the two extremes represented by a light switch.

SEVEN COLORS...OR SIX? The distribution of colors across the spectrum is as follows: red-orange-yellow-green-blue-violet. The reasons for this arrangement, explained below in the context of the electromagnetic spectrum, were unknown to Newton. Not only did he live in an age that had almost no understanding of electromagnetism, but he was also a product of the era called the Enlightenment, when intellectuals (scientists included) viewed the world as a highly rational, ordered mechanism. His Enlightenment viewpoint undoubtedly influenced his interpretation of the spectrum as a set of seven colors, just as there are seven notes on the musical scale.

In addition to the six basic colors listed above, Newton identified a seventh, indigo, between blue and violet. In fact, there is a noticeable band of color between blue and violet, but this is because one color fades into another. With a spectrum, there is a blurring of lines between one color and the next: for instance, orange exists at a certain point along the spectrum, as does yellow, but between them is a nearly unlimited number of orange-yellow and yellow-orange gradations.

Indigo itself is not really a distinct color—just a deep, purplish blue. But its inclusion in the listing of colors on the spectrum has given generations of students a handy mnemonic (memo-

ISAAC NEWTON. *(The Bettmann Archive. Reproduced by permission.)*

rization) device: the name "ROY G. BIV." These letters form an acrostic (a word constructed from the first letters of other words) for the colors of the spectrum. Incidentally, there is something arbitrary even in the idea of six colors, or for that matter seven musical notes: in both cases, there exists a very large gradation of shades, yet also in both cases, the divisions used were chosen for practical purposes.

WAVES, PARTICLES, AND OTHER QUESTIONS CONCERNING LIGHT

THE WAVE-PARTICLE CONTROVERSY BEGINS. Newton subscribed to the corpuscular theory of light: the idea that light travels as a stream of particles. On the other hand, Dutch physicist and astronomer Christiaan Huygens (1629-1695) maintained that light travels in waves. During the century that followed, adherents of particle theory did intellectual battle with proponents of wave theory. "Battle" is not too strong a word, because the conflict was heated, and had a nationalistic element. Reflecting both the burgeoning awareness of the nation-state among Europeans, as well as Britons' sense of their own island as an entity separate from the European continent, particle theory had its strongest defenders in Newton's homeland, while continental scientists generally accepted wave theory.

According to Huygens, the appearance of the spectrum, as well as the phenomena of reflection and refraction, indicated that light was a wave. Newton responded by furnishing complex mathematical calculations which showed that particles could exhibit the behaviors of reflection and refraction as well. Furthermore, Newton challenged, if light were really a wave, it should be able to bend around corners. Yet, in 1660, an experiment by Italian physicist Francesco Grimaldi (1618-1663) proved that light could do just that. Passing a beam of light through a narrow aperture, or opening, Grimaldi observed a phenomenon called diffraction, or the bending of light.

In view of the nationalistic character that the wave-particle debate assumed, it was ironic that the physicist whose work struck a particularly forceful blow against corpuscular theory was himself an Englishman: Thomas Young (1773-1829), who in 1801 demonstrated interference in light. Directing a light beam through two closely spaced pinholes onto a screen, Young reasoned that if light truly were made of particles, the beams would project two distinct points onto the screen. Instead, what he saw was a pattern of interference—a wave phenomenon.

THE QUESTION OF A MEDIUM. As the nineteenth century progressed, evidence in favor of wave theory grew. Experiments in 1850 by Jean Bernard Leon Foucault (1819-1868)—famous for his pendulum —showed that light traveled faster in air than through water. Based on studies of wave motion up to that time, Foucault's work added substance to the view of light as a wave.

Foucault also measured the speed of light in a vacuum, a speed which he calculated to within 1% of its value as it is known today: 186,000 mi (299,339 km) per second. An understanding of just how fast light traveled, however, caused a nagging question dating back to the days of Newton and Huygens to resurface: how did light travel?

All types of waves known to that time traveled through some sort of medium: for instance, sound waves were propagated through air, water, or some other type of matter. If light was a wave, as Huygens said, then it, too, must have some medium. Huygens and his followers proposed a weak theory by suggesting the existence of an invisible substance called ether, which existed throughout the universe and which carried light.

Ether, of course, was really no answer at all. There was no evidence that it existed, and to many scientists, it was merely a concept invented to shore up an otherwise convincing argument. Then, in 1872, Scottish physicist James Clerk Maxwell (1831-1879) proposed a solution that must have surprised many scientists. The "medium" through which light travels, Maxwell proposed, was no medium at all; rather, the energy in light is transferred by means of radiation, which requires no medium.

ELECTROMAGNETISM

Maxwell brought together a number of concepts developed by his predecessors, sorting these out and adding to them. His work led to the identification of a "new" fundamental interaction, in addition to that associated with gravity. This was the mode of particle interaction associated with electromagnetic force.

The particulars of electromagnetic force, waves, and radiation are a subject unto themselves—really, many subjects. As for the electromagnetic spectrum, it is treated at some length in an essay elsewhere in this volume, and the reader is encouraged to review that essay to gain a greater understanding of light and its place in the spectrum.

In addition, some awareness of wave motion and related phenomena would also be of great value, and, for this purpose, other essays are recommended. In the present context, a number of topics relating to these larger subjects will be handled in short order, with a minimum of explanation, to enable a more speedy transition to the subject of principal importance here: light.

ELECTROMAGNETIC WAVES. There is, of course, no obvious connection between light and the electromagnetic force observed in electrical and magnetic interactions. Yet, light is an example of an electromagnetic wave, and is part of the electromagnetic spectrum. The breakthrough in establishing the electromagnetic quality of light can be attributed both to Maxwell and German physicist Heinrich Rudolf Hertz (1857-1894).

In his *Electricity and Magnetism* (1873), Maxwell suggested that electromagnetic force might have aspects of a wave phenomenon, and his experiments indicated that electromagnetic waves should travel at exactly the same speed as light. This appeared to be more than just a coincidence, and his findings led him to theorize that the electromagnetic interaction included not only electricity and magnetism, but light as well. Some time later, Hertz proved Maxwell's hypothesis by showing that electromagnetic waves obeyed the same laws of reflection, refraction, and diffraction as light.

Hertz also discovered the photoelectric effect, the process by which certain metals acquire an electrical potential when exposed to light. He could not explain this behavior, and, indeed, there was nothing in wave theory that could account for it. Strangely, after more than a century in which acceptance of wave theory had grown, he had encountered something that apparently supported what Newton had said long before: that light traveled in particles rather than waves.

THE WAVE-PARTICLE DEBATE REVISITED

One of the modern physicists whose name is most closely associated with the subject of light is Albert Einstein (1879-1955). In the course of proving that matter is convertible to energy, as he did with the theory of relativity, Einstein predicted that this could be illustrated by accelerating to speeds close to that of light. (Conversely, he also showed that it is impossible for matter to reach the speed of light, because to do so would—as he proved mathematically—result in the matter acquiring an infinite amount of mass, which, of course, is impossible.)

Much of Einstein's work was influenced by that of German physicist Max Planck (1858-1947), father of quantum theory. Quantum theory and quantum mechanics are, of course, far too complicated to explain in any depth here. It is enough to say that they called into question everything physicists thought they knew, based on Newton's theories of classical mechanics. In particular, quantum mechanics showed that, at the subatomic level, particles behave in ways not just different from, but opposite to, the behavior of larger physical objects in the observable world. When a quantity is "quantized," its values or properties at the atomic or subatomic level are separate from one another—meaning that some-

thing can both be one thing and its opposite, depending on how it is viewed.

Interpreting Planck's observations, Einstein in a 1905 paper on the photoelectric effect maintained that light is quantized—that it appears in "bundles" of energy that have characteristics both of waves and of particles. Though light travels in waves, as Einstein showed, these waves sometimes behave as particles, which is the case with the photoelectric effect. Nearly two decades later, American physicist Arthur Holly Compton (1892-1962) confirmed Einstein's findings and gave a name to the "particles" of light: photons.

LIGHT'S PLACE IN THE ELECTRO-MAGNETIC SPECTRUM

The electromagnetic spectrum is the complete range of electromagnetic waves on a continuous distribution from a very low range of frequencies and energy levels, with a correspondingly long wavelength, to a very high range of frequencies and energy levels, with a correspondingly short wavelength. Included on the electromagnetic spectrum are radio waves and microwaves; infrared, visible, and ultraviolet light; x rays, and gamma rays. As discussed earlier, concerning the visible color spectrum, each of these occupies a definite place on the spectrum, but the divisions between them are not firm: in keeping with the nature of a spectrum, one band simply "blurs" into another.

Of principal concern here is an area near the middle of the electromagnetic spectrum. Actually, the very middle of the spectrum lies within the broad area of infrared light, which has frequencies ranging from 10^{12} to just over 10^{14} Hz, with wavelengths of approximately 10^{-1} to 10^{-3} centimeters. Even at this point, the light waves are oscillating at a rate between 1 and 100 trillion times a second, and the wavelengths are from 1 millimeter to 0.01 millimeters. Yet, over the breadth of the electromagnetic spectrum, wavelengths get much shorter, and frequencies much greater.

Infrared lies just below visible light in frequency, which is easy to remember because of the name: red is the lowest in frequency of all the colors, as discussed below. Similarly, ultraviolet lies beyond the highest-frequency color, violet. Neither infrared nor ultraviolet can be seen, yet we experience them as heat. In the case of ultraviolet (UV) light, the rays are so powerful that exposure to even the minuscule levels of UV radiation that enter Earth's atmosphere can cause skin cancer.

Ultraviolet light occupies a much narrower band than infrared, in the area of about 10^{15} to 10^{16} Hz—in other words, oscillations between 1 and 10 quadrillion times a second. Wavelengths in this region are from just above 10^{-6} to about 10^{-7} centimeters. These are often measured in terms of a nanometer (nm)—equal to one-millionth of a millimeter—meaning that the wavelength range is from above 100 down to about 10 nm.

Between infrared and ultraviolet light is the region of visible light: the six colors that make up much of the world we know. Each has a specific range and frequency, and together they occupy an extremely narrow band of the electromagnetic spectrum: from $4.3 \cdot 10^{14}$ to $7.5 \cdot 10^{14}$ Hz in frequency, and from 700 down to 400 nm in wavelength. To compare its frequency range to that of the entire spectrum, for instance, is the same as comparing 3.2 to 100 billion.

REAL-LIFE APPLICATIONS

Colors

Unlike many of the topics addressed by physics, color is far from abstract. Numerous expressions in daily life describe the relationship between energy and color: "red hot," for instance, or "blue with cold." In fact, however, red—with a smaller frequency and a longer wavelength than blue—actually has less energy; therefore, blue objects should be hotter.

The phenomenon of the red shift, discovered in 1923 by American astronomer Edwin Hubble (1889-1953), provides a clue to this apparent contradiction. As Hubble observed, the light waves from distant galaxies are shifted to the red end, and he reasoned that this must mean those galaxies are moving away from the Milky Way, the galaxy in which Earth is located.

To generalize from what Hubble observed, when something shows red, it is moving away from the observer. The laws of thermodynamics state that where heat is involved, the movement is always away from an area of high temperature and toward an area of low temperature. Heated molecules that reflect red light are, thus, to use a colloquialism, "showing their tail end" as they

move toward an area of low temperature. By contrast, molecules of low temperature reflect bluish or purple light because the tendency of heat is to move toward them.

There are other reasons, aside from heat, that some objects tend to be red and others blue—or another color. Chemical factors may be involved: atoms of neon, for example, can be made to vibrate at a particular wavelength, producing a specific color. In any case, the color that an object reflects is precisely the color that it does not absorb: thus, if something is red, that means it has absorbed every color of the spectrum but red.

WHY IS THE SKY BLUE? The placement of colors on the electromagnetic spectrum provides an answer to that age-old question posed by generations of children to their parents: "Why is the sky blue?" Electromagnetic radiation is scattered as it enters the atmosphere, but all forms of radiation are not scattered equally. Those having shorter wavelengths—that is, toward the blue end of the spectrum—tend to scatter more than those with longer wavelengths, on the red and orange end.

Yet the longer-wavelength light becomes visible at sunset, when the Sun's light enters the atmosphere at an angle. In addition, the dim quality of evening light means that it is easiest to see light of longer wavelengths. This effect is known as Rayleigh scattering, after English physicist John William Strutt, Lord Rayleigh (1842-1919), who discovered it in 1871. Thanks to Rayleigh's discovery, there is an explanation not only for the question of why the sky is blue, but why sunsets are red, orange, and gold.

RAINBOWS. On the subject of color as children perceive it, many a child has been fascinated by a rainbow, seeing in them something magical. It is easy to understand why children perceive these beautiful phenomena this way, and why people have invented stories such as that of the pot of gold at the end of a rainbow. In fact, a rainbow, like many other "magical" aspects of daily life, can be explained in terms of physics.

A rainbow, in fact, is simply an illustration of the visible light spectrum. Rain drops perform the role of tiny prisms, dispersing white sunlight, much as scientists before Newton had learned to do. But if there is a pot of gold at the end of the rainbow, it would be impossible to find. In order for a rainbow to be seen, it must be viewed from a specific perspective: the observer must be in a position between the sunlight and the raindrops.

Sunlight strikes raindrops in such a way that they are refracted, then reflected back at an angle so that they represent the entire visible light spectrum. Though they are beautiful to see, rainbows are neither magical nor impossible to reproduce artificially. Such rainbows can be produced, for instance, in the spiral of small water droplets emerging from a water hose, viewed when one's back is to the Sun.

PERCEPTION OF LIGHT AND COLOR

People literally live and die for colors: the colors of a flag, for instance, present a rallying point for soldiers, and different colors are assigned specific political meanings. Blue, both in the American and French flags, typically stands for liberty. Red can symbolize the blood shed by patriots, or it can mean some version of fraternity or brotherhood. Such is the case with the red of the French tricolor (red-white-blue); likewise, the red in the flag of the former Soviet Union and other Communist countries stood for the alleged international brotherhood of all working peoples. In Islamic countries, by contrast, green stands for the unity of all Muslims.

These are just a few examples, drawn from a specific realm—politics—illustrating the meanings that people ascribe to colors. Similarly, people find meanings in images presented to them by light itself. In his *Republic*, the ancient Greek philosopher Plato (c. 427-347 B.C.) offered a complex parable, intended to illustrate the difference between reality and illusion, concerning a group of slaves who do not recognize the difference between sunlight and the light of a torch in a cave. Modern writers have noted the similarities between Plato's cave and a phenomenon which the ancient philosopher could hardly have imagined: a movie theatre, in which an artificial light projects images—images that people sometimes perceive as being all too real—onto a screen.

People refer to "tricks of the light," as, for instance, when one seems to see an image in a fire. One particularly well-known "trick of the light," a mirage, is discussed below, but there are also manmade illusions created by light, shapes, and images. An optical illusion is something that produces a false impression in the brain, causing one to believe that something is as it appears, when, in fact, it is not. When two lines of equal length are placed side by side, but one has arrows pointed outward at either end while the other line has arrows pointing inward, it appears that the line with the inward-pointing arrows are shorter.

This is an example of the ways in which human perception plays a role in what people see. That topic, of course, goes far beyond physics and into the realms of psychology and the social sciences. Nonetheless, it is worthwhile to consider, from a physical standpoint, how humans see what they see—and sometimes see things that are not there.

A MIRAGE. Because they can be demonstrated in light waves as well as in sound waves, diffraction and interference are discussed in separate essays. As for refraction, or the bending of light waves, this phenomenon can be seen in the familiar example of a mirage. While driving down a road on a hot day, one may observe that there are pools of water up ahead, but by the time one approaches them, they disappear.

Of course, the pools were never there; light itself has created an optical illusion of sorts. As light moves from one material to another, it bends with a different angle of refraction, and, though, in this instance, it is traveling entirely through air, it is moving through regions of differing temperature. Light waves travel faster through warm air than through cool air, and, thus, when the light enters the area over the heated surface of the asphalt, it experiences refraction. The waves are thus bent, creating the impression of a reflection, which suggests to the observer that there is water up ahead.

HOW THE EYE SEES COLOR. White, as noted earlier, is the combination of all colors; black is the absence of color. Where ink, dye, or other forms of artificial pigmentation are concerned, of course, black is a "real" color, but in terms of light, it is not. In the same way, the experience of coldness is real, yet "cold" does not exist as a physical phenomenon: it is simply the absence of heat.

The mixture of pigmentation is an entirely different matter from the mixture of light. In artificial pigmentation, the primary colors—the three colors which, when mixed, yield the remainder of the shades on the rainbow—are red, blue, and yellow. Red mixed with blue creates purple, blue mixed with yellow makes green, and red mixed with yellow yields orange. Black and white are usually created by using natural substances of that color—chalk for white, for instance, or various oxides for black. For light, on the other hand, blue and red are primary colors, but the third primary color is green, not yellow. From these three primary colors, all other shades of the visible spectrum can be made.

The mechanism of the human eye responds to the three primary colors of the visible light spectrum: thus, the eye's retina is equipped with tiny cones that respond to red, blue, and green light. The cones respond to bright light; other structures called rods respond to dim light, and the pupil regulates the amount of light that enters the eye.

The eye responds with maximum sensitivity to light at the middle of the visible color spectrum—specifically, green light with a wavelength of about 555 nm. The optimal wavelength for maximum sensitivity in dim light is around 510 nm, on the blue end. It is difficult for the eye to recognize red light, at the far end of the spectrum, against a dark background. However, this can be an advantage in situations of relative darkness, which is why red light is often used to maintain vision for sailors, amateur astronomers, and the military on night maneuvers. Because there is not much difference between the darkness and the red light, the eye adjusts and is able to see beyond the red light into the darkness. A bright yellow or white light in such situations, on the other hand, would minimize visibility in areas beyond the light.

ARTIFICIAL LIGHT

PREHISTORIC LIGHTING TECHNOLOGY.
Prehistoric humans did not know it, but they were making use of electromagnetic radiation when they lit and warmed their caves with light from a fire. Though it would seem that warmth was more essential to human survival than artificial light, in fact, it is likely that both functions emerged at about the same time: once humans began using fire for warmth, it would have been a relatively short time before they comprehended the power of fire to drive out both darkness and the fierce creatures (for instance, bears) that came with it.

These distant forebears advanced to the fashioning of portable lighting technology in the form of torches or rudimentary lamps. Torches were probably made by binding together resinous material from trees, while lamps were made either from stones with natural depressions, or from soft rocks—for example, soapstone or steatite—into which depressions were carved by using harder material. Most of the many hundreds of lamps found by archaeologists at sites in southwestern France are made of either limestone or sandstone. Limestone was a particularly good choice, since it conducts heat poorly; lamps made of sandstone, a good conductor of heat, usually had carved handles to protect the hands of the user.

ARTIFICIAL LIGHT IN PRE-MODERN TIMES.
The history of lighting is generally divided into four periods, each of which overlap, and which together illustrate the slow pace of change in illumination technology. First was the primitive, a period encompassing the torches and lamps of prehistoric human beings—though, in fact, French peasants used the same lighting methods depicted in nearby cave paintings until World War I.

Next came the classical stage, the world of Greece and Rome. Earlier civilizations, such as that of Egypt, belong to the primitive era in lighting—before the relatively widespread adoption of the candle and of vegetable oil as fuel. Third was the medieval stage, which saw the development of metal lamps. Last came the modern or invention stage, which began with the creation of the glass lantern chimney by Leonardo da Vinci (1452-1519) in 1490, culminated with Thomas Edison's (1847-1931) first practical incandescent bulb in 1879, and continues today.

At various times, ancient peoples used the fat of seals, horses, cattle, and fish as fuel for lamps. (Whale oil, by contrast, entered widespread use only during the nineteenth century.) Primitive humans sometimes used entire animals—for example, the storm petrel, a bird heavy in fat—to provide light. Even without such cruel excesses, however, animal fat made for a smoky, dangerous, foul-smelling fire.

The use of vegetable oils, a much more efficient medium for lighting, did not take hold until Greek, and especially, Roman times. Animal oils remained in use, however, among the poor, whose homes often reeked with the odor of castor oil or fish oil. Because virtually all fuels came from edible sources, times of famine usually meant times of darkness as well.

The candle, as well as the use of vegetable oils, dates back to earliest antiquity, but the use of candles only became common among the richest citizens of Rome. Because it used animal fat, the candle was apparently a return to an earlier stage, but its hardened tallow actually represented a much safer, more stable fuel than lamp oil.

INCANDESCENT LIGHT. Lighting technology in the period from about 1500 to the late nineteenth century involved a number of improvements, but in one respect, little had progressed since prehistoric times: people were still burning fuel to provide illumination. This all changed with the invention of the incandescent bulb, which, though it is credited to Edison, was the product of experimentation that took place throughout the nineteenth century. As early as 1802, British scientist Sir Humphry Davy (1778-1829) showed that electricity running through thin strips of metal could heat them enough to cause them to give off light—that is, electromagnetic radiation.

Edison, in fact, was just one of several inventors in the 1870s attempting to develop a practical incandescent lamp. His innovation lay in his understanding of the parameters necessary for developing such a lamp—in particular, decreasing the electrical resistance in the lamp filament (the part that is heated) so that less energy would be required to light it. On October 19, 1879, using low-resistance filaments of carbon or platinum, combined with a high-resistance carbon filament in a vacuum-sealed glass container, Edison produced the first practical lightbulb.

Much has changed in the design of lightbulbs during the decades following Edison's ingenious invention, of course, but his design provided the foundation. There is just one problem with incandescent light, however—a problem inherent in the definition and derivation of the word incandescent, which comes from a Latin root meaning "to become hot." The efficiency of a light is determined by the ratio of light, or usable energy, to heat—which, except in the case of a campfire, is typically not a desirable form of energy where lighting is concerned.

Amazingly, only about 10% of the energy output from a typical incandescent light bulb is in the form of visible light; the rest comes through the infrared region of the spectrum, producing heat rather than light that people can use. The visible light tends to be in the red and yellow end of the spectrum—closer to infrared—but a blue-tinted bulb helps to absorb some of the red and yellow, providing a color balance. This, however, only further diminishes the total light output, and, hence, in many applications today, fluorescent light takes the place of incandescent light.

LASERS

A laser is an extremely focused, extremely narrow, and extremely powerful beam of light. Actually, the term laser is an acronym, standing for *Light Amplification* by *Simulated Emission of Radiation.* Simulated emission involves bringing a large number of atoms into what is called an "excited state." Generally, most atoms are in a ground state, and are less active in their movements, but the energy source that activates a laser brings about population inversion, a reversal of the ratios, such that the majority of atoms within the active medium are in an excited rather than a ground state. To visualize this, picture a popcorn popper, with the excited atoms being the popping kernels, and the ground-state atoms the ones remaining unpopped. As the atoms become excited, and the excited atoms outnumber the ground ones, they start to cause a multiplication of the resident photons. This is simulated emission.

A laser consists of three components: an optical cavity, an energy source, and an active medium. To continue the popcorn analogy, the "popper" itself—the chamber which holds the laser—is the optical cavity, which, in the case of a laser, involves two mirrors facing one another. One of these mirrors fully reflects light, whereas the other is a partly reflecting mirror. The light not reflected by the second mirror escapes as a highly focused beam. As with the popcorn popper, the power source involves electricity, and the active medium is analogous to the oil in a conventional popper.

TYPES OF LASERS. There are four types of lasers: solid-state, semiconductor, gas, and dye. Solid-state lasers are generally very large and extremely powerful. Having a crystal or glass housing, they have been implemented in nuclear energy research, and in various areas of industry. Whereas solid-state lasers can be as long as a city block, semiconductor lasers can be smaller than

the head of a pin. Semiconductor lasers (involving materials such as arsenic that conduct electricity, but do not do so as efficiently as the metals typically used as conductors) are applied for the intricate work of making compact discs and computer microchips.

Gas lasers contain carbon dioxide or other gases, activated by electricity in much the same way the gas in a neon sign is activated. Among their applications are eye surgery, printing, and scanning. Finally, dye lasers, as their name suggests, use different colored dyes. (Laser light itself, unlike ordinary light, is monochromatic.) Dye lasers can be used for medical research, or for fun—as in the case of laser light shows held at parks in the summertime.

LASER APPLICATIONS. Laser beams have a number of other useful functions, for instance, the production of compact discs (CDs). Lasers etch information onto a surface, and because of the light beam's qualities, can record far more information in much less space

KEY TERMS

APERTURE: An opening.

DIFFRACTION: The bending of waves around obstacles, or the spreading of waves by passing them through an aperture.

DIFFUSION: A process by which the concentration or density of something is decreased.

ELECTROMAGNETIC SPECTRUM: The complete range of electromagnetic waves on a continuous distribution from a very low range of frequencies and energy levels, with a correspondingly long wavelength, to a very high range of frequencies and energy levels, with a correspondingly short wavelength. Included on the electromagnetic spectrum are long-wave and short-wave radio; microwaves; infrared, visible, and ultraviolet light; x rays, and gamma rays.

ELECTROMAGNETIC WAVE: A transverse wave with electric and magnetic fields that emanate from it. The directions of these fields are perpendicular to one another, and both are perpendicular to the line of propagation for the wave itself.

FREQUENCY: The number of waves passing through a given point during the interval of one second. The higher the frequency, the shorter the wavelength.

HERTZ: A unit for measuring frequency, named after nineteenth—century German physicist Heinrich Rudolf Hertz (1857-1894).

INCIDENCE: Contact between a ray—for example, a light ray—and a surface. Types of incidence include reflection and refraction.

MEDIUM: A substance through which light travels, such as air, water, or glass.

Because light moves by radiation, it does not require a medium, and, in fact, movement through a medium slows the speed of light somewhat.

OPTICS: An area of physics that studies the production and propagation of light.

PHOTOELECTRIC EFFECT: The phenomenon whereby certain metals acquire an electrical potential when exposed to light.

PHOTON: A particle of electromagnetic radiation—for example, light—carrying a specific amount of energy, measured in electron volts (eV).

PRISM: A three-dimensional glass shape used for the diffusion of light rays.

PROPAGATION: The act or state of traveling from one place to another.

RADIATION: The transfer of energy by means of electromagnetic waves, which require no physical medium (for example, water or air) for the transfer. Earth receives the Sun's energy (including its light), via the electromagnetic spectrum, by means of radiation.

RAY: In geometry, a ray is that part of a straight line that extends in one direction only. The term "ray" is used to describe the directed line made by light as it moves through space.

REFLECTION: A type of incidence whereby a light ray is returned toward its source rather than being absorbed at the interface.

REFRACTION: The bending of a light ray that occurs when it passes through a dense medium, such as water or glass.

SPECTRUM: The continuous distribution of properties in an ordered arrangement across an unbroken range. Examples of spectra (the plural of "spectrum") include the colors of visible light, or the electromagnetic spectrum of which visible light is a part.

TRANSVERSE WAVE: A wave in which the vibration or motion is perpendi-cular to the direction in which the wave is moving.

VACUUM: An area of space devoid of matter, including air.

WAVELENGTH: The distance between a crest and the adjacent crest, or the trough and an adjacent trough, of a wave. The shorter the wavelength, the higher the frequency.

than the old-fashioned ways of producing phonograph records.

Lasers used in the production of CD-ROM (Read-Only Memory) disks are able to condense huge amounts of information—a set of encyclopedias or the New York metropolitan phone book—onto a disk one can hold in the palm of one's hand. Laser etching is also used to create digital videodiscs (DVDs) and holograms. Another way that lasers affect everyday life is in the field of fiber optics, which uses pulses of laser light to send information on glass strands.

Before the advent of fiber-optic communications, telephone calls were relayed on thick bundles of copper wire; with the appearance of this new technology, a glass wire no thicker than a human hair now carries thousands of conversations. Lasers are also used in scanners, such as the price-code checkers at supermarkets and various kinds of tags that prevent thefts of books from libraries or clothing items from stores. In an industrial setting, heating lasers can drill through solid metal, or in an operating room, lasers can remove gallstones or cataracts. Lasers are also used for guiding missiles, and to help building contractors ensure that walls and floors and ceilings are in proper alignment.

WHERE TO LEARN MORE

Burton, Jane and Kim Taylor. *The Nature and Science of Colors.* Milwaukee, WI: Gareth Stevens Publishing, 1998.

Glover, David. *Color and Light.* New York: Dorling Kindersley Publishing, 2001.

Kalman, Bobbie and April Fast. *Cosmic Light Shows.* New York: Crabtree Publishing, 1999.

Kurtus, Ron. *"Visible Light"* (Web site). <http://www.school-for-champions.com/science/light.html> (May 2, 2001).

"Light Waves and Color." The Physics Classroom (Web site). <http://www.glenbrook.k12.il.us/gbssci/phys/Class/light/lighttoc.html> (May 2, 2001).

Miller-Schroeder, Patricia. *The Science and Light of Color.* Milwaukee, WI: Gareth Stevens Publishing, 2000.

Nassau, Kurt. *Experimenting with Color.* New York: F. Watts, 1997.

Riley, Peter D. *Light and Color.* New York: F. Watts, 1999.

Taylor, Helen Suzanne. *A Rainbow Is a Circle: And Other Facts About Color.* Brookfield, CT: Copper Beech Books, 1999.

"Visible Light Waves" NASA: National Aeronautics and Space Administration (Web site). <http://imagers.gsfc.nasa.gov/ems/visible.html> (May 2, 2001).

LUMINESCENCE

CONCEPT

Luminescence is the generation of light without heat. There are two principal varieties of luminescence, fluorescence and phosphorescence, distinguished by the delay in reaction to external electromagnetic radiation. The ancients observed phosphorescence in the form of a glow emitted by the oceans at night, and confused this phenomenon with the burning of the chemical phosphor, but, in fact, phosphorescence has nothing at all to do with burning. Likewise, fluorescence, as applied today in fluorescent lighting, involves no heat—thus creating a form of lighting more efficient than that which comes from incandescent bulbs.

HOW IT WORKS

RADIATION

Elsewhere in this volume, the term "radiation" has been used to describe the transfer of energy in the form of heat. In fact, radiation can also be described, in a more general sense, as anything that travels in a stream, whether that stream be composed of subatomic particles or electromagnetic waves.

Many people think of radiation purely in terms of the harmful effects produced by radioactive materials—those subject to a form of decay brought about by the emission of high-energy particles or radiation, including alpha particles, beta particles, or gamma rays. These high-energy forms of radiation are called ionizing radiation, because they are capable of literally ripping through some types of atoms, removing electrons and leaving behind a string of ions.

Ionizing radiation can indeed cause a great deal of damage to matter—including the matter in a human body. Even radiation produced by parts of the electromagnetic spectrum possessing far less energy than gamma rays can be detrimental, as will be discussed below. In general, however, there is nothing inherently dangerous about radiation: indeed, without the radiation transmitted to Earth via the Sun's electromagnetic spectrum, life simply could not exist.

THE ELECTROMAGNETIC SPECTRUM

The electromagnetic spectrum is the complete range of electromagnetic waves on a continuous distribution from those with very low frequencies and energy levels, along with correspondingly long wavelengths, to those with very high frequencies and energy levels, with correspondingly short wavelengths.

An electromagnetic wave is transverse, meaning that even as it moves forward, it oscillates in a direction perpendicular to the line of propagation. An electromagnetic wave can thus be defined as a transverse wave with mutually perpendicular electrical and magnetic fields that emanate from it. Though their shape is akin to that of waves on the ocean, electromagnetic waves travel much, much faster than any waves that human eyes can see. Their speed of propagation in a vacuum is equal to that of light: 186,000 mi (299,339 km) per second.

PARTS OF THE ELECTROMAGNETIC SPECTRUM. Included on the electromagnetic spectrum are radio waves and microwaves; infrared, visible, and ultraviolet light; x rays, and gamma rays. Though each occu-

MODERN UNDERSTANDING OF LUMINESCENCE OWES MUCH TO MARIE CURIE.

pies a definite place on the spectrum, the divisions between them are not firm: as befits the nature of a spectrum, one simply "blurs" into another.

Though the Sun sends its energy to Earth in the form of light and heat from the electromagnetic spectrum, not everything within the spectrum is either "bright." The "bright" area of the spectrum—that is, the band of visible light—is incredibly small, equal to about 3.2 parts in 100 billion. This is like comparing a distance of 16 ft (4.8 m) to the distance between Earth and the Sun: 93 million miles ($1.497 \cdot 10^9$ km).

When electromagnetic waves of almost any frequency are asborbed in matter, their energy can be converted to heat. Whether or not this happens depends on the absorption mechanism. However, the realm of "heat" as it is most experienced in daily life is much smaller, encompassing infrared, visible, and ultraviolet light. Below this frequency range are various types of radio waves, and above it are ultra high-energy x rays and gamma rays. Some of the heat experienced in a nuclear explosion comes from the absorption of gamma rays emitted in the nuclear reaction.

FREQUENCY AND WAVELENGTH RANGE. There is nothing arbitrary about the order in which the different types of electromagnetic waves are listed above: this is their order in terms of frequency (measured in Hertz, or Hz) and energy levels, which are directly related. This ordering also represents the reverse order (that is, from longer to shorter) for wavelength, which is inversely related to frequency.

Extremely low-energy, long-wavelength radio waves have frequencies of around 10^2 Hz, while the highest-energy, shortest-wavelength gamma rays can have frequencies of up to 10^{25} Hz. This means that these gamma rays are oscillating at the rate of 10 trillion trillion times a second!

The wavelengths of very low-energy, low-frequency radio waves can be extremely long: 10^8 centimeters, equal to 1 million meters or about 621 miles. Precisely because these wavelengths are so very long, they are hard to apply for any practical use: ordinary radio waves of the kind used for actual radio broadcasts are closer to 10^5 cm (about 328 ft).

At the opposite end of the spectrum are gamma rays with wavelengths of less than 10^{-15} centimeters—in other words, a decimal point followed by 14 zeroes and a 1. There is literally nothing in the observable world that can be compared to this figure, equal to one-trillionth of a centimeter. Even the angstrom—a unit so small it is used to measure the diameter of an atom—is 10 million times as large.

EMISSION AND ABSORPTION

The electromagnetic spectrum is not the only spectrum: physicists, as well as people who are not scientifically trained, often speak of the color spectrum for visible light. The reader is encouraged to study the essays on both subjects to gain a greater understanding of each. In the present context, however, two other types of spectra (the plural of "spectrum") are of interest: emission and absorption spectra.

Emission occurs when internal energy from one system is transformed into energy that is carried away from that system by electromagnetic radiation. An emission spectrum for any given system shows the range of electromagnetic radiation it emits. When an atom has energy transferred to it, either by collisions or as a result of exposure to radiation, it is said to be experiencing excitation, or to be "excited." Excited atoms

COMPARED TO STANDARD INCANDESCENT LIGHT BULBS (SHOWN ON THE LEFT), NEWER FLUORESCENT BULBS (SHOWN ON THE RIGHT) USE FAR LESS ELECTRICITY AND LAST MUCH LONGER. *(Photograph by Roger Ressmeyer/Corbis. Reproduced by permission.)*

will emit light of a given frequency as they relax back to their normal state. Atoms of neon, for instance, can be excited in such a way that they emit light at wavelengths corresponding to the color red, a property that finds application in neon signs.

As its name suggests, absorption has a reciprocal relationship with emission: it is the result of any process wherein the energy transmitted to a system via electromagnetic radiation is added to the internal energy of that system. Each material has a unique absorption spectrum, which makes it possible to identify that material using a device called a spectrometer. In the phenomenon of luminescence, certain materials absorb electromagnetic radiation and proceed to emit that radiation in ways that distinguish the materials as either fluorescent or phosphorescent.

REAL-LIFE APPLICATIONS

EARLY OBSERVATIONS OF LUMINESCENCE

For the most part, prior to the nineteenth century, scientists had little concept of light without

heat. Even what premodern observers called "phosphorescence" was not phosphorescence as the term is used in modern science. Instead, the word was used to describe light given off in a fiery reaction that occurs when the element phosphorus is exposed to air.

There are, however, examples of luminescence in nature that had been observed from ancient times onward—for instance, the phosphorescent glow of the ocean, visible at night under certain conditions. At one time this, too, was mistakenly associated with phosphorus, which was supposedly burning in the water. In fact, the ocean's phosphorescence comes neither from phosphorus nor water, but from living creatures called dinoflagellates. This is an example of a phenomenon known as bioluminescence—fireflies are another example—discussed below.

Modern understanding of luminescence owes much to Polish-French physicist and chemist Marie Curie (1867-1934). Operating in fields that had been dominated by men since the birth of the physical sciences, Curie distinguished herself with a number of achievements, becoming the first scientist in history to receive two Nobel prizes (physics in 1903 and chemistry in 1911). While working on her doctoral thesis,

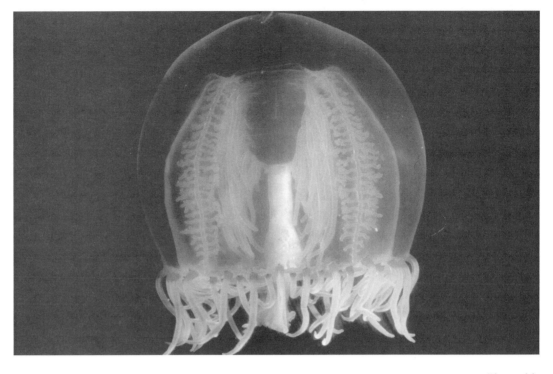

LIKE MANY MARINE CREATURES, JELLYFISH PRODUCE THEIR OWN LIGHT THROUGH PHOSPHORESCENCE. *(Photograph by Mark A. Johnson. The Stock Market. Reproduced by permission.)*

Curie noted that calcium fluoride glows when exposed to a radioactive material known as radium.

Curie—who also coined the term "radioactivity"—helped spark a revolution in science and technology. As a result of her work and the discoveries of others who followed, interest in luminescence and luminescent devices grew. Today, luminescence is applied in a number of devices around the household, most notably in television screens and fluorescent lights.

FLUORESCENCE

As indicated in the introduction to this essay, the difference between the two principal types of luminescence relates to the timing of their reactions to electromagnetic radiation. Fluorescence is a type of luminescence whereby a substance absorbs radiation and almost instantly begins to re-emit the radiation. (Actually, the delay is 10^{-6} seconds, or a millionth of a second.) Fluorescent luminescence stops within 10^{-5} seconds after the energy source is removed; thus, it comes to an end almost as quickly as it begins.

Usually, the wavelength of the re-emitted radiation is longer than the wavelength of the radiation the substance absorbed. British mathematician and physicist George Gabriel Stokes (1819-1903), who coined the term "fluorescence," first discovered this difference in wavelength. However, in a special type of fluorescence known as resonance fluorescence, the wavelengths are the same. Applications of resonance include its use in analyzing the flow of gases in a wind tunnel.

BLACK LIGHTS AND FLUORESCENCE. A "black light," so called because it emits an eerie bluish-purple glow, is actually an ultraviolet lamp, and it brings out vibrant colors in fluorescent materials. For this reason, it is useful in detecting art forgeries: newer paint tends to fluoresce when exposed to ultraviolet light, whereas older paint does not. Thus, if a forger is trying to pass off a painting as the work of an Old Master, the ultraviolet lamp will prove whether the artwork is genuine or not.

Another example of ultraviolet light and fluorescent materials is the "black-light" poster, commonly associated with the psychedelic rock music of the late 1960s and early 1970s. Under ordinary visible light, a black-light poster does not look particularly remarkable, but when exposed to ultraviolet light in an environment in which visible light rays are not propagated (that is, a darkened room), it presents a dazzling array of colors. Yet, because they are fluorescent, the moment the black light is turned off, the colors of the poster cease to glow. Thus, the poster, like the light itself, can be turned "on" and "off," simply by activating or deactivating the ultraviolet lamp.

RUBIES AND LASERS. Fluorescence has applications far beyond catching art forgers or enhancing the experience of hearing a Jimi Hendrix album. In 1960, American physicist Theodore Harold Maiman developed the first laser using a ruby, a gem that exhibits fluorescent characteristics. A laser is a very narrow, highly focused, and extremely powerful beam of light used for everything from etching data on a surface to performing eye surgery.

Crystalline in structure, a ruby is a solid that includes the element chromium, which gives the gem its characteristic reddish color. A ruby exposed to blue light will absorb the radiation and go into an excited state. After losing some of the absorbed energy to internal vibrations, the ruby passes through a state known as metastable before dropping to what is known as the ground state, the lowest energy level for an atom or molecule. At that point, it begins emitting radiation on the red end of the spectrum.

The ratio between the intensity of a ruby's emitted fluorescence and that of its absorbed radiation is very high, and, thus, a ruby is described as having a high level of fluorescent efficiency. This made it an ideal material for Maiman's purposes. In building his laser, he used a ruby cylinder which emitted radiation that was both coherent, or all in a single direction, and monochromatic, or all of a single wavelength. The laser beam, as Maiman discovered, could travel for thousands of miles with very little dispersion—and its intensity could be concentrated on a small, highly energized pinpoint of space.

FLUORESCENT LIGHTS. By far the most common application of fluorescence in daily life is in the fluorescent light bulb, of which there are more than 1.5 billion operating in the United States. Fluorescent light stands in contrast to incandescent, or heat-producing, electrical light. First developed successfully by Thomas Edison (1847-1931) in 1879, the incandescent lamp quite literally transformed human life, making possible a degree of activity after dark that would have been impractical in the age of gas lamps. Yet, incandescent lighting is highly

inefficient compared to fluorescent light: in an incandescent bulb, fully 90% of the energy output is wasted on heat, which comes through the infrared region.

A fluorescent bulb consuming the same amount of power as an incandescent bulb will produce three to five times more light, and it does this by using a phosphor, a chemical that glows when exposed to electromagnetic energy. (The term "phosphor" should not be confused with phosphorescence: phosphors are used in both fluorescent and phosphorescent applications.) The phosphor, which coats the inside surface of a fluorescent lamp, absorbs ultraviolet light emitted by excited mercury atoms. It then re-emits the ultraviolet light, but at longer wavelengths—as visible light. Thanks to the phosphor, a fluorescent lamp gives off much more light than an incandescent one, and does so without producing heat.

PHOSPHORESCENCE. In contrast to the nearly instantaneous "on-off" of fluorescence, phosphorescence involves a delayed emission of radiation following absorption. The delay may take as much as several minutes, but phosphorescence continues to appear after the energy source has been removed. The hands and numbers of a watch that glows in the dark, as well as any number of other items, are coated with phosphorescent materials.

Television tubes also use phosphorescence. The tube itself is coated with phosphor, and a narrow beam of electrons causes excitation in a small portion of the phosphor. The phosphor then emits red, green, or blue light—the primary colors of light—and continues to do so even after the electron beam has moved on to another region of phosphor on the tube. As it scans across the tube, the electron beam is turned rapidly on and off, creating an image made up of thousands of glowing, colored dots.

PHOSPHORESCENCE IN SEA CREATURES. As noted above, one of the first examples of luminescence ever observed was the phosphorescent effect sometimes visible on the surface of the ocean at night—an effect that scientists now know is caused by materials in the bodies of organisms known as dinoflagellates. Inside the body of a dinoflagellate are the substances luciferase and luciferin, which chemically react with oxygen in the air above the water to produce light with minimal heat levels. Though

KEY TERMS

ABSORPTION: The result of any process wherein the energy transmitted to a system via electromagnetic radiation is added to the internal energy of that system. Each material has a unique absorption spectrum, which makes it possible to identify that material using a device called a spectrometer. (Compare absorption to emission.)

ELECTROMAGNETIC SPECTRUM: The complete range of electromagnetic waves on a continuous distribution from a very low range of frequencies and energy levels, with a correspondingly long wavelength, to a very high range of frequencies and energy levels, with a correspondingly short wavelength. Included on the electromagnetic spectrum are long-wave and short-wave radio; microwaves; infrared, visible, and ultraviolet light; x rays, and gamma rays.

ELECTROMAGNETIC WAVE: A transverse wave with electric and magnetic fields that emanate from it. These waves are propagated by means of radiation.

EMISSION: The result of a process that occurs when internal energy from one system is transformed into energy that is carried away from it by electromagnetic radiation. An emission spectrum for any given system shows the range of electromagnetic radiation it emits. (Compare emission to absorption.)

EXCITATION: The transfer of energy to an atom, either by collisions or due to radiation.

FLUORESCENCE: A type of luminescence whereby a substance absorbs radiation and begins to re-emit the radiation 10^{-6} seconds after absorption. Usually the wavelength of emission is longer than the wavelength of the radiation the substance absorbed. Fluorescent luminescence stops within 10^{-5} seconds after the energy source is removed.

FREQUENCY: The number of waves passing through a given point during the interval of one second. The higher the frequency, the shorter the wavelength.

dinoflagellates are microscopic creatures, in large numbers they produce a visible glow.

Nor are dinoflagellates the only bioluminescent organisms in the ocean. Jellyfish, as well as various species of worms, shrimp, and squid, all produce their own light through phosphorescence. This is particularly useful for creatures living in what is known as the mesopelagic zone, a range of depth from about 650 to 3,000 ft (200-1,000 m) below the ocean surface, where little light can penetrate.

One interesting bioluminescent sea creature is the cypridina. Resembling a clam, the cypridina mixes its luciferin and luciferase with sea water to create a bright bluish glow. When dried to a powder, a dead cypridina can continue to produce light, if mixed with water. Japanese soldiers in World War II used the powder of cypridina to illuminate maps at night, providing themselves with sufficient reading light without exposing themselves to enemy fire.

PROCESSES THAT CREATE LUMINESCENCE

The phenomenon of bioluminescence actually goes beyond the frontiers of physics, into chemistry and biology. In fact, it is a subset of chemiluminescence, or luminescence produced by chemical reactions. Chemiluminescence is, in

KEY TERMS CONTINUED

HERTZ: A unit for measuring frequency, named after ninetenth-century German physicist Heinrich Rudolf Hertz (1857-1894).

LUMINESCENCE: The generation of light without heat. There are two principal varieties of luminescence, fluorescence and phosphorescence.

PHOSPHORESCENCE: A type of luminescence involving a delayed emission of radiation following absorption. The delay may take as much as several minutes, but phosphorescence continues to appear after the energy source has been removed.

PROPAGATION: The act or state of travelling from one place to another.

RADIATION: In a general sense, radiation can refer to anything that travels in a stream, whether that stream be composed of subatomic particles or electromagnetic waves.

RADIOACTIVE: A term describing materials which are subject to a form of decay brought about by the emission of high-energy particles or radiation, including alpha particles, beta particles, or gamma rays.

SPECTRUM: The continuous distribution of properties in an ordered arrangement across an unbroken range. Examples of spectra (the plural of "spectrum") include the colors of visible light, the electromagnetic spectrum of which visible light is a part, as well as emission and absorption spectra.

TRANSVERSE WAVE: A wave in which the vibration or motion is perpendicular to the direction in which the wave is moving.

VACUUM: An area of space devoid of matter, including air.

WAVELENGTH: The distance between a crest and the adjacent crest, or the trough and an adjacent trough, of a wave. The shorter the wavelength, the higher the frequency.

---

turn, one of several processes that can create luminescence.

Many of the types of luminescence discussed above are described under the heading of electroluminescence, or luminescence involving electromagnetic energy. Another process is triboluminescence, in which friction creates light. Though this type of friction can produce a fire, it is not to be confused with the heat-causing friction that occurs when flint and steel are struck together.

Yet another physical process used to create luminescence is sonoluminescence, in which light is produced from the energy transmitted by sound waves. Sonoluminescence is one of the fields at the cutting edge in physics today, and research in this area reveals that extremely high levels of energy may be produced in small areas for very short periods of time.

WHERE TO LEARN MORE

Birch, Beverley. *Marie Curie: Pioneer in the Study of Radiation.* Milwaukee, WI: Gareth Stevens Children's Books, 1990.

Evans, Neville. *The Science of a Light Bulb.* Austin, TX: Raintree Steck-Vaughn Publishers, 2000.

"*Luminescence.*" *Slider.com* (Web site). <http://www.slider.com/enc/32000/luminescence.html> (May 5, 2001).

"*Luminescence.*" *Xrefer* (Web site). <http://www.xrefer.com/entry/642646> (May 5, 2001).

Macaulay, David. *The New Way Things Work.* Boston: Houghton Mifflin, 1998.

Pettigrew, Mark. *Radiation.* New York: Gloucester Press, 1986.

Simon, Hilda. *Living Lanterns: Luminescence in Animals.* Illustrated by the author. New York: Viking Press, 1971.

Skurzynski, Gloria. *Waves: The Electromagnetic Universe.* Washington, D.C.: National Geographic Society, 1996.

Suplee, Curt. *Everyday Science Explained.* Washington, D.C.: National Geographic Society, 1996.

"UV-Vis Luminescence Spectroscopy" (Web site). <http://www.shu.ac.uk/virtual_campus/courses/241/lumin1.html> (May 5, 2001).

GENERAL
SUBJECT INDEX

*Boldface type indicates main entry page numbers. Italic
type indicates photo and illustration page numbers.*

laws of motion, 62
Handlebars (bicycles), 109
Harmonic motion
    damping, 266
    Doppler effect, 301
    electromagnetism, 343
    frequency, 271–274, 273–274
    resonance, 280
    wave motion, 255
    See also Oscillations; Wave motion
Harmonics (acoustics)
    frequency, 274, 276
    interference, 289
Harvard University, 211
HCFCs. *See* Hydrochlorofluorocarbons
Hearts, 324
Heat, **227–235**
    calorimetry, 230–231
    engines, 231–232
    friction, 56
    glossary, 233*t*–234*t*
    laws of thermodynamics, 232, 234–235
    luminescence, 366
    measurement, 219, 229
    specific heat, 230
    thermal energy, 218–219, 227, 236–237
    transfer, 228–229, 348
    See also Thermodynamics
Heat capacity, 219, 230
Heat conduction, 220, 228
Heat engines, 231–232
Heat transfer, 220–221, 348
    conduction, 228
    convection, 228–229
    radiation, 229
    See also Heat
Heaviside layer, 346
Heaviside, Oliver, 346
Helical gears, 163
Heliocentric universe
    frame of reference, 9–10
    gravity, 70–71
    laws of motion, 61–62, 65
Helium
    balloons and dirigibles, 126
    liquefaction, 198, 200
Helmets, bicycle, 109
Helsinki University of Technology Low Temperature Laboratory, 223
Henry's law, 187
Hero of Alexandria, 120, 163, 231
Herodotus, 164
Herschel, William, 349
Hertz, Heinrich Rudolf
    electromagnetism, 341–342
    light, 356–357

wave properties, 256–257
Hertz (unit of measure), 256–257
Hindenburg (dirigible), 105, 126, 128
Hockey, 54–55
    See also Ice skating
Holograms, *295, 296,* 298–300
Holographic memory, 300
Holographic optical elements, 299
Hooke, Robert, 148, 239
Hooke's law, 148
Hooke's scale (temperature), 239
Hoover bugles, 117
Horizontal motion. *See* Projectile motion
Horsepower (unit of measure), 173
Hot-air balloons, 126, *186,* 188
    See also Balloons
Hot extrusion of metals, 151
Houot, Georges, 125
Hubble, Edwin, 307, 358
Hubble Space Telescope, *342, 350*
Human cannonballs, *70*
Humason, Milton, 307
Huygens, Christiaan
    pendulum clocks, 267, 274
    wave theory of light, 296, 343, 355–356
Hydraulic presses, 167–169
    fluid mechanics, 98–99
    origins, 160
    Pascal's principle, 142–143
Hydraulic rams. *See* Hydraulic presses
Hydrochlorofluorocarbons (HCFCs), 188
Hydrodynamica (Bernoulli), 113, 195
Hydrodynamics, 96–97
    See also Dynamics
Hydroelectric dams
    conservation of energy, 176
    fluid mechanics, 101
Hydrogen
    balloons and airships, 105, 126–128
    infrared astronomy, 350
    liquefaction, 200
Hydrogen bombs, 177–179
Hypersonic flight. *See* Supersonic flight
Hypersound, 313
Hypoteneuse. *See* Trigonometry

I

I-hsing, 267
ICBMs. *See* Intercontinental ballistic missiles (ICBMs)
Ice
    characteristics, 208
    floating characteristics, *204*
    thermal expansion, *247,* 248–249, 249